**Natural Flavors, Fragrances, and Perfumes**

# Natural Flavors, Fragrances, and Perfumes

Chemistry, Production, and Sensory Approach

*Edited by Sreeraj Gopi, Nimisha Pulikkal Sukumaran, Joby Jacob, and Sabu Thomas*

WILEY-VCH

**Editors**

***Dr. Sreeraj Gopi***
Chemical Faculty
Gdansk University of Technology
Gdansk
Poland
Overseas Expert Professor - Shanghai University
China

***Mrs Nimisha Pulikkal Sukumaran***
Aurea Biolabs (P) Ltd
R&D Center
Kolenchery
Cochin
682311 Kerala
India

***Dr. Joby Jacob***
Aurea Biolabs (P) Ltd
R&D Center
Kolenchery
Cochin
682311 Kerala
India

***Prof. Sabu Thomas***
Mahatma Gandhi University
Center for Nanoscience and Nanotechnology
Priyadarshini Hills
686-560 Kottayam, Kerala
India



**Library of Congress Card No.:** applied for

**British Library Cataloguing-in-Publication Data**
A catalogue record for this book is available from the British Library.

**Bibliographic information published by the Deutsche Nationalbibliothek**
The Deutsche Nationalbibliothek lists this publication in the Deutsche Nationalbibliografie; detailed bibliographic data are available on the Internet at <http://dnb.d-nb.de>.



**Print ISBN:** 978-3-527-34708-7
**ePDF ISBN:** 978-3-527-82479-3
**ePub ISBN:** 978-3-527-82480-9
**oBook ISBN:** 978-3-527-82481-6

**Typesetting** Straive, Chennai, India

# Contents

# Preface

It has been a while since a book was put together to address the natural flavors in industrial formulations, including concise treatments of the relation between sophisticated methods and better comprehension of consumer perception and behavior using a multidisciplinary metabolic engineering methodology by connecting with fields. It appears, however, that a gap has arisen between the new advances in basic information and the direct application to product situations, with a critical requirement for scientific data. Besides, flavor innovation is a developing powerful field, which keeps on expanding its applications from its foundations in food and beverage to incorporate categories as different as personal care products and household products. But, consumer perceptions of naturalness and perceived healthiness influence their product choices and increase the preference for products labeled as natural. Consequently, keeping in mind the negative health perceptions brought about by many factors, including artificial flavors, ingredient unfamiliarity, advancements in processing, and so on, researchers all over the world have recently focused on flavors from natural resources with much success.

The idea of compiling the book *Natural Flavors, Fragrances, and Perfumes: Chemistry, Production, and Sensory Approach* took seed in mid-2018. The urge to foster such a book originated from the perspective of growing in this area, giving special emphasis on fundamental and applied research findings, and more persuasively in a food manufacturing business, to enhance product quality, and extend the shelf life of the products and improve process efficiency. This book is a systematic, complete, sequential compilation of information about the engineering of flavors and flavor products in the area of its application to provide insights to help guide development and in commercial strategy. Indeed, the goal of this book is to bring together some of the core knowledge in the field to provide a practical and wide-ranging guide for molecular biologists, new product researchers, and scientists involved in the commercial development of natural flavors and flavor-related compounds, and their use in applications as varied as pharmaceuticals, nutraceuticals, or fragrances.

This book is divided into 5 parts, including 13 chapters encompassing from natural product diversity and synthesis of flavor compounds to all different kinds of applications in the food and fragrance industries. It begins with a part devoted to the natural product diversity and its biomolecular aspects. Chapters 2 and 3 present advances in the strategies of metabolic pathway engineering and the biogenesis of plant-derived aroma compounds. The next part of the book (Chapters 4–6) focuses on flavor technology and the different kinds of flavor-delivering systems, signatures, and biochemistry of beverages. Chapters 7 and 8 are oriented to specific examples and applications of seasonings, herbs, spices, and resinoids to get a better understanding of their chemistry and uses. The last chapters are focused on regulatory aspects, sensory science and its challenges, advances in *in silico* approaches, and sensory perceptions.

A book like this is impossible without the support and effort of the contributors who have taken the time to submit their manuscripts during the pandemic period. Our editors wish to thank the authors who have generously contributed material to this book. They are experts in their fields and have provided valuable information and insights into the naturally derived flavors. The contributors include Valsaraj T.V.; Akhila Nair; Józef T. Haponiuk; Sumith K.; Anjali Anil; Neha Naijo Areekal; Sneha George; Irene Mary Peter; Roshin Thankachan, Maurício Bonatto Machado de Castilhos; Ana Paula Garrido de Queiroga; Lia Lúcia Sabino; Jorge Roberto dos Santos Júnior; Jorge Alejandro Santiago-Urbina; Hipócrates Nolasco-Cancino; Francisco Ruíz-Terán; Vanildo Luiz Del Bianchi; Daniel Jan Strub; Maria Strub; Nicolas Baldovini; Vaishak Ramachandran; Anirudh Jayakumar; Constantina Tzia; Virginia Giannou; Tryfon Kekes; Charikleia Chranioti; Maria Katsouli; Nirosha Pulikkal; and Dhanesh Haridas.

25 August 2022

*Sreeraj Gopi*
*Nimisha P. Sukumaran*
*Joby Jacob*
*Sabu Thomas*
*Kochi, India*

# Part I

# Biodiversity

# 1

# Natural Product Diversity and its Biomolecular Aspects in Flavors and Fragrances

*Themanamveedu Valsaraj, Akhila Nair, and Joby Jacob*

*Aurea Biolabs (P) Ltd., R&D Centre, Kolenchery, Cochin, Kerala, 682311, India*

## 1.1 Introduction

In pharmaceutical, food, cosmetic, and nutraceutical industries, flavors and fragrances play a vital role. The natural selection method or processes facilitate unique as well as wide chemical diversity with optimal interactions with other biological macromolecules. Moreover, since a millennium, it is observed that the introduction of continental and conventional selective breeding efforts has resulted in land race, elite cultivars that could not only adapt to globally diverse habitats but also ensure vivid quality and productivity in flavors and fragrances worldwide. However, unraveling the genomic basis of these vivid adaptations remains indecipherable. For example, the world's oldest and most popular caffeine-containing beverage, the tea, comes along with immense medicinal, economic, and cultural virtues. Constant research will definitely pave way for a diverse metabolic, functional, and genomic refinement for the evaluation of their biosynthetic pathways [1]. Although it is well recognized that the differential accumulation of the three major characteristic constituents in tea tree leaves largely determines the quality of tea, little genomic information is currently available. Sequencing of the tea tree genome would facilitate to uncover the molecular mechanisms underlying secondary metabolic biosynthesis with the promise to improve breeding efficiency and thus develop better tea cultivars with even higher quality. The development of tea clones with more desirable quality traits and enhanced stress resistance becomes a necessity. Strategizing such crop improvement procedures based on miRNAs requires a detailed understanding of the miRNA–mRNA modules associated with stress tolerance and quality in tea plants [2].

*Natural Flavours, Fragrances, and Perfumes: Chemistry, Production, and Sensory Approach,*
First Edition. Edited by Sreeraj Gopi, Nimisha Pulikkal Sukumaran, Joby Jacob, and Sabu Thomas.

Biosynthesis of aroma compounds involves metabolic pathways in which the main precursors are fatty acids and amino acids, and the main products are aldehydes, alcohols, and esters. Some enzymes are crucial in the production of volatile compounds, such as lipoxygenase, alcohol dehydrogenase, and alcohol acyltransferase. Composition and concentration of volatiles in apples may be altered by pre- and postharvest factors that cause a decline in apple flavor [3]. Among the volatile aroma compounds produced by ripe apples, esters account for the majority. For example, among the volatile aroma compounds of Golden Delicious and Starking Delicious, esters account for 80% [4]. This chapter discusses the genetic resources and plant breeding, agricultural diversity, conservation of agrobiodiversity, and the economically useful natural products used as flavors and fragrances.

## 1.2 Genetic Resources and Plant Breeding

From time immemorial, the breeding and domestication of plant varieties and/or species for flavor, aroma, and other characteristics have been a constant and ongoing process. Novel heterogeneity in concentration and combination of secondary metabolites has been a constant source to develop new varieties of flavors and fragrances. These variations in the composition of secondary metabolites are affected by human preferences and domestication in flavors and aromas [5]. Moreover, the need for a higher nutrition crop or fruit variety in terms of sustainable agriculture has put into limelight the genomic breeding approaches inclusive of marker-assisted selection, backcrossing, haplotype breeding, and genomic prediction methods in synergy with artificial intelligence and machine learning to increase the speed of these breeding approaches. Figure 1.1 depicts an example of the use of an integrated framework of genomic resources [7].

## 1.3 Agricultural Diversification

Globally, in an agricultural system, aromatic plants are those with aromatic compounds. These aromatic plants synthesize secondary metabolites to produce essential oils, which provide relief from biotic and environmental stress. In addition, these essential oils are used in diverse applications like flavors, perfumes, and fragrance, which will provide economic returns to farmers and manufacturing industries. The increasing interest of research scholars worldwide encourages agricultural activities like proper land utilization as well as focuses on economic returns for aromatic crops. The ecological applications of these aromatic plants in agricultural systems lie in soil erosion control, carbon sequestration, phytoremediation, utilization of low-quality water, pest and disease management, and augmentation of soil properties [8]. The sensory evaluation or validation of spices depends on dominant attributes like color, aroma, and pungency, which is heavily influenced by the varieties, primary processing cultivation, and the processed products.

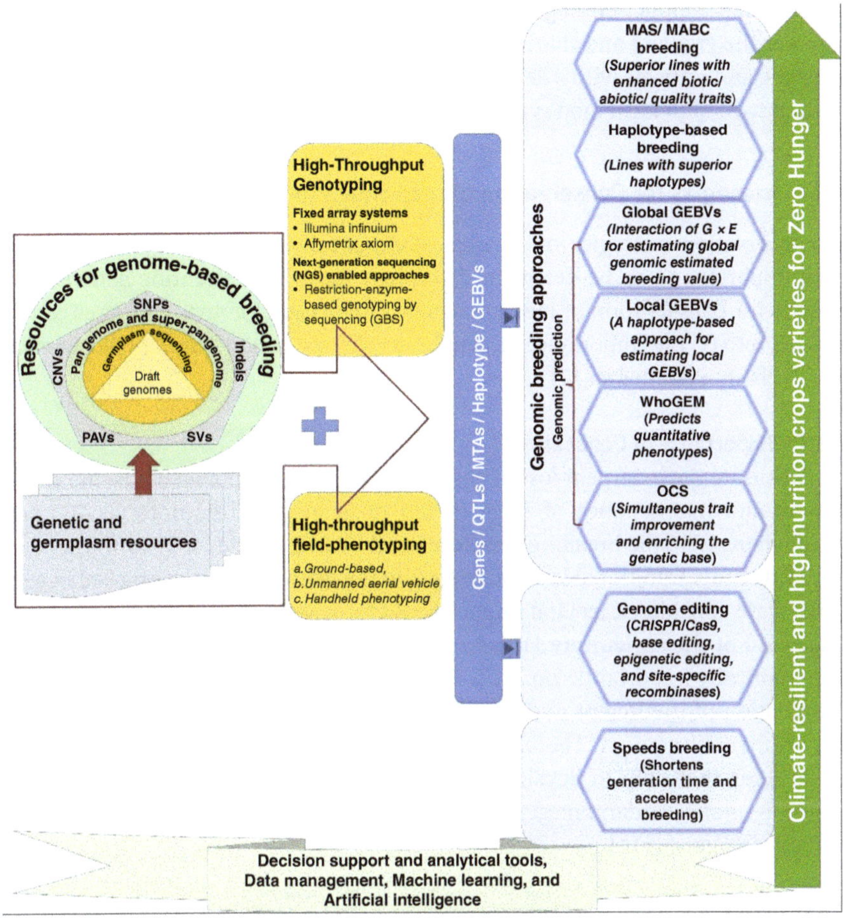


**Figure 1.1** A unified framework of using genomic resources for genomic breeding to tailor climate resilient and high nutrition crops. Source: Adapted from Ashry et al. [6].

## 1.4 Conservation of Agrobiodiversity

The natural product diversity witness difficulties at economic level due to denial of traditional collective seed ownership, make people helpless to grow, harvest, and channel sufficient surplus food. There are many internationally acclaimed reciprocated responses that work in favor of the intellectual property law of farmers. Furthermore, the United States of America have designed vivid grassroot agricultural and biodiverse conservation projects to regulate the open pollinated seeds within fraternities of similar interest. This project involves exploration of the functions of pollinated seeds and focuses on various other strategies for agricultural biodiversity conservation. There are research projects that collectively disseminate and document open-pollinated seed around Appalachian Mountains and Ozark highlands of southeastern United States. The research methods involve an anthropology team who can conduct ethnographic interviews and make participant observation that cover the growing and sharing of seed varieties with local farmers, seed savers, gardeners, and activists with a definite aim of constructing more integrated, sustainable, and sovereign local food systems.

### 1.4.1 Strategies for Conservation of Medicinal Plants

The conservation strategies of near-to-extinct species of medicinal and food or aroma importance can be determined through social and scientific actions. The strategies for conservation of medicinal plants to be used for vivid purposes can be classified into (i) the importance of genebanks, (ii) molecular-based phylogenetics, and (iii) chemosystematics [9].

#### 1.4.1.1 Importance of Genebanks

To compensate the emergent loss of genetic diversity in medicinal crops, the establishment and maintenance of large *ex situ* plant genetic resources (PGRs) was started where systemic breeding was developed by using genetically uniform cultivars to substitute traditional land races around the world. The seeds stocked in genebanks were considered as vital due to the fact it gave an insight of the historical background of the agriculture [10]. To illustrate, *Elettaria cardamomum*, which is an economically important crop, faces limitation in its genomic analysis because of the limitation of inefficient nucleic acid extraction due to its high polysaccharide and polyphenolic content. Therefore, genebanks provide an extraction protocol for nucleic acids that help to develop genetic markers for cardamom, perform gene expression, clone cardamom genes, analyze small RNAs, and clone cardamom-infecting viral genes [11].

### 1.4.2 Molecule-Based Phylogenetics

Cryptic diversity is often not recognized due to the incapability of recognizing the distinguishable morphological traits and because of inability to quantify the chemical communication systems. For certain plants or animals, species-level taxonomy is obstructed because of its distortion upon preservation and morphological plasticity. The morphological characteristics to differentiate likely related species using these methods become difficult, but recent advances in morphological characteristic-based studies imply several differences in the phenotypes. The revisions in taxonomic as well as molecular-based phylogenetic studies have proved to be promising to garner information related to large species groups with different genera [12].

### 1.4.3 Metabolomic-Based Phylogeny or Chemosystematics

The initial part of the last century witnessed the evolution of metabolomics-based phylogeny or chemosystematics that eventually gained its popularity in the 1970s [13]. However, these studies centered on the intrafamily classification at the species level as well as the measurement of particular components of single biochemical families, especially alkaloids, in accordance with the technologies of that period. For example, the chemical systematics of the family Rutaceae and the order Rutales received immense research attention. The authenticity of chemosystematic classification was proved by comparison with the phylogeny determined by molecular polymorphism analyses.

## 1.5 Economically Important Natural Products Used in Flavors and Fragrances

The economically most important plants serving the purpose of flavors and fragrances are cardamom, cinnamon, cocoa, fenugreek, marigold, nutmeg, vanilla, paprika, rosemary, davana oil, olibanum carterii/serrata, lavender, vetiver, and so on (Tables 1.1 and 1.2)

### 1.5.1 Flavors

#### 1.5.1.1 Cardamom

Cardamom (*Elettaria cardamomum*) has been associated with numerous pharmacological properties [14]. This aromatic plant is one of the most expensive species in the world. India provides the most favorable warm humid climate with loamy-rich organic soil, well-distributed rainfall, and unique cultivation and processing methods that result in unique aroma, flavor, size, and green color. It has been used in culinary, confectionary, sweets, and medicines since time immemorial [33].

The diverse metabolites impose restriction to provide a standard method for RNA isolation for the available plants. The polysaccharide and polyphenol content of cardamom tissues obstruct the RNA extraction procedure. However, the combination of commercial kits as well as conventional cetyl trimethylammonium bromide (CTAB) method yields RNA with higher yield, good purity, and good integrity. The total RNA isolated from this approach was found compatible for small RNA analysis and transcriptome through next-generation sequencing platforms [34].

#### 1.5.1.2 Cinnamon

Globally, cinnamon is a valuable source as an antioxidant compound [35]. Cinnamic aldehyde is widely used as a flavoring agent in foods and dentifrices [36] due to their antibacterial effects of cinnamon essential oil, cinnamon extracts, and pure compounds against different oral pathogens [37]. It is known for its aroma and essence in culinary, perfumes, and medicinal products. The active constituents of cinnamon, which are found in its essential oils, are cinnamaldehyde and trans-cinnamaldehyde that constitute its immense biological activities and fragrances. Cinnamon bark contains catechins and procyanidins. The components of procyanidins include both procyanidin A-type and B-type linkages. These procyanidins extracted from cinnamon and berries also possess antioxidant activities [36].

**Table 1.1** Economically important plants and their ingredients.

| Plant source | Active ingredients | Economic importance |
|---|---|---|
| Cardamom [14] | 1,8-cineole, α-pinene, α-terpineol, linalool, linalyl acetate and nerolidol, and the ester constituent α-terpinyl acetate | Antioxidant, antimicrobial, antibacterial, anti-inflammatory |
| Cinnamon [15] | Cinnamaldehyde, trans-cinnamic acid, *o*-methoxycinnamaldehyde, eugenol, and monoterpenoids | Antioxidant, anti-inflammatory, antidiabetic, antimicrobial, anticancer, lipid-lowering, and cardiovascular-disease-lowering compound |
| Cocoa [16] | Theobromine, serotonin, anandamide, phenylethylamine | Antioxidation, anticancer, anti-microbial, anti-inflammation, anti-diabetes, cardioprotective, physical improvement, antiphotoaging, antidepression, and blood glucose regulation |
| Fenugreek [17] | Carbohydrates, proteins, lipids, alkaloids, flavonoids, fibers, saponins, steroidal saponins | Hypocholesterolemia, lactation aid, antibacterial, gastric stimulant, for anorexia, antidiabetic agent, galactagogue, hepatoprotective effect, and anticancer |
| Marigold [18] | Carotenoids, flavonoids, saponins, sterols, phenolic acids | Anti-inflammatory, anti-edematous activity, antioxidant, antibacterial and antifungal activity, immunostimulant activity, genotoxic and antigen-toxic activity, spasmogenic and spasmolytic activity, hepatoprotective activity |
| Nutmeg [19] | Terpenes (α-pinene, *p*-cymene, sabinene, camphene, myrcene, and γ-terpinene), terpene derivatives (terpinol, geraniol, and linalool), and phenylpropanes | Antioxidant activity, immuno-modulatory and radioprotective activities, anti-carcinogenic and hepatoprotective activity, anti-inflammatory activity, antimicrobial activity |
| Paprika [20] | Carotenoids | Antioxidant activity, antifungal activity<br>Anti-inflammatory properties |
| Rosemary [21] | Triterpenes, phenolic diterpenes rosmarinic acid, carnosic acid, rosmanol, carnosol, ursolic acid, betulinic acid | Anti-depressant, antispasmodic antimicrobial, anti-inflammatory, anti-tumorigenic |
| Black pepper [22] | Piperine, alkaloids, amides, essential volatile oils | Anti-inflammatory, antioxidant improves blood sugar control, antithyroid activity |
| Oregano [23] | Carvacrol, β-fenchyl alcohol, thymol, and γ-terpinene | Antioxidant, antimicrobial |
| Annatto [24] | Carotenoids, Terpenoids, and terpenes | Anti-inflammatory, antibacterial, anti-endemic, and antifungal activity |
| Turmeric [25] | Curcumin | Anti-inflammatory, antioxidant, antimicrobial |

**Table 1.2** Important aromatic plants and their bioactives.

| Fragrance source | Active ingredients | Properties |
|---|---|---|
| Davana oil [26] | Davanone, bicyclogermacrene, linalool, caryophyllene oxide, phytol | Antioxidant, antibacterial |
| Olibanum Carterii/ Serrata [27] | Boswellic acid | Anti-inflammatory, antifungal activity, antibacterial activity, antioxidant activity, anti-arthritic activity |
| Lavender [28] | Linalool, linalyl acetate, 1,8-cineole, β-ocimene terpinen-4-ol, and camphor | Antifungal activity, antibacterial activity, ascaricidal activity |
| Vetiver [29] | cedr-8-en-13-ol, β-guaiene Cycloisolongifolene | Antifungal activity, antibacterial activity, antioxidant activity |
| Red Sandalwood [30] | Carbohydrates, flavonoids, terpenoids, phenolic compounds, alkaloids, saponins, tannins, glycosides | Antibacterial activity, hepatoprotective, hypolipidemic activity, angiogenesis, and wound-healing activity |
| Lemongrass [31] | Myrcene, limonene, citral, geraniol, citronellol, geranyl acetate, neral, nerol | Anticancer and chemopreventive activity, anti-inflammatory, antifungal activity, antibacterial activity, antioxidant, allelopathic, insect repellent, anthelmintic activities |
| Elemi [32] | Elemol, limonene, elemicin, coumarins, furans, phenols | Hepatoprotective activities, antifungal activity, antibacterial activity, antioxidant, analgesic Antidiabetic |
| Myrrh [6] | Myrrhol, Myrrhin | Antihealing, antiseptic, antimicrobial |

#### 1.5.1.3 Cocoa

Cocoa seeds are a valuable food as well as a wellness product. These fruits are the main source of chocolate that is relished all over the world. The major producer of cocoa is Brazil [38]. The protein fractions of cocoa have an influence on the sensory and bioactive potential of cocoa products. The possible modifications during ripening, maturation, and post-harvest processing (drying, roasting, fermentation, and alkalization); composition of the phenolic compounds; and modifications in manufacturing processes are well documented [39, 40].

The phenolic compounds of cocoa contain antiradical and antioxidant properties with different biological properties like protection against cardiovascular diseases. Clovamide, a minor component of cocoa, is effective against oxidative stress induced in the rat cardiomyoblast cell line as compared to rosmarinic acid, other bio-isosteric forms, and epicatechin. All these three components were analyzed with DNA fragmentation, annexin V positivity, and caspase release as well as activation and were found to be effective to inhibit the production of reactive oxygen species and apoptosis [40].

#### 1.5.1.4 Fenugreek

*Trigonella foenum-graecum* (fenugreek), belonging to the family Fabaceae, is a legume cultivated as a semiarid crop in India, Canada, Northern Africa, and Mediterranean region. This spice is known worldwide to enrich the sensory quality of foods and has high nutraceutical value. The presence of alkaloids, steroid saponin, and fiber in *Trigonella* seeds shows antidiabetic activity. Other bioactive compounds like trigonelline, orientin, isoorientin, isovitexin, and vitexin were quantified by high-performance liquid chromatography (HPLC). Other compounds identified by ultrahigh-performance liquid chromatography-hybrid electrospray triple quadrupole linear ion trap mass spectrometry were trigonelline, pinitol, isoorientin, sarsapogenin, and isovitexin [41]. The bioassay-guided isolation revealed 1 new pterocarpan and 12 known pterocarpans. These pterocarpans are important in terms of nutritional value as functional foods, foods, or antioxidants [18]. Moreover, fenugreek gum, a natural galactomannan, originates from the endosperm of *Trigonella foenum-graecum* seeds. This gum is composed of (1 → 4)-β-D-mannose (Man) backbone attached to a single α-D-galactose (Gal) group at the O-6 position with a Gal/Man ratio of 1:1 or in few cases of 1:2. It has been over a decade that fenugreek gums are used in food and pharmaceutical industry as a stabilizing and thickening agent [13, 42].

#### 1.5.1.5 Marigold

*Tagetes* (marigold) is mainly found in America and is also cultivated in Europe, Asia, and Africa. Many species of this plant, including *T. erecta*, *T. patula*, *T. minuta*, and *T. tenuifolia*, are researched as medicinal plants. The major bioactive components of marigold are carotenoids, which are lipophilic pigments and well recognized as health-promoting agents. Although the native profile of carotenoids is not much studied because of the difficult analysis of carotenoid esters, it is observed that the hydroxyl carotenoids are found in both esterified and free form in numerous plant matrices. These carotenoids are used as supplements or marigold with no saponification process. The marigold petals contain lutein esters as major compounds. 18 caratenoids, 20 monoesters, 30 diesters (zeaxanthin, auroxanthin, violaxanthin, zeinoxanthin, and β-cryptoxanthin) were identified [43]. Various parts of the plants of *Tagetes* species are used to treat dental, stomach, and digestive disorders as well as anxiety and depression. These plants are also used for their fungicidal, bactericidal, insecticidal, anti-inflammatory, antioxidant, and enzyme inhibitory properties. They likewise find applications as a food additive and for their antimicrobial activities [43].

#### 1.5.1.6 Nutmeg

Nutmeg is a plant found in tropical regions and its seeds are known for the unique flavor, nutritive value, and medicinal properties [44]. Occurrences of misuse have been accounted for nutmeg, including a family zest produced by crushing the seeds of *Myristica fragrans*, inferable from its stimulating properties [45]. A review of tangible metabolite dispersion in nutmeg has given a most far-reaching guide of its tactile metabolites. The key flavoring agent myristicin (40% in organic products) and 53 volatiles were differentiated in various classes, namely fragrant ethers, monoterpenes, and sesquiterpenes. In any case, monoterpene hydrocarbons are considered significant unstable structures in seeds [44].

#### 1.5.1.7 Vanilla

Beginning from the locale of Mexico, *Vanilla planifolia* is a blooming climbing orchid that is internationally appreciated for the "vanilla" flavor created from its cases. The plant is distributed everywhere, particularly in tropical countries, for example, Madagascar, Uganda, Papua New Guinea, Indonesia, India, and islands like Comoros, Mayotte, Tahiti, and La Réunion. In addition to producing vanillin, the significant compound of vanilla concentrate, vanilla cases are the main wellspring of the more mind-boggling vanilla flavor. The photosynthesis process of *V. planifolia* also known as "Crassulacean Acid Metabolism (CAM) plant" involves the intake of carbon dioxide during the evening and its eventual stockpiling in cell vacuoles as malate. The very next day, during daytime, malate is secreted from the vacuoles, and carbon dioxide is produced by malate decarboxylation that enters the Calvin cycle and is utilized as a substrate for Rubisco to produce sugars and different carbohydrates for the plant. Albeit many investigations have been conducted on the natural chemistry of vanilla beans and the vanillin biosynthetic pathway, very little work has been performed on vanilla leaf metabolites [46]. Sun et al. detailed the presence of *p*-ethoxymethylphenol, *p*-butoxymethylphenol, vanillin, *p*-hydroxy-2-methoxycinnamaldehyde, and 3,4-dihydroxyphenylacetic compounds in the ethanol concentrate of vanilla leaves as well as stems. Tokoro et al. (1990) detailed the presence of bis[4-(*b*-D-glucopyranosyloxy)-benzyl]-2-isopropyltartrate (glucoside A) and bis[4-(*b*-D-glucopyranosyloxy)-benzyl]-2-(2-butyl)-tartrate (glucoside B) in vanilla leaves and stems. [47].

During the restoring process of Hainan vanilla beans, the key vanilla flavors, vanillin antecedents, and principal catalysts are removed. During handling, vanillin content increased, while glucovanillin content decreased, and vanillic content is found in beans; however, this content is decreased in drying beans. Both p-hydroxybenzaldehyde and p-hydroxybenzoic compounds show the highest content in restored beans. The ferulic compound is fundamentally produced in dry beans and is decreased in restored beans. The content of the *p*-coumaric compound is increased during the restoration process. During the relieving stage Vanillyl liquor in drying beans (0.22%) is subjected to hydrolysis of glucoside, that then changes. Besides, the enzymatic action of β-glucosidase is not observed in whitened and perspiring beans. But after drying, peroxidase activity reduces by 94% during relieving in restored beans. Polyphenol oxidase activity is low in early stages, while cellulose activity in handled beans is higher than in green beans, apart from restored beans. This study unfolds the biosynthesis pathway of vanillin [48].

Vanillin is the main flavor compound in the vanilla unit. *V. planifolia* vanillin synthase (VpVAN) catalyzes the transformation of ferulic compound and ferulic glucoside into vanillin and vanillin glucoside, respectively. Desorption electrospray ionization-mass spectrometry imaging (DESI-MSI) of vanilla case areas shows that vanillin glucoside is especially restricted inside the mesocarp and placental laminae, while vanillin is especially limited inside the mesocarp. Immunoblotting, a neutralizer intended for visualizing the C-terminal arrangement of VpVAN. Through this neutralizer it is seen that VpVAN is a developed structure (25 kDa) and contingent upon the tissue and seclusion system. In addition, limited quantities of the youthful natural structure (40 kDa) and putative oligomers (50, 75, and 100 kDa) are also seen. The VpVAN protein is confined inside chloroplasts and detached chloroplasts named phenyloplasts, as determined during the course of unit improvement. Detached chloroplasts were shown to convert [14C] phenylalanine and [14C] cinnamic compounds into [14C] vanillin glucoside, demonstrating that the whole vanillin biosynthetically changes phenylalanine completely to vanillin glucoside, which is available in the chloroplast [47].

#### 1.5.1.8 Paprika

Red pepper and its dietary items contain varieties of carotenoids, which might add to the carotenoids, for example, of human blood and tissues. The yellow-orange shades of stew pepper natural products are essentially because of the accumulation of α-and β-carotene, zeaxanthin, lutein, and β-cryptoxanthin. Carotenoids, for example, capsanthin, capsorubin, and capsanthin-5,6-epoxide present red tones. [49]. Capsaicin is the essential bioactive substance in red chili peppers, which delivers the sharp flavor. Capsicum natural products are famous overall and are utilized in cuisines around the globe. With its various varieties, structures, and uses, the capsicum spice adds to the whole range of tangible experience variety as finely ground paprika powder or concentrate in sausages, goulash, cheddar, and snacks; both sharpness and variety as the numerous forms of chilies utilized in Mexican, African, Indian, and southeast Asian foods; variety, smell, and gentle sharpness as the new green chilies utilized in a considerable quantity in developing nations; and appearance, variety, fragrance, and surface as a new natural product in servings of mixed greens and as a cured and canned item [50]. Sharpness as a particular gustatory insight, alongside different characteristics are accordingly seen for carotenoids, volatiles, as well as total capsaicinoids [51].

Capsicum species produce organic products that incorporate and collect carotenoid shades, which are responsible for the yellow, orange, and red tones of the natural products. The methods for measuring the content of carotenoids, chlorophyll, polyphenols, tannins, and flavonoids in red paprika (RP) were developed in Korea, which include light treatment of water (W) and ethanolic (Et) extracts under high tension sodium (HPS) and light emitting plasma (LEP) lights (RPControl, RPHPS, and RPLEP). The results of this study showed that of all the compounds, chlorophyll and carotenes were the most noteworthy in RPHPS ($10.50 \pm 1.02$ and $33.90 \pm 3.26\,\mu g\,g^{-1}$ dry weight [DW]). The paprika ethanolic extracts show lower values in their bioactivity than the water extracts. The cytotoxicity capacities of all polyphenols in paprika are accounted together. The paprika tests can be utilized as an aid to cell reinforcements [52].

#### 1.5.1.9 Rosemary

Rosemary oil (RO) is famous in the Mediterranean region as a culinary added substance, which protects sensitive organs like the liver, cerebrum, and heart. Rosemary (*Rosmarinus officinalis*) are broadly utilized in the food, nutraceutical, and restorative regions. Their major bioactive show antioxidant, anti-inflammatory, antimicrobial, antitumorigenic, and chemopreventive properties [53]. The active molecules obtained from rosemary oil are 1,8-cineol (15–20%), camphor (15–25%), borneol (16–20%), α-pinene (25%), and acetic acid derivative (up to 7%); moreover, the oil contains minor amounts of β-pinene, linalool, camphene, sabinene, myrcene, α-phellandrene, α-terpinene, limonene, *p*-cymene, terpinolene, thujene, terpinen-4-ol, α-terpineol, caryophyllene, methyl chavicol, and thymol. The distillation fraction generally contains α-thujene, α-pinene, camphene, β-pinene, and 1,8-cineol, while camphor and bornyl acetic acid derivatives comprising the major part were extracted later after refining. Rosemary leaves and flowering tops have numerous culinary purposes: mutton preparations, lamb roast, marinades, baked fish, bouquet garni, rice, soups, mixed greens, sporadic use with egg preparations, dumplings, apples, summer wine cups, and fruits cordials and use in vinegar and oil [54].

The new and dried leaves of rosemary are used in Mediterranean foods as they have a harsh astringent taste and fragrance. Dried and powdered leaves are added to cooked meat, fish, poultry soup, stews, sauces, dressings, jelly, and sticks. The leaves are additionally utilized in pork preparations. At the time of consumption, the leaves emit a unique mustard smell. Rosemary is the best spice with a wide scope of purposes in food handling. In Europe and the USA, rosemary is economically accessible for use as a cancer prevention agent; however, it is not actually recorded as a normal additive or as an antioxidant, particularly in Europe [55]. Rosemary has broad applications in subsiding warmed flavor [56]. The bioactives in rosemary are carnosic acid, 12-methoxy carnosic acid, and carnosol as well as the diterpenes, like, epirosmarinol, isorosmanol, rosmaridiphenol, rosmariquinone, and rosmarinic acid [57].

The antioxidant properties of rosemary are attributed to its capacity to savage superoxide radicals, lipid antioxidation, metal chelating, and so forth. Essential oils and extracts of rosemary can be utilized to stabilize fats, oils, and fat-containing food varieties, for example, margarine, against oxidation and rancidity and to settle matured meat items [58, 59]. A helpful audit by Yanishlieva-Maslarova and Heinonen examines the antioxidant properties of rosemary and sage, covering their chemistry, including properties, extraction, and application [55]. Commercial antioxidants, deodorized liquid of rosemary, either as mono-home grown or as polyherbal (rosemary, thyme, sage, and oreganum) details, for example, "Herbor 025" and "spice Cocktail," are found [60]. Rosemary oil likewise has applications in preservation of raw meat. The expansion of rosemary oleoresin to ground chicken affected crude meat appearance during storage and on the kind of the cooked meat [61]. The antioxidative property of rosemary oleoresin is well known [21]. However, the addition of rosemary oleoresin to ground chicken affected crude meat appearance during storage and even cooked meat [61].

### 1.5.2 Fragrances

There are countless fragrances which are used in numerous foods, savories, cosmetics, and perfumes, including davana, olibanaum, lavender, and vetiver, as depicted in Table 1.2.

#### 1.5.2.1 Davana Oil

Davana (*Artemisia pallens*) is an annual branched herb. The plant grows knee-high and looks much like a small fern. It is most fragrant at maturity, when it produces tiny flowers that are rich in essential oils. India is the sole source of *A. pallens*. The crop can be found in southern India, namely Tamil Nadu and Karnataka, with the outskirts of Bangalore producing the largest volumes [62]. Davana's reddish brown essential oil boasts a potent, exquisite scent. Its characteristic is shockingly reminiscent of old spirits like cognac, with sharp, dry fruit notes, and a full body with thick honey herbaceous notes. Davanone, a sesquiterpene ketone, is the major component and quality driving factor of davana oil [63]. High-quality material usually boasts of a davanone content at or above 50%. Davana is first and foremost a flavor ingredient. Fine fragrance perfumer Ilias Ermenidis (Firmenich) began using davana with Givenchy Pour Homme and Givenchy Pour Homme Blue Label. Since that time, it became one of his favorite ingredients in men's fragrances. In rose formulations, davana is used to round the fragrance and to tame the minty metallic edge of geranium; it is used more for its low-dosage effect than for its characteristic itself [64].

#### 1.5.2.2 Olibanum Carterii/Serrata

Frankincense, or olibanum, is a characteristic oleo-gum-tar made out of around 5–9% medicinal balm, 65–85% liquor solvent resins, and the leftover water-dissolvable gums. Frankincense is harvested as tears or as the size of a pea to that of a walnut. It is light yellow or pale golden in variety. The scent of frankincense is depicted as new balsamic, dry and resinous, marginally green smell with a natural product top note and a diffusive note of green, unripe apple strip. Frankincense is utilized by perfumers as an outright (by liquor extraction), oil, or resinoid (by hydrocarbon extraction) [65]. It is utilized in Oriental bases, ambers, "powder" scents, flower aromas, citrus colognes, flavor mixes, violet fragrances, male scents, and so on. It mixes well with flavor oils, labdanum, mimosa, neroli, muguet bases, woody notes, and other balsamic notes. The main bioactives of frankincense is ß-boswellic acid, one of the super dynamic parts of frankincense. These are a portion of the synthetic mixtures present in frankincense: · Acid sap (56%), dissolvable in liquor and having the recipe $C_{20}H_{32}O_4$ · Gum (like Gum Arabic) 30–36% · 3-Acetyl-beta-boswellic acid (*Boswellia sacra*) · Alpha-boswellic acid (*Boswellia sacra*) · 4-O-methyl-glucuronic acid (*Boswellia sacra*) [66]·

Incensole acetic acid derivative · Phellandrene. Olibanum is portrayed by a balsamic-fiery, somewhat lemon, and regular aroma of incense, with a marginally conifer-like feeling. It is utilized in aroma as well as beauty care products and pharmaceutical industry [67].

#### 1.5.2.3 Lavender

Lavender (*Lavandula angustifolia*) is a spice found in northern Africa and the Mediterranean mountains, and its extract is frequently used in medicinal ointments. It is likewise developed for the production creation of its natural oil, which comes from the refining of the bloom spikes of specific lavender species. The lavender oil has cosmetic purposes, and its few therapeutic uses are additionally accepted. The biological advantages of utilizing lavender are utilized to treat hair loss, wound, anxiety, and fungal infections. Lavender is not utilized to treat high blood pressure, menstrual pain, eczema, nausea, and other different circumstances. This spice is not approved by the Food and Drug Administration (FDA) and should not be assumed as a replacement of endorsed and recommended medications [67]. This class Lavandula is generally rich in phenolic constituents, with 19 flavones and 8 anthocyanins (Harbourne and Williams, 2002). The family normally contains different glycosides of hypolaetin and scutellarein. Triterpenoids incorporate ursolic acid [68]. The spice lavender is widely used in skin products, hair shampoos and as aroma agents. It can be bought from the pharmacy counter [69]. A few variants of lavenders are utilized to add flavors to prepared products and food varieties. This spice likewise contains numerous therapeutic properties. Lavender smells are fit for modifying depressive states and may show that the utilization of scents is useful in diminishing uneasiness in dental patients [70]

#### 1.5.2.4 Vetiver

*Vetiveria zizanioides* is a kind of spiky sinewy grass with rhizome-like roots of 2 m or more in length. Vetiver oil is among the best, thick light brown, and most positive perfumery fixings. Generally appreciated for its trademark refined and sweet woody amber smell, it shows up in over 33% of all aromas. It has a rich green-woody natural and nut-like aroma. Less fortunate grades produced in China and Java by nearby ranchers with crude gear are oftentimes more obscure in variety and have smoky back notes. It is an exceptionally valued oil and has broad usage in fine perfumery for its richness and profundity. Vetiver is likewise utilized for the partition of its main alcohol – vetiverol – for an even cleaner note. Vetiverol may in this way be acetylated for the development of vetiveryl acetate, which with its marginally more grounded velvety, fruity, green-woody subtleties is indispensable in the base notes of "haute couture" fragrances [71]. An eleven-step chemical synthesis, with a novel asymmetric organocatalytic Mukaiyama–Michael addition revealed that (+)-2-epi-ziza-6(13)en-3-one is the active smelling principle of vetiver oil. The trademark semi-pheromone-like vetiver odor was obtained by synthetic blend of 2-epi-ziza-6(13)en-3-one utilizing a remarkable imidophosphoimidate (IDPi)-catalyzed Mukaiyama–Michael expansion of a silyl ketene acetal to cyclopent-2-en-1-one.

## 1.6 Conclusion

The industry of flavor and fragrance is highly priced with huge and rapid changes in the technological aspects related to breeding, conservation, and analytical techniques. These sensory aspects are either synthetic or natural. The important natural flavors, including cardamom, cinnamon, cocoa, fenugreek, marigold, nutmeg, vanilla, paprika, and rosemary, are discussed briefly. In addition, davana oil, olibanum carterii, lavender, and vetiver are also described. However, more comprehensive studies need to be conducted to understand the potential of these products and improve their quality and cost in terms of flavor and fragrances. This would definitely boost the savory industries in terms of economic aspects to become the most sought-after ones among pharmaceutics and nutraceutics.

## Acknowledgment

The authors of this chapter take this chance as an esteemed privilege to show their gratitude to the management of Aurea Biolabs (Pvt.) Ltd., Cochin, India, who were an invariable source of support to put pen to paper and encouraged us during the entire course of this chapter. We also grab this golden opportunity to express deep appreciation to our colleagues for their valuable support that accelerated the successful completion of this task.

## Declaration of Interest

The authors profess no declaration of interest in this present chapter.

## References

1 Xia, E.H., Zhang, H.B., Sheng, J. et al. (2017). The tea tree genome provides insights into tea flavor and independent evolution of caffeine biosynthesis. *Mol. Plant* 10 (6): 866–877, ISSN 1674-2052.

2 Krishnatreya, D.B., Agarwala, N., Gill, S.S., and Bandyopadhyay, T. (2021). Understanding the role of miRNAs for improvement of tea quality and stress tolerance. *J. Biotechnol.* 328: 34–46. https://doi.org/10.1016/j.jbiotec.2020.12.019.

3 Miguel, E.D. et al. (2016). Biochemistry of apple aroma: a review. *Food Technol. Biotechnol.* 54 (4): 375–394.

4 Song, J. and Bangerth, F. (1996). The effect of harvest date on aroma compound production from "Golden Delicious" apple fruit and relationship to respiration and ethylene production. *Postharvest Biol. Technol.* 8 (4): 259–269. https://doi.org/10.1016/0925-5214(96)00020-8.

**5** Frydman, A., Liberman, R., Huhman, D.V. et al. (2013). The molecular and enzymatic basis of bitter/non-bitter flavor of citrus fruit: evolution of branch-forming rhamnosyltransferases under domestication. *Plant J.* 73: 166–178. https://doi.org/10.1111/tpj.12030.

**6** El Ashry, E.S.H., Rashed, N., Salama, O.M., and Saleh, A. (2003). Components, therapeutic value and uses of myrrh. *Pharmazie* 58 (3): 163–168.

**7** Thudi, M., Palakurthi, R., Schnable, J.C. et al. (2021). Genomic resources in plant breeding for sustainable agriculture. *J. Plant Physiol.* 257: 153351, ISSN 0176-1617, https://doi.org/10.1016/j.jplph.2020.153351.

**8** Rao, P.V. and Gan, S.H. (2014). Cinnamon: a multifaceted medicinal plant. *Evid. Based Complement Alternat Med.* 2014: 642942. https://doi.org/10.1155/2014/642942.

**9** Breseghello, F. and Coelho, A.S.G. (2013). Traditional and modern plant breeding methods with examples in rice (*Oryza sativa* L.). *J. Agric. Food. Chem.* 61 (35): 8277–8286. https://doi.org/10.1021/jf305531j.

**10** Mascher, M., Schreiber, M., Scholz, U. et al. (2019). Genebank genomics bridges the gap between the conservation of crop diversity and plant breeding. *Nat. Genet.* 51: 1076–1081. https://doi.org/10.1038/s41588-019-0443-6.

**11** Palani, S.N., Elangovan, S., Menon, A. et al. (2019). An efficient nucleic acids extraction protocol for *Elettaria cardamomum. Biocatal. Agri. Biotechnol.* 17: 207–212, ISSN 1878-8181, https://doi.org/10.1016/j.bcab.2018.11.026.

**12** Amor, M.D., Norman, M.D., Roura, A. et al. (2017). Morphological assessment of the *Octopus vulgaris* species complex evaluated in light of molecular-based phylogenetic inferences. *R. Swedish Acad. Sci.* 46 (3): 275–288.

**13** Wu, Z., Cai, Y.S., Yuan, R. et al. (2020). Bioactive pterocarpans from *Trigonella foenum-graecum* L. *Food Chem.* 30 (313): 126092. https://doi.org/10.1016/j.foodchem.2019.126092.

**14** Souissi, M., Azelmat, J., Chaieb, K., and Grenier, D. (2020). Antibacterial and anti-inflammatory activities of cardamom (*Elettaria cardamomum*) extracts: potential therapeutic benefits for periodontal infections. *Anaerobe* 61: 102089. https://doi.org/10.1016/j.anaerobe.2019.102089.

**15** Charles, D.J. (2012). Cinnamon. In: *Antioxidant Properties of Spices, Herbs and Other Sources*, 231–243. https://doi.org/10.1007/978-1-4614-4310-0_19. New York, NY: Springer.

**16** Cova, I., Leta, V., Mariani, C. et al. (2019). Exploring cocoa properties: is theobromine a cognitive modulator? *Psychopharmacology* 236 (2): 561–572. https://doi.org/10.1007/s00213-019-5172-0.

**17** Helambe, S. and Payal, D.R. (2012). Fenugreek (*Trigonella foenum-graecum* L.): an overview. *Int. J. Curr. Pharm. Rev. Res.* 2 (4): 169–187.

**18** Salarbashi, D., Bazeli, J., and Fahmideh-Rad, E. (2019). Fenugreek seed gum: biological properties, chemical modifications, and structural analysis – a review. *Int. J. Biol. Macromol.* 138: 386–393. https://doi.org/10.1016/j.ijbiomac.2019.07.006., PMID: 31276725.

**19** Nelofer, J., Andrabi, K.I., and John, R. (2017). Calendula officinalis-an important medicinal plant with potential biological properties. *Proc. Indian Nat. Sci. Acad.* 83.

**20** Namra, N. (2016). Nutmeg: a review on uses and biological properties. *Int. J. Chem. Biochem. Sci.* 9: 107–110.

**21** Saikumar, B. (2012). Rosemary. In: *Handbook of Herbs and Spices* (ed. K.V. Peter), 452–468.

**22** Ravindran, P.N. and Kallupurackal, J.A. (2012). Black pepper. In: *Handbook of Herbs and Spices* (ed. K.V. Peter), 86–115. https://doi.org/10.1533/9780857095671.86.

**23** Kintzios, S.E. (2012). Oregano. In: *Handbook of Herbs and Spices* (ed. K.V. Peter), 417–436. https://doi.org/10.1533/9780857095688.417.

**24** Islam, S., Rather, L.J., and Mohammad, F. (2016). Phytochemistry, biological activities and potential of annatto in natural colorant production for industrial applications – a review. *J. Adv. Res.* 7 (3): 499–514. https://doi.org/10.1016/j.jare.2015.11.002.

**25** Muriel, M.T., Stangarlin, J.R., Kuhn, O.J. et al. (2017). Biological properties of turmeric. *Sci. Agrar. Paran.* 16 (1): 1–12.

**26** Singh, S., Kurmi, A., Swamy Gowda, M.R., and Singh, M. (2022). Chemical investigation, quality assessment, and antimicrobial activity of davana (*Artemisia pallens* Wall. ex DC) essential oil collected from different locations in India. *J. Essent. Oil Res.* 1–13.

**27** Yasiry, A., Mustafa, A.R., and Kiczorowska, B. (2016). Frankincense-therapeutic properties. *Adv. Hyg. Exp. Med.*/Postepy Higieny i Medycyny Doswiadczalnej 70.

**28** Cavanagh, H.M.A. and Wilkinson, J.M. (2002). Biological activities of lavender essential oil. *Phytother. Res.* 16 (4): 301–308. https://doi.org/10.1002/ptr.1103.

**29** Khusminder Kaur, C., Kaushal, S., and Sandhu, A.K. (2015). Chemical composition and biological properties of *Chrysopogon zizanioides* (L.) Roberty syn. *Vetiveria zizanioides* (L.) Nash-A review. *Indian J. Nat. Prod. Res. (IJNPR)* [Formerly Natural Product Radiance (NPR)] 6 (4): 251–260.

**30** Saradamma, B., Reddyvari, H., Nallanchakravarthula, V., and Vaddi, D.R. (2016). Therapeutic potential of *Pterocarpus santalinus* L.: an update. *Pharmacogn. Rev.* 10 (19): 43.

**31** Haque, A.N.M., Ahsanul, R.R., and Naebe, M. (2018). Lemongrass (Cymbopogon): a review on its structure, properties, applications and recent developments. *Cellulose* 25 (10): 5455–5477.

**32** Mogana, R. and Wiart, C. (2011). *Canarium* L.: a phytochemical and pharmacological review. *J. Pharm. Res.* 4 (8): 2482–2489.

**33** Parthasarathy, V.A. and Prasath, D. (2012). Chapter 8 – Cardamom. In: *Woodhead Publishing Series in Food Science, Technology and Nutrition, Handbook of Herbs and Spices*, 2e (ed. K.V. Peter), 131–170. Woodhead Publishing, ISBN 9780857090393, https://doi.org/10.1533/9780857095671.131.

**34** Nadiya, F., Anjali, N., Gangaprasad, A., and Sabu, K.K. (2015). High-quality RNA extraction from small cardamom tissues rich in polysaccharides and polyphenols. *Anal. Biochem.* 485: 25–27, ISSN 0003-2697, https://doi.org/10.1016/j.ab.2015.05.017.

**35** Almatroodi, S.A., Alsahli, M.A., Almatroudi, A. et al. (2020). Cinnamon and its active compounds: a potential candidate in disease and tumour management through modulating various genes activity. *Gene Rep.* 21: 100966, ISSN 2452-0144, https://doi.org/10.1016/j.genrep.2020.100966.

**36** Isaac-Renton, M., Li, M.K., and Parsons, L.M. (2015). Cinnamon spice and everything not nice: many features of intraoral allergy to cinnamic aldehyde. *Dermatitis* 26 (3): 116–121. doi: 10.1097/DER.0000000000000112.

**37** Yanakiev, S. (2020). Effects of cinnamon (*Cinnamomum* spp.) in dentistry: a review. *Molecules* 25 (18): 4184. https://doi.org/10.3390/molecules25184184. PMID: 32932678.

**38** De Araujo, Q.R., Gattward, J.N., Almoosawi, S. et al. (2016). Cocoa and human health: from head to foot--a review. *Crit. Rev. Food Sci. Nutr.* 56 (1): 1–12. https://doi.org/10.1080/10408398.2012.657921.

**39** Rawel, H.M., Huschek, G., Sagu, S.T., and Homann, T. (2019). Cocoa bean proteins-characterization, changes and modifications due to ripening and post-harvest processing. *Nutrients* 11 (2): 428. https://doi.org/10.3390/nu11020428. . PMID: 30791360.

**40** Zamperone, A., Pietronave, S., Colangelo, D. et al. (2014). Protective effects of clovamide against $H_2O_2$-induced stress in rat cardiomyoblasts H9c2 cell line. *Food Funct.* 5 (10): 2542–2551. https://doi.org/10.1039/c4fo00195h.

**41** Dugo, L., Tripodo, G., Santi, L., and Fanali, C. (2018). Cocoa polyphenols: chemistry, bioavailability and effects on cardiovascular performance. *Curr. Med. Chem.* 25 (37): 4903–4917. https://doi.org/10.2174/0929867323666160919094339.

**42** Zhang, Z., Wang, H., Chen, T. et al. (2019). Synthesis and structure characterization of sulfated galactomannan from fenugreek gum. *Int. J. Biol. Macromol.* 125: 1184–1191. https://doi.org/10.1016/j.ijbiomac.2018.09.113. , PMID: 30244132.

**43** Salehi, B., Valussi, M., Morais-Braga, M.F.B. et al. (2018). *Tagetes* spp. Essential oils and other extracts: chemical characterization and biological activity. *Molecules* 23 (11): 2847. https://doi.org/10.3390/molecules23112847. PMID: 30388858.

**44** Z. Cao Z, Xia W, Zhang X, Yuan H, Guan D, Gao L. (2020). Hepatotoxicity of nutmeg: a pilot study based on metabolomics. *Biomed. Pharmacother.* 131: 110780. https://doi.org/10.1016/j.biopha.2020.110780. , PMID: 33152938.

**45** Faragab, M.A., Mohsena, E., El Nasser, A., and El-Gendy, G. (2018). Sensory metabolites profiling in *Myristica fragrans* (Nutmeg) organs and in response to roasting as analysed via chemometric tools. *LWT* 97, 684–692: 0023–6438. https://doi.org/10.1016/j.lwt.2018.08.002.

**46** Palama, T.L., Fock, I., Choi, Y.H. et al. (2010). Biological variation of *Vanilla planifolia* leaf metabolome. *Phytochemistry* 71 (5–6): 567–573. https://doi.org/10.1016/j.phytochem.2009.12.011.

**47** Cai, Y., Gu, F., Hong, Y. et al. (2019). Metabolite transformation and enzyme activities of hainan vanilla beans during curing to improve flavor formation. *Molecules* 24 (15): 2781. https://doi.org/10.3390/molecules24152781. , PMID: 31370187.

**48** Gallage, N.J., Jørgensen, K., Janfelt, C. et al. (2018). The intracellular localization of the vanillin biosynthetic machinery in pods of *Vanilla planifolia*. *Plant Cell Physiol.* 59 (2): 304–318. https://doi.org/10.1093/pcp/pcx185. , PMID: 29186560.

**49** Pérez-Gálvez, A., Martin, H.D., Sies, H., and Stahl, W. (2003). Incorporation of carotenoids from paprika oleoresin into human chylomicrons. *Br. J. Nutr.* 89 (6): 787–793. https://doi.org/10.1079/BJN2003842. , PMID: 12828795.

**50** Gómez-García, R.M. and Ochoa-Alejo, N. (2013). Biochemistry and molecular biology of carotenoid biosynthesis in chili peppers (*Capsicum* spp.). *Int. J. Mol. Sci.* 14 (9): 19025–19053. https://doi.org/10.3390/ijms140919025 , PMID: 24065101.

**51** Govindarajan, V.S., Rajalakshmi, D., and Chand, N. (1987). Capsicum--production, technology, chemistry, and quality. Part IV. Evaluation of quality. *Crit. Rev. Food Sci. Nutr.* 25 (3): 185–282. https://doi.org/10.1080/10408398709527453. , PMID: 3297498.

**52** Lu, M., Chen, C., Lan, Y. et al. (2020). Capsaicin-the major bioactive ingredient of chili peppers: bio-efficacy and delivery systems. *Food Funct.* 11 (4): 2848–2860. https://doi.org/10.1039/d0fo00351d. , PMID: 32246759.

**53** Onçalves, G.A., Corrêa, R.C.G., Barros, L. et al. Effects of in vitro gastrointestinal digestion and colonic fermentation on a rosemary (*Rosmarinus officinalis* L) extract rich in rosmarinic acid. *Food Chem.* 271.

**54** A. Bonar A. (1994). *Herbs – A Complete Guide to their Cultivation and Use.* London: Tiger Books International.

**55** Maslarova, Y. and Heinonen (2001). Sources of natural antioxidants: vegetables, fruits, herbs, spices and teas. In: *Antioxidants in Food, Practical Applications*, Woodhead Publishing Series in Food Science, Technology and Nutrition (ed. J. Pokorny, N. Yanishlieva and M. Gordon), 210–263. USA: Woodhead Publishing.

**56** Valenzuela, A., Nieto, S., and Aceites, G.Y. (1996). Synthetic and natural antioxidants: food quality protectors. *Semantic Scholar* https://doi.org/10.3989/GYA.1996.V47.I3.859.

**57** Richheimer, S.L., Bernart, M.W., King, G.A. et al. (1996). Antioxidant activity of lipid-soluble phenolic diterpenes from rosemary. *JAOCS* 73: 507–514.

**58** Zegarska, Z., Amarowicz, R., Karmac, M., and Raflowski, R. (1996). Anti-oxidative effect of rosemary ethanolic extract on butter. *Milchwissenchaft* 51: 195–198.

**59** Pokorny, J., Rehbolya, Z., and Janitz, W. (1998). Extracts from rosemary and sage as natural anti-oxidants for fats and oils. *Czech. J. Food Sci.* 16: 227–234.

**60** Aruoma, O.I., Spencer, J.P.E., Rossi, R. et al. (1996). An evaluation of the antioxidant and antiviral action of extracts of rosemary and provençal herbs. *Food Chem. Toxicol.* 34: 449–456.

**61** Rašković, A., Milanović, I., Pavlović, N. et al. Antioxidant activity of rosemary (*Rosmarinus officinalis* L.) essential oil and its hepatoprotective potential. *BMC Complementary and Alternative Medicine* https://doi.org/10.1186/1472-6882-14-225.

**62** Mallavarapu, G.R., Kulkarni, R.N., Baskaran, K. et al. (1999). Influence of plant growth stage on the essential oil content and composition in Davana (*Artemisia pallens* wall.). *J. Agric. Food. Chem.* 47 (1): 254–258. https://doi.org/10.1021/jf980624c.

**63** Hellivan, P.J. (2011). Davana oil, Perfumer & Flavorist

**64** Misra, L.N., Raghunath, A.C., and Thakur, S. (1991). Fragrant components of oil from *Artemisia pallens*. *Phytochemistry* 30 (2): 549–552. https://doi.org/10.1016/0031-9422(91)83725-Z.

**65** Holmes, P. (1998–1999). Frankincense oil: the rainbow bridge. *Int. J. Aromather.* 9: 156–161.

**66** Byler, K.G. and Setzer, W.N. (2018). Protein targets of frankincense: a reverse docking analysis of terpenoids from Boswellia oleo-gum resins. *Medicines* 5: 96.

**67** Sharma, L., Chandra, M., and Ajmera, P. (2019). Health benefits of lavender (*Lavandula angustifolia*). *Int. J. Phys. Nutr. Phys. Edu.* 1274–1277.

**68** Lis-Balchin, M.T. (2012). *Handbook of Herbs and Spices*, Woodhead Publishing Series in Food Science, Technology and Nutrition, 2e (ed. K.V. Peter), 329–347. USA: Woodhead Publishing https://doi.org/10.1533/9780857095688.329.

**69** Kim, N.S. and Lee, D.S. (2002). Comparison of different extraction methods for the analysis of fragrances from Lavandula species by gas chromatography mass spectrometry. *J. Chromatogr. A* 982: 31–47.

**70** Lehrner, J., Marwinski, G., and Deecke, L. (2005). Ambient odors of orange and lavender reduce anxiety and improve mood in a dental office, psychology. *Physiol. Behav.* .

**71** Dowthwaite, S.V. and Rajani, S. (2000). Vetiver: perfumer's liquid gold. In: *Proceedings of ICV-2 held in Cha-am*, Phetchaburi, Thailand, 18–22 Jan. 2000, 478–481.

# Part II

# Commercial Biotechnology Pathways, and Their Applications to Industrial Sustainability

# 2

# Biogenesis of Plant-Derived Flavor Compounds

*Nimisha P. Sukumaran and Sreeraj Gopi*

*Aurea Biolabs (P) Ltd, R&D Centre, Kolenchery, Cochin, Kerala, 682311, India*

## 2.1 Introduction

Plant-derived flavor compounds, with their great chemical diversity, varied biological functions, and with importance for human health and food sensory quality, have been of great significance in food, textile, pharmaceutical, cosmetic, and perfume industries. Furthermore, it has become evident that there is a global shift in food consumption patterns along with increased consumer wakefulness to food-sensory qualities, taste, and aroma, which therefore have become important factors affecting end user choices [1]. Moreover, they play an important role in consumer preferences for many products, as there exist a wide variety of tastes for each group of consumers. In other words, consumers who are interested in moving toward healthier and sustainable foods in addition to being economically advantageous often look for something completely new, stable over time with improved flavor qualities [2]. Therefore, the foremost purpose of these industries is to offer good, stable flavors to their customers, in addition to being economically profitable. The industry is therefore constantly researching new, more efficient, and cheaper ways of extracting plants for its delicate flavor ingredients. To achieve a novel flavor profile, they need to research deep into the biogenesis or formation of these compounds inside a plant cell. Although a fair amount of work has been performed on volatile analysis, ascribing specific importance or assigning volatile markers with quality attributes is extremely dubious.

Flavor is well defined as an intermingled but unitary experience, which includes sensations of taste, smell, and pressure, and often cutaneous sensations such as warmth, color, or mild pain. Flavors can be broadly classified into taste (nonvolatile compounds) and aroma (volatile compounds) (Figure 2.1). Aroma compounds are perceived mainly with the nose, whereas taste receptors exist in the mouth and are impacted when the food is chewed, where it is first released into the saliva before they are transferred to the mouth's headspace and the nose [3, 4]. Thus, in food

*Natural Flavours, Fragrances, and Perfumes: Chemistry, Production, and Sensory Approach,*
First Edition. Edited by Sreeraj Gopi, Nimisha Pulikkal Sukumaran, Joby Jacob, and Sabu Thomas.

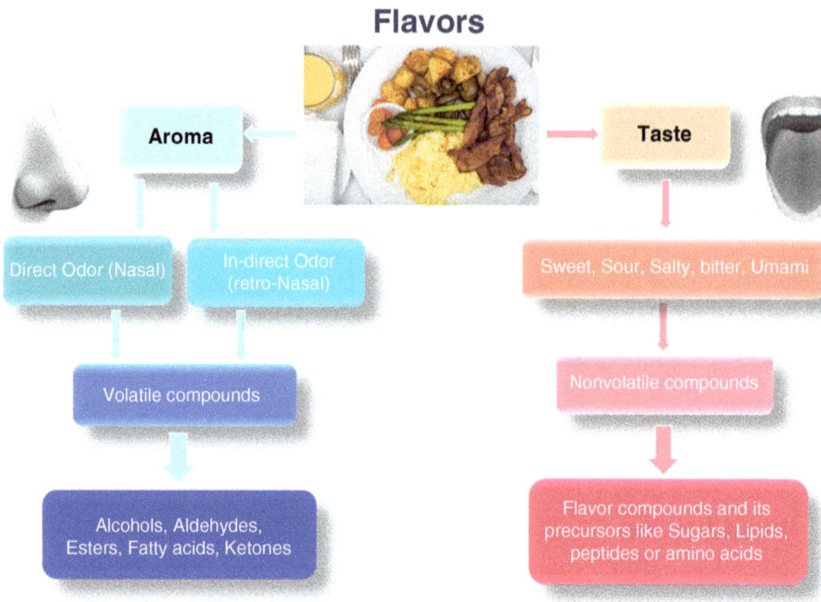


**Figure 2.1** Classification of flavors. Source: (b) lavenda/Adobe Stock.

matrices, aroma compounds have to be released from the food during the eating process and must reach the olfactory receptors. In turn, the release of volatile-molecule phase highly depends on their interaction with non-volatile compounds present in the food matrix, such as proteins, carbohydrates, and lipids (i.e. primary metabolites) [5].

Generally, the food flavor quality is governed by some features, such as chemical reactivity flavor; the environment like availability of light and atmospheric oxygen; and the matrix system and its constituents (carbohydrate, protein, transition metal, fat, radical, and other polymers) [2], which limits the stability of flavor profile. Similarly, the extraction tool or with which the flavor metabolites are extracted also reflects the quality characteristics of sensory profile. For instance, comparison of extraction techniques for essential oils from white tea reported variation in alcohols, aldehydes, ketones, esters, heterocyclics and alkanes, affected the sweet, roasted and woody aroma, which significantly declined [6]. Thus, a comprehensive evaluation of metabolic contents to characterize the biosynthetic pathway through a bottom-up strategy would help to identify and establish the flavor signatures. Moreover, an important goal for improvement of flavors is by targeting specific metabolites and/or downregulation of some other target metabolites. This is mainly because, like in almost all production chains, reducing the number of intermediates and/or metabolites can provide yield and reduce costs. The same holds factual for use of plants as programmable biochemical workhorses for the synthesis of flavor-active molecules. Further, the synthesis of novel compounds directly from the plant with the use of metabolic engineering is the best strategy to produce bulk commodity products at low cost.

For all these reasons, intensive efforts have been devoted to the dissection of biogenesis of plant-derived flavor-active compounds. In particular, the recent application of transcriptomic and metabolomics profiling data make it possible to study and facilitate our understanding of metabolic make-up processes and may also unravel novel ways to provide an alternative production system for plant-derived flavor compounds for commercial exploitation. In this chapter, we bring together recent and traditional findings on the biogenesis of flavor compounds in plants. First, we review the metabolic diversity in plant species and then discuss how the metabolite are formed, which determines the flavor patterns.

## 2.2 Primary and Secondary Flavor Compounds

With over 400 000 extant species of vascular plants on the Earth, the metabolic diversity within the plant kingdom is commonly being identified to comprise between 200 000 and 1 million metabolites. This arsenal of compounds being expected within a single species is thought to contain an upward of 5000 metabolites [7]. In fact, there exists a high level of qualitative and quantitative variations of metabolism within a plant species, which makes it even more complex [3]. The vast and versatile flavor effects of plants basically depend on their diversity in their chemical scaffolds. For instance, primary and secondary metabolites determine the aroma and flavor quality of grapevine berries, must, and wine, which is chiefly influenced by environmental and viticultural inputs and their complex interactions [8]. In the same way, the accumulation of primary metabolites like ethanol, lactic acid, and acetic acid and secondary metabolites like 2-methyl-1-butanol, isoamyl acetate, and ethyl acetate in cocoa beans is the key for flavor development [9]. Generally, the phytochemical constituents of plants fall into two categories based on their role in basic metabolic processes, namely primary and secondary metabolites [7].

### 2.2.1 Primary Metabolites

Primary metabolites consist of a large group of compounds such as glycolysis and tricarboxylic acid (TCA) cycle intermediates, amino acids, proteins, purines, pyrimidine bases, polysaccharides, and fatty acids. They are rarely used as therapeutic compounds but commonly used as nutritional supplements or flavoring agents [10]. Figure 2.2 depicts the examples of plant primary flavor-active metabolites with their structures arranged by major classes.

#### 2.2.1.1 Organic Acids

Organic acids are a primary metabolite having an important role in flavor "with" or/ and "without" added organic acids. For instance, Cocoas from South East Asia and the South Pacific tended to be more acidic than West African beans in terms of both chemical and sensory characteristics. This was due to the presence of greater concentrations of lactic and acetic acids in cocoas from the former regions which were considered to be largely responsible for higher acid flavor scores, but with lower

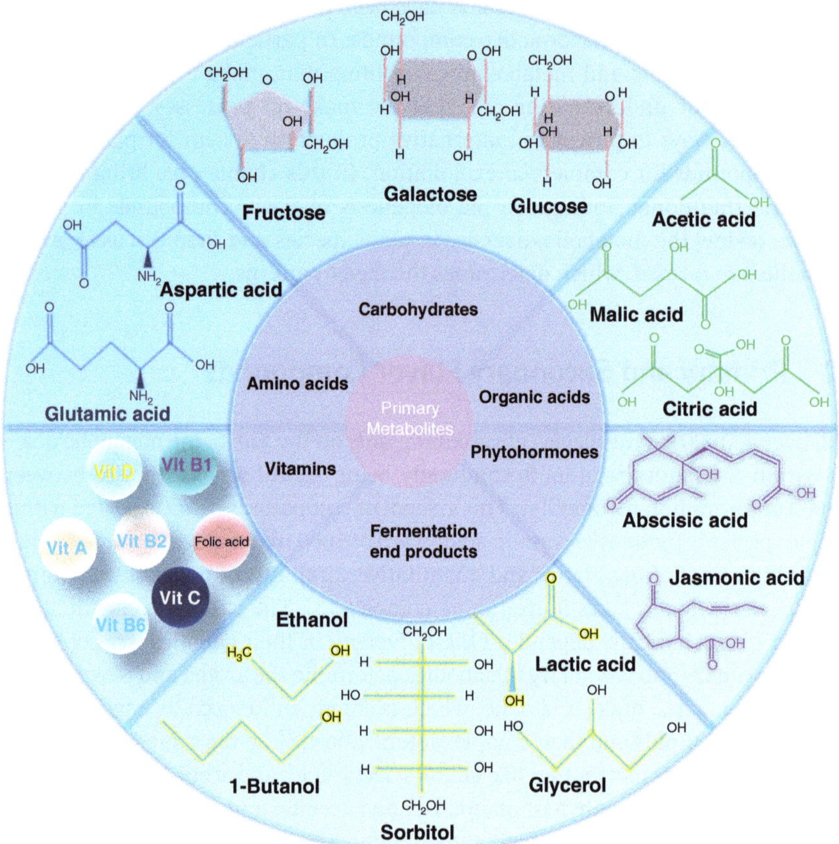


**Figure 2.2** Examples of plant primary flavor-active metabolites with their structures arranged by major classes.

levels of citric acid and oxalic acids [11]. Studies on the identification of volatile organic acids of bread and their influence on its taste, have revealed the role of organic acids in bread aroma. Remarkably, organic acids were found to be acetic acid being the major component followed distantly by propionic acid, isobutyric, butyric, isovaleric, valeric, and caproic acids [12]. Similarly, naturally ripened banana has more aroma than acetylene- and ethephon-treated banana, with the highest number of volatile compounds and high level of esters (65%). Malic acid, citric acid, and oxalic acids were significantly low in treated bananas than in naturally ripened bananas. This is because, though ethephon and acetylene could trigger the ripening process, the lower levels of organic acids and sugars in flesh of banana would eventually lead to poor aroma profile [13]. Aroma profile of cheddar cheese had acetic acid, butanoic acid, dimethyl trisulfide, methional, hexanal, (*E*)-2-nonenal, acetoin, 1-octen-3-one, δ-dodecalactone, furaneol, hexanoic acid, heptanal, and ethyl caproate [14]. The nonhybrid Raf tomato, one of the most tasteful commercial tomatoes that is traditionally grown in this Spanish region, contains high fructose to glucose and fructose to citric acid ratios. Moreover, the content of malic acid during ripening also remains high for Raf tomato. These combined features are responsible of the unique characteristic and exceptional taste of this traditional line of tomato [15].

#### 2.2.1.2 Phytohormones

Phytohormones like ethylene are considered as an indicator of flavor quality evaluation, for instance, the volatile profile of Fuji apples has change in ethylene production. Ethylene is considered to be the most influential factor in converting starch to sugar and forming aromatic substances and plays an important role in the synthesis of esters as well [16]. A metabolite analysis carried on peach aroma, suggested a crucial association between peach-like volatiles and "ethylene production and modulation of FA levels" [17]. Also, though unclear, ethylene has a pivotal role in the fruit aroma biosynthetic pathways and the volatile aroma formation during climacteric fruit ripening [18, 19]. Pointedly, it is now known that ethylene regulated ripening related genes is found to have involvement in sugar metabolism and subsequent volatile aroma formation [16]. Similarly, abscisic acid (ABA), another phytohormone, which is derived from the carotenoid pathway, has a major effect on the aroma volatile profile of the fruit. For instance, ABA promoted the accumulation of bioactive components and the antioxidant capacity via the regulation of gene expression during tomato ripening [20]. As important upstream signals, phytohormones regulate the plant volatiles' biosynthesis under various stresses. For instance, the formation of some characteristic aromas during the manufacturing process of oolong tea (postharvest stage) is due to the defense responses of tea leaves to stress. Jasmonic acid (JA) and ABA levels enhanced during the manufacturing processes (enzyme-active stage) of oolong tea [21].

#### 2.2.1.3 Vitamins

Effect of folic acid fortification on the sensory characteristics of lemon yogurt was studied to develop dairy products with new flavors that has added health benefits, which further helps the dairy industry increase sales of products as well as provide consumers with products they enjoy [22]. Vitamin concentrates with vitamins A and D are used for fortification of fluid milk. Although many of the degradation components of vitamins A and D have an important role in flavor/fragrance applications, they may also be source(s) of off-flavor(s) in vitamin fortified milk due to their heat, oxygen, and the light sensitivity. It is very important for the dairy industry to understand how vitamin concentrates can impact flavor and flavor stability of fluid milk [23]. Sunlight flavor, where milk proteins form a loosely bound complex with riboflavin, which greatly depends on the presence of tryptophan in the protein. Precisely, the loosely bound complex absorb light, undergoes an electronic transition to an excited state, thereby dissociation of the complex and decreased concentrations of riboflavin and tryptophan [24]. Thiamine-Derived Taste Enhancer, which was found to be formed by a Maillard-type reaction of thiamine and cysteine, has great potential in commercial meat like process flavors [25]. Rather riboflavin is removed from beer for the related light-struck off-flavor formation [26].

#### 2.2.1.4 Amino Acids

Umami, the "savory" taste, is characterized by the presence of amino acids like aspartic and glutamic acids, where glutamate, has been extensively used in the food supply as a sodium salt (monosodium glutamate; MSG) or via glutamate-rich ingredients (e.g. hydrolyzed yeast extract) to enhance flavor of savory products (soups, meats, cheese, etc.) [27]. The chicken nuggets prepared using spent meat enriched with milkfat and potato mash had characteristic umami taste is due to the presence of umami-taste amino acids (aspartic and glutamic acids) along with higher 5′-nucleotides (GMP, AMP, ADP) and. Besides the presence of abundant volatiles like trimethyldodecane, camphene, 5-ethyl-2,2,3-trimethylheptane, 3,6-dimethylundecane, 2,2,4-trimethylheptane, and α-pinene [28]. The taste-active amino acids found in lemon fruits were increased slightly both in the fruit peel and pulp after postharvest ethephon degreening treatment, which is done as an effective and economical approach to improve the peel color and increase the internal quality of lemon fruits. Though individual organic acid and sugar fluctuated after ethephon treatment, the volatile compounds in fruit peel were obviously increased about one time as compared to the untreated fresh fruits [29].

#### 2.2.1.5 Fermentation Products

Modulated wine characteristics with sequential inoculations with nine non-Saccharomyces yeasts belonging to 6 species and two mixed inoculum of *T. delbrueckii* and *Lactobacillus thermotolerans* tended to produce higher ethanal and glycerol and lower volatile acidity [30]. Wine yeasts play a pivotal role in the final aroma profiles of wines. Yet, mixed-culture fermentations using non-*Saccharomyces* yeast and genetic modifications of *Saccharomyces cerevisiae* have all been shown to greatly enhance the chemical composition and sensory profile of wines [31]. The main volatile compounds found in apple ciders were esters and higher alcohols, followed by aldehydes and fatty acids. Precisely, principal component analysis (PCA) showed that floral and fruity (fresh apple, banana, and pear) odors were highly associated with sweet taste and opposed to the more complex aroma attributes (yeasty, lactic, chemical, moldy, black pepper, and earthy) and sour taste. Ciders with marked levels of acetate esters were characterized by cooked/fresh apple, citrus and tropical fruit odors. Flavor profiling of apple ciders from the UK and Scandinavian region [32]. Figure 2.3. Glycerol has favorable impacts on the quality of beer, wine, and spirit. Large surveys showed that obvious higher level of glycerol was present in wines of superior grade than that of ordinary. Nevertheless, flavor profile is unique for each wine. Individual wine, as well as other beverage, has debating benefits from glycerol [33].

#### 2.2.1.6 Proteins, Lipids, and Carbohydrates

Primary metabolites like proteins, lipids, and carbohydrates play primary roles in flavor development, as they include various compounds which can develop into key flavor precursors when heated. The compounds that elicit various tastes and odors have different thresholds for perception. For instance, sweet flavor in meat is derived from sugars, amino acids, and organic acids presence, while sour flavors arise after

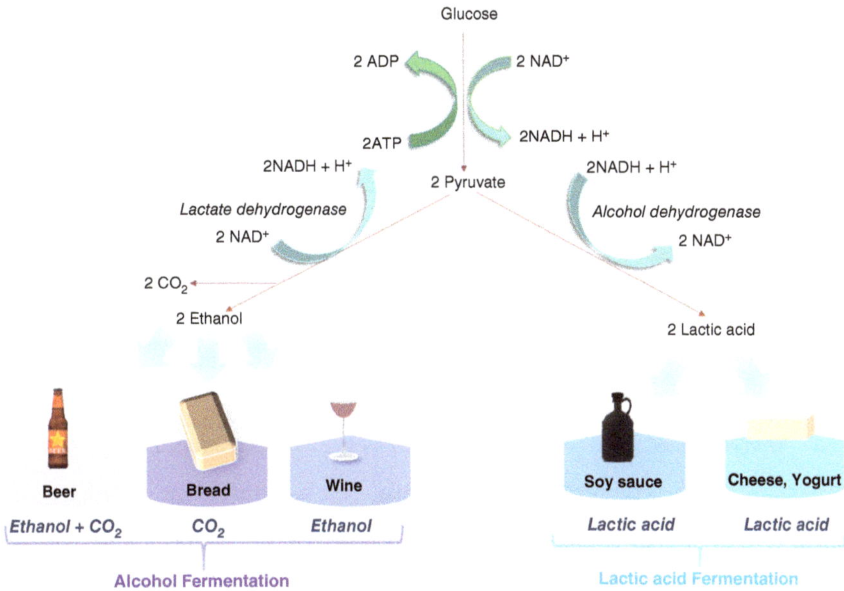


**Figure 2.3** Types of fermentation and their use in food industries.

the amino acids are coupled with organic acids. Inorganic salts and sodium salts of glutamate and aspartate generate saltiness and the bitterness is likely due to hypoxanthine, anserine and carnosine as well as some amino acids. Moreover, some flavor compounds can be regarded as "acceptable flavor," such as the fried meat flavor originating from the Maillard reaction between amino acids and carbohydrates and from the degradation of lipids [34]. In fruits, the flavor compounds change during postharvest, mainly due to enzymatic ripening reactions; with the general trends for the taste compounds being increase of sweetness, due to accumulation of glucose and fructose (reflected in an increase of total soluble solids), and decrease of sourness, due to degradation of organic acids (reflected in a decrease of titratable acidity) [35]. Similarly, medium-chain fatty acid (MCFA) ethyl esters, as yeast metabolites, considerably contribute to the fruity aroma of foods and beverages [36].

### 2.2.2 Secondary Metabolites

Secondary metabolites, derived from primary metabolites with diverse physiological activities, have great structural diversity and are counted as distinctive traits of plants [37]. More than 200 000 chemical structures have been isolated and described so far, and they are the products of subsidiary pathways for instance, the shikimic acid pathway [7]. Alternatively, secondary metabolites are defined as a heterogeneous group of natural metabolic products that are not essential for vegetative growth of the producing organisms, but considered as differentiation compounds conferring adaptive roles. Yet, multitude of secondary metabolite secretions are harvested

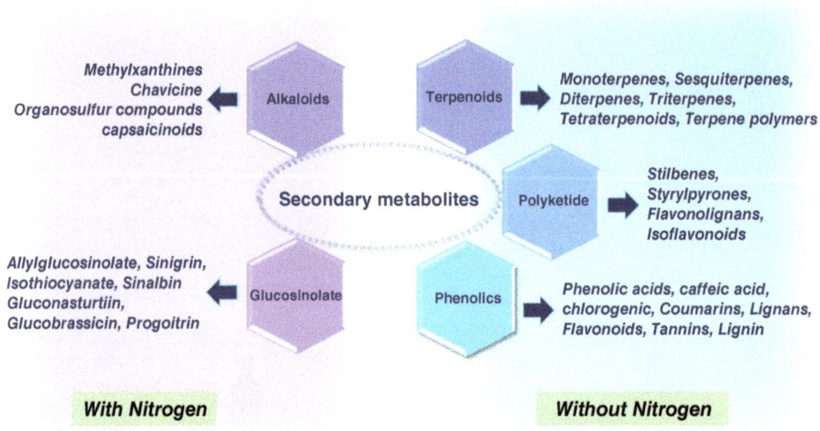


**Figure 2.4** Classification of secondary metabolites with or without nitrogen with examples.

to improve their health, widen the pyramid of healthy nutrition, enhancing agricultural productivity. Secondary metabolites can be divided into secondary metabolites with and without nitrogen in their structures as depicted in Figure 2.4 [7].

### 2.2.3 Secondary Metabolites with Nitrogen

#### 2.2.3.1 Alkaloids

Alkaloids, a group of biologically important amine structures with diverse matrix containing nitrogen atom, are either derived from amino acids or from transamination [38]. In pure form most alkaloids are colorless, nonvolatile, crystalline solids, with bitter taste. Despite the supposed universality of bitter taste denial, many commonly consumed foods and beverages such as fruits, tea, coffee, chocolate, and alcohol have bitterness as a major sensory attribute which, in the overall taste profile of a food, is often appreciated by the consumers [39]. Among the alkaloids, caffeine is perhaps the most popular due to its use in beverages can positively affect flavor profiles and the experience of beverage consumption [40]. Similarly, piperine, an alkaloid, which is the major bio-active component of pepper, imparts pungency and biting taste. Though other alkaloids, including piperanine, piperettine, piperylin A, piperolein B, and pipericine, all possessing some degree of pungency in the pepper extract studied, piperine is considered as the principal compound as it constitutes around 98% of the total alkaloids in pepper [41]. Likewise, allicin (diallyl-dithiosulfinate), an organosulfur compound obtained from garlic, is the most important alkaloid that imparts flavor and owe to the economic importance of garlic [42]. As well, capsaicinoids are the pungent compounds that are responsible for the characteristic taste of hot peppers, which are spicy or savory food additives widely used in numerous countries around the world [43].

#### 2.2.3.2 Glucosinolates

Among nitrogen containing secondary plant metabolites, the class of alkaloids is the largest with over 27 000 structures, followed by 700 non-protein amino acids [37]. Cyanogenic glucosides, glucosinolates, and amines are less diverse, with fewer than 100 structures in each class. Glucosinolates occur in various edible plants such as cabbage (white cabbage, Chinese cabbage, broccoli), Brussels sprouts, watercress, horseradish, capers, and radishes where the breakdown products often contribute a significant part of the distinctive taste [43].

### 2.2.4 Secondary Metabolites Without Nitrogen

The largest group of secondary metabolites do not contain nitrogen in their structures; among them are terpenoids (mono-, sesqui-, di-, tri-, and tetraterpenes, saponins, iridoid glucosides) the most frequent, followed by the extensively distributed phenolics (phenylpropanoids, flavonoids, catechins, tannins, lignans, coumarins, furanocoumarins, anthraquinones) [37].

#### 2.2.4.1 Terpenoids

Herbs and higher plants containing terpenoids and their oxygenated derivatives have been used as fragrances and flavors for centuries [44]. With more than 40 000 known molecules, terpenes are the largest family of natural plant compounds, of which many are volatiles that contribute to the flavor and taste [45]. They are usually the main constituents of essential oils of most plants and offer a wide variety of pleasant scents from flowery to fruity, to woody or balsamic notes. For this reason, terpenoids constitute a very important class of compounds in flavor and fragrance industries, which has broad market prospects and economic benefits [46]. Some examples of terpenoids that are used in flavor industries are given in the Table 2.1. For instance, hoppy aroma of wine is due to the transformations of flavor-active compounds like terpenes, sulfur compounds and oxygenated compounds during fermentation [53].

**Table 2.1** Terpenoids and their examples used in flavor and fragrance industries.

| | Type of classification | Compound | Sources | Uses | References |
|---|---|---|---|---|---|
| Terpenoids | Monoterpenes | Camphor | Camphor Laurel | Fragrance | [47] |
| | Sesquiterpenes | β-caryophyllene | Ylang Ylang | Fragrance | [48] |
| | Diterpenes | Cafestol and Kahweol | Coffee beans | Beverage | [49] |
| | Triterpenes, | Oleanolic acid | Olive oil | Seasoning | [50] |
| | Tetraterpenoids | Lutein | Marigold flowers | Food colors | [51] |
| | Polyterpenes | Chicle | Sapodilla tree | Natural gum | [52] |

**Table 2.2** Flavor and Fragrance industry relevant phenolic compounds.

| Compounds | Types | Plant source | Applications | References |
|---|---|---|---|---|
| Anthocyanin | Flavonoids | Grapes | Red wine | [59] |
| Punicalagin | Tannins | Pomegranates | Beverage | [60] |
| Benzenoid lactone | Coumarins | Cinnamon | Fragrance substance in cosmetic preparations | [61] |
| Carnosic acid, rosmarinic acid | Phenolic acids | Rosemary leaves | Flavoring agent | [62] |
| Caffeic acid, ferulic acid, gallic acid | Phenolic acids | Coffee | Beverage | [63] |

Similarly, terpene's presence even at very low concentrations, can affect the flavor profiles of olive oil [45]. As well, degradation of terpenes, leads to a reduced formation of reaction products from the lipoxygenase pathway and less pronounced fruitiness and flavor [54].

#### 2.2.4.2 Phenolics

Phenolic compounds are chemically defined as compounds containing hydroxylated aromatic rings, the hydroxy group being attached directly to the phenyl, substituted phenyl, or other aryl group [55]. Phenolics have been applied in the food industry as biopreservatives, have been widely studied for improving the shelf life of perishable products [56]. They are a group of metabolites derived from the secondary pathways which comprise flavonoids, tannins, coumarins, lignans, and phenolic acids, compounds naturally found in fruits, vegetables, cereals, roots, and leaves among other plant products [57]. Phenolic compounds are closely associated with the flavor, color, foam, colloidal, and sensory properties of beer and contribute to berry and wine quality [8]. For instance, phenolic compounds in beer mainly originate from the raw material of brewing, i.e. barley malt and hop, and include simple phenols, phenolic acids, hydroxycinnamic acid derivatives and flavonoids [58]. Some flavor and fragrance industry relevant phenolic compounds are briefed with examples in Table 2.2.

## 2.3 Mechanistic Pathways of Flavor Formation

Taste is mainly determined by saccharides and organic acids, which form the background flavor, while aroma profiles are quite complicated, including many volatile compounds that differ depending on various conditions. The taste component of the fruits and vegetables flavor for instance is positively correlated with sugars (sweet taste), organic acids (sour taste), phenolic compounds (bitter taste), tannins, and capsaicinoids (chemical feelings), while the odor/aroma component is linked to volatile compounds (esters, terpenes, alcohols, aldehydes, etc.) [35]. Similarly, vegetable flavor comes from the interplay of sensory factors, like secondary metabolites, namely low-molecular-weight organic compounds, which are often restricted to a limited number of plant families or species [64].

### 2.3.1 Primary Metabolites

Central carbon metabolism, also known as primary metabolism, contributes to the synthesis of intermediate compounds that act as precursors for plant secondary metabolism. Primary metabolites such as amino acids, organic acids, or nucleosides are commonly used as nutritional supplements or flavoring agents, they are rarely used as therapeutic compounds [10]. For example, free amino acids produced in the carbohydrate pools are converted to various flavor compounds through amino acid catabolism. The branched-chain amino acids (valine, leucine, and isoleucine), the aromatic amino acids (tyrosine, tryptophan, and phenylalanine), and the sulfur-containing amino acids (methionine and cysteine) are the main amino acid sources for flavor compounds [65].

The fruity or corn chip flavor profiles in beer were associated beer purines/pyrimidines, volatile ketones, amines, and phenolics, and malt lipids, saccharides, phenols, amines, and alkaloids [66]. The pictorial representation of pathways responsible for the biosynthesis of major classes of primary metabolites is given as Figure 2.5. Moreover, secondary metabolites derived essentially from the modification of primary metabolites by different pathways, for instance, the shikimate pathway is considered the major pathway that forms the basic building block for a wide range of phenolic and flavonoid compounds [67].

### 2.3.2 Secondary Metabolites

Primary metabolism products derived from glycolysis, the TCA cycle, or the shikimate pathway often serve as precursors for the synthesis of the tens of thousands of secondary metabolites. Precisely, secondary metabolic pathways originate from different nodes of core primary metabolic pathway, suggesting that emergent enzymatic activities against primary metabolites yielded new compounds that were able to increase plant adaptation to particular environments and were gradually converted into specialized metabolites [68]. These low-molecular-weight flavor molecule substances derived from the fatty acid, amino acid and carbohydrate pools constitute the heterogenous group of molecules called secondary metabolites with saturated and unsaturated, straight-chain, branched-chain and cyclic structures bearing various functional groups (e.g. alcohols, aldehydes, ketones, esters, and ethers) and also nitrogen and sulfur [69]. Usually they are derived from the phenylpropanoid, isopropanoid, alkaloid and fatty acid pathways. Umami taste and characteristic aroma for instance in fermented soybean paste is due to acids, alcohols, aldehydes, terpenes, and sulfur-containing compounds [70]. Though the secondary metabolite biosynthetic pathways are too numerous and is difficult to elucidate, there are a few typical pathways involved in the biosynthesis of major classes of these compounds [67]. An overview of secondary pathway leading to flavor is depicted in Figure 2.6.

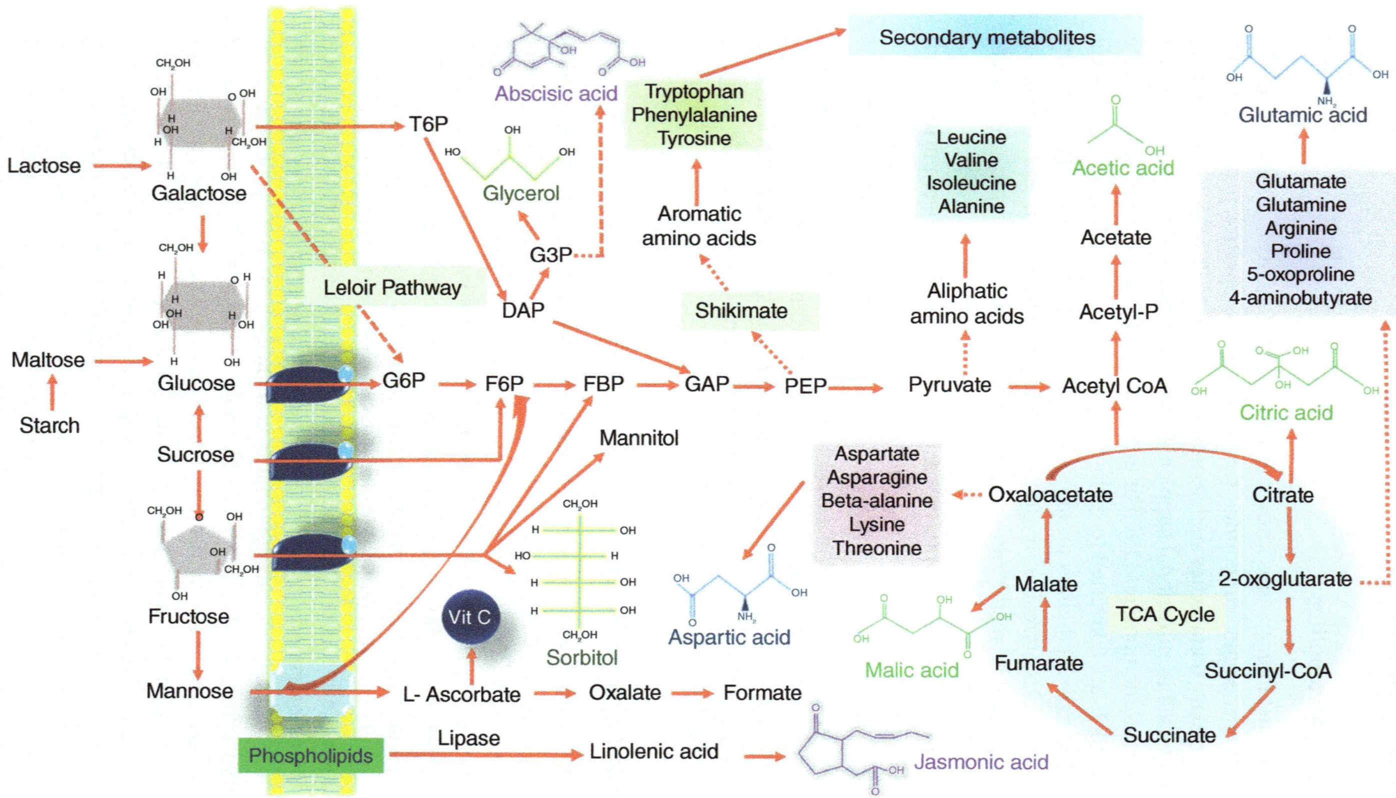


**Figure 2.5** **Schematic representation of pathways responsible for biosynthesis of major classes of primary metabolites** G6P; Glucose-6-phosphate, F6P; Fructose-6-phosphate, FBP; Fructose-1,6-bisphosphate, GAP; Glyceraldehyde 3-phosphate, PEP; Phosphoenolpyruvate, T6P; Tagatose 6-phosphate, DAP; Dihydroacetone-P, TCA; Citric Acid Cycle or Krebs cycle, G3P; Glycerol-3-phosphate.

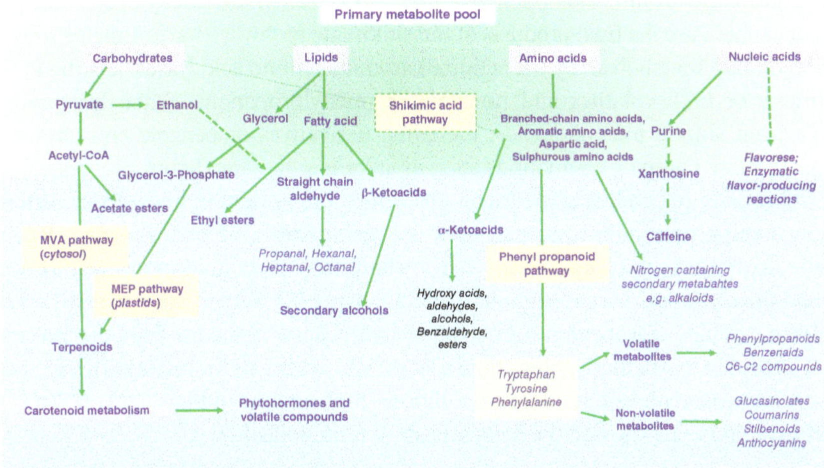


**Figure 2.6 Scheme of secondary metabolite pathways leading to flavor.** MVA; mevalonate, MEP; methylerythritol-4-phosphate, the shikimate pathway connects central carbon metabolism (glycolysis and pentose phosphate pathways) with the biosynthesis of aromatic amino acids and derived products.

#### 2.3.2.1 Purine Metabolism

A purine is an aromatic heterocycle composed of carbon and nitrogen, which constitute a number of important biomolecules such as ATP, GTP, cyclic AMP, NADH, and coenzyme A. Purines have an $NH_2$ group and oxo groups that exhibit keto-enol and amine-imine tautomerism, although amino and oxo forms predominate in physiological conditions [71]. Caffeine, for instance a purine alkaloid, is a natural component of coffee, tea, and cocoa, and the biosynthetic pathway of caffeine has been elucidated and has been studied extensively. Xanthosine, which is derived from purine metabolites, is the first committed intermediate in caffeine biosynthesis [72]. Similarly, glutamate and 5′-inosinate, which are the taste principles in traditional Japanese soup stock, are derived from purine metabolism [73].

#### 2.3.2.2 Aminoacid Metabolism

Free amino acids produced in the carbohydrate pools are converted to various flavor compounds through amino acid catabolism. The branched-chain amino acids (valine, leucine, and isoleucine), the aromatic amino acids (tyrosine, tryptophan, and phenylalanine), and the sulfur-containing amino acids (methionine and cysteine) are the main amino acid sources for flavor compounds. The conversion of these amino acids into flavors proceeds via two distinct routes: transamination and elimination. The transamination route is initiated by aminotransferases that convert amino acids into their corresponding α-keto acids. The α-keto acids are then further converted into aldehydes, alcohols, and esters, which are important aroma compounds. It was shown that branched-chain amino acids, aromatic amino acids, and methionine are catabolized via the transamination route. The elimination route has been described for methionine where activity by carbon-sulfur lyases results in the release of methanethiol [65].

The aromatic amino acids phenylalanine and tyrosine, are phenylpropanoid precursors, which are the final products of the shikimate pathway, that is directed toward the secondary metabolism by the action of aromatic amino acid lyases, leading to the synthesis of both volatiles and non-volatile phenylpropanoids. Phenylpropanoids range from simple phenolic acids, including derivatives of benzoic and cinnamic acids, to more complex compounds such as stilbenes, lignans (lignin precursors), or the ubiquitous and well-studied flavonoids, which are present in many fruits. Briefly, phenylalanine ammonia lyase catalyzes the deamination of phenylalanine to cinnamic acid and is the gateway enzyme to the phenylpropanoid pathway, directing carbon flow from primary to secondary metabolism [68]. Terpenoids can also be synthesized through an isopentenyl diphosphate pathway, which arose from the intermediate substrate, particularly, mevalonic acid (MVA) via the mevalonate pathway and a methylerythritol phosphate/deoxy-D-xylulose 5-phosphate pathway (MEP/DOXP) pathway. Figure 2.5 showed the biosynthesis of terpenoids [67].

#### 2.3.2.3 Carotenoid Metabolism

Carotenoids are lipophilic secondary metabolites derived from the isoprenoid pathway and provide precursors for the biosynthesis of the phytohormones, ABA, and strigolactones [74]. Briefly, in carotenoid biosynthesis, the first step is the condensation of two geranylgeranyl diphosphate to produce phytoene that is catalyzed by phytoene synthase. From phytoene, lycopene is formed by a series of desaturation and isomerization reactions, which after cyclization gives rise to α-carotene and β-carotene in the presence of lycopene ε-cyclase and lycopene β-cyclase. Then, oxidative cleavage of carotenoid molecules by carotenoid cleavage dioxygenases and non-enzymatic cleavage between the C9 and C10 position, yield to apocarotenoid formation, including phytohormones and volatile compounds such as α- and β-ionone, 6-methyl-5-hepten-2-one, or geranylacetone that play an important role in the aroma of fruits [68].

#### 2.3.2.4 Fatty Acid Metabolism

Fatty acids are metabolized through several biosynthetic pathways to generate volatile compounds, thereby act as the major precursors of aroma volatiles. β-Oxidation and lipoxygenase action are the main pathways for the formation of aldehydes, alcohols, esters, ketones, and acids from lipids. Butyl acetate, hexyl acetate, 2-methylbutyl acetate, and ethyl 2-methylbutyrate are the most important esters in ripe fruits. The flavor properties of cheese for instance is directly influenced by free fatty acids. Briefly, lipolysis in dairy products is supported by the activity of lipases which catalyze the triglyceride hydrolysis, with the consequent production of medium-chain (chain lengths up to 10 carbon atoms) and long-chain (chain lengths over than 10 carbon atoms) free fatty acids, di- and mono-glycerides, and glycerol. VOCs that contribute to the formation of cheese deriving from the catabolism of FFAs, include methyl ketones, secondary alcohols, straight-chain aldehydes, lactones, esters, and *S*-thioesters [75].

#### 2.3.2.5 Carbohydrate Metabolism

Carbohydrate metabolism is one of the key biologic processes supporting human life because of their roles in energy production and various biosynthetic pathways [76]. Pyruvic acid has been established as the key intermediate substance in the metabolism of carbohydrates, and almost all six-, five-, and four-carbon compounds are converted initially to pyruvate, from which further catabolic or synthetic reactions proceed [77]. Carbohydrates are the major constituents of fruits and vegetables. Yet, there are only two groups of volatiles that are directly derived from carbohydrates: (i) terpenoid compounds and (ii) furanones [78]. The fundamental for terpene biosynthesis is the isoprene units. The shikimate pathway which involves multiple isoprene units ($C_5H_8$) linked together in a head-to-tail pattern can synthesize terpenoids according to the number of isoprene units incorporated in the molecular skeleton [67]. Furanones, which are found in some food products, represent one of the dominating class of flavor compounds formed as sugar degradation products by Maillard reaction induced by heat treatments [79].

#### 2.3.2.6 Organic Acid Metabolism

Organic acids are intermediates in the degradation pathways of amino acids, fats, and carbohydrates that affect fruits and vegetables' organoleptic properties such as color, flavor, and aroma [80]. For instance, organic acids are the second most abundant soluble solids component in fruit juices, and are typically present at about 1% of the total weight of a fruit juice. Citric and malic acids are the primary organic acids found in fruit juices [81]. Organic acids are either derived from main metabolic pathways, such as the tricarboxylic acid cycle (TCA) and glycolysis (For e.g. citric, lactic, itaconic, and malic acids), or sometimes derived from direct oxidation of glucose, where it is produced by one or two enzymatic steps from glucose (For e.g. gluconic and acetic acids) [82].

## 2.4 Conclusion

In flavor industries, flavor signatures are regarded as one of the most important aspects of quality and a key factor for quality grading and customer satisfaction. Thus, the concept of estimating flavor quality has been evolved and needs further research into chemo diversity of plant-derived flavor compounds, which need consolidation or discovering, one of the interim prospects would concern the investigation of the complex metabolic pathways. For this, key molecules have to be found after targeted and untargeted metabolic profiling. Once, the key molecules responsible for volatile biosynthesis within a cell are exploited, the relationship between the perceived aroma of a mixture of volatiles, its chemical composition, and the aromas of its components can be elucidated.

## References

1 Roger Harker, F., Anne Gunson, F., and Jaeger, S.R. (2003). The case for fruit quality: an interpretive review of consumer attitudes, and preferences for apples. *PostharvestBiol. Technol.* 28(3):333–347,ISSN0925-5214,https://doi.org/10.1016/S0925-5214(02)00215-6.

2 Lopes, D.B., Speranza, P., and Macedo, G.A. (2016). Chapter 18 – A new approach for flavor and aroma encapsulation. In: *In Nanotechnology in the Agri-Food Industry, Novel Approaches of Nanotechnology in Food* (ed. A.M. Grumezescu), 623–661. Academic Press, ISBN 9780128043080, https://doi.org/10.1016/B978-0-12-804308-0.00018-2.

3 Barrett, D.M., Beaulieu, J.C., and Shewfelt, R. (2010). Color, flavor, texture, and nutritional quality of fresh-cut fruits and vegetables: desirable levels, instrumental and sensory measurement, and the effects of processing. *Crit. Rev. Food Sci. Nutr.* 50 (5): 369–389. https://doi.org /10.1080/10408391003626322.

4 de Roos, K.B. (2006). Chapter 11 – Modelling aroma interactions in food matrices. In: *Woodhead Publishing Series in Food Science, Technology and Nutrition, Flavour in Food* (ed. A. Voilley and P. Etiévant), 229–259, ISBN 9781855739604. Woodhead Publishing https://doi.org/10.1533/9781845691400.2.229.

5 Guichard, E. (2015). Chapter 13 – Interaction of aroma compounds with food matrices. In: *Woodhead Publishing Series in Food Science, Technology and Nutrition, Flavour Development, Analysis and Perception in Food and Beverages* (ed. J.K. Parker, J.S. Elmore and L. Methven), 273–295, ISBN 9781782421030. Woodhead Publishing https://doi.org/10.1016/B978-1-78242-103-0.00013-8.

6 Huang, C., Zhou, S., Tong, Y. et al. (2021). Simultaneous nitrogen-blow distillation extraction: a novel approach for aroma extraction of white tea. *LWT* 138: 110675, ISSN 0023-6438, https://doi.org/10.1016/j.lwt.2020.110675.

7 Hussein, R.A. and El-Anssary, A.A. (2018). Plants secondary metabolites: the key drivers of the pharmacological actions of medicinal plants. In: *Herbal Medicine* (ed. P.F. Builders). IntechOpen https://doi.org/10.5772/intechopen.76139.

8 Wang, S., Alseekh, S., Fernie, A.R., and Luo, J. (2019). The structure and function of major plant metabolite modifications. *Mol. Plant* 12 (7): 899–919, ISSN 1674-2052, https://doi.org/10.1016/j.molp.2019.06.001.

9 Viesser, J.A., de Melo Pereira, G.V., de Carvalho Neto, D.P. et al. (2021). Co-culturing fructophilic lactic acid bacteria and yeast enhanced sugar metabolism and aroma formation during cocoa beans fermentation. *Int. J. Food Microbiol.* 339: 109015, ISSN 0168-1605, https://doi.org/10.1016/j.ijfoodmicro.2020.109015.

10 Dowd, C.J. and Kelley, B. (2011). Chapter 2.56 – Purification process design and the influence of product and technology platforms. In: *Comprehensive Biotechnology*, 2e (ed. M. Moo-Young), 799–810, ISBN 9780080885049. Academic Press https://doi.org/10.1016/B978-0-08-088504-9.00137-9.

11 Holm, C.S., Aston, J.W., and Douglas, K. (1993). The effects of the organic acids in cocoa on the flavour of chocolate. *J. Sci. Food Agric.* 61: 65–71. https://doi.org/10.1002/jsfa.2740610111.

12 Calvel, R. (2001). Organic acids. In: *The Taste of Bread* (ed. J.J. MacGuire). Boston, MA: Springer https://doi.org/10.1007/978-1-4757-6809-1_5.

13 Maduwanthi, S.D.T. and Marapana, R.A.U.J. (2019). Comparative study on aroma volatiles, organic acids, and sugars of ambul banana (*Musa acuminata*, AAB) treated with induced ripening agents. *J. Food Qual.* 2019: 7653154, 9 p. https://doi.org/10.1155/2019/7653154.

**14** Wang, J., Yang, Z.J., Xu, L.Y. et al. (2021). Key aroma compounds identified in Cheddar cheese with different ripening times by aroma extr ct dilution analysis, odor activity value, aroma recombination, and omission. *J. Dairy Sci.* 104 (2): 1576–1590, , ISSN 0022-0302. https://doi.org/10.3168/jds.2020-18757.

**15** Pérez, E.M.S., López, J.G., Iglesias, M.J. et al. (2011). HRMAS-nuclear magnetic resonance spectroscopy characterization of tomato "flavor varieties" from Almería (Spain). *Food Res. Int.* 44 (10): 3212–3221, , ISSN 0963-9969. https://doi.org/10.1016/j.foodres.2011.08.012.

**16** Qi, W., Wang, H., Zhou, Z. et al. (2020). Ethylene emission as a potential indicator of Fuji apple flavor quality evaluation under low temperature. *Hortic. Plant J.* 6 (4): 231–239, , ISSN 2468-0141. https://doi.org/10.1016/j.hpj.2020.03.007.

**17** Cai, H., Han, S., Jiang, L. et al. (2019). 1-MCP treatment affects peach fruit aroma metabolism as revealed by transcriptomics and metabolite analyses. *Food Res. Int.* 122: 573–584, , ISSN 0963-9969. https://doi.org/10.1016/j.foodres.2019.01.026.

**18** Zhu, H.L., Zhu, B.Z., Fu, D.Q. et al. (2005). Role of ethylene in the biosynthetic pathways of aroma volatiles in ripening fruit. *Russ. J. Plant Physiol.* 52: 691–695. https://doi.org/10.1007/s11183-005-0103-9.

**19** Bauchot, A.D., Mottram, D.S., and John, P. (2002). Aroma volatiles in fruits in which ethylene production is depressed by antisense technology. In: *Analysis of Taste and Aroma. Molecular Methods of Plant Analysis*, vol. 21 (ed. J.F. Jackson and H.F. Linskens). Berlin, Heidelberg: Springer https://doi.org/10.1007/978-3-662-04857-3_8.

**20** Tao, X., Wu, Q., Aalim, H. et al. (2020). Effects of exogenous abscisic acid on bioactive components and antioxidant capacity of postharvest tomato during ripening. *Molecules* 25 (6): 1346. https://doi.org/10.3390/molecules25061346.

**21** Zeng, L., Wang, X., Liao, Y. et al. (2019). Formation of and changes in phytohormone levels in response to stress during the manufacturing process of oolong tea (*Camellia sinensis*). *Postharvest Biol. Technol.* 157: 110974, ISSN 0925-5214, https://doi.org/10.1016/j.postharvbio.2019.110974.

**22** Boeneke, C.A. and Aryana, K.J. (2008). Effect of folic acid fortification on the characteristics of lemon yogurt. *LWT Food Sci. Technol.* 41 (7): 1335–1343, , ISSN 0023-6438. https://doi.org/10.1016/j.lwt.2007.08.010.

**23** Yeh, E.B., Barbano, D.M., and Drake, M. (2017). Vitamin fortification of fluid milk. *J. Food Sci.* 82: 856–864. https://doi.org/10.1111/1750-3841.13648.

**24** Aurand, L.W., Singleton, J.A., and Matrone, G. (1964). Sunlight flavor in milk. II. Complex formation between milk proteins and Riboflavin1, 2. *J. Dairy Sci.* 47 (8): 827–830, ISSN 0022-0302, https://doi.org/10.3168/jds.S0022-0302(64)88784-1.

**25** Brehm, L., Jünger, M., Frank, O., and Hofmann, T. (2019). Discovery of a thiamine-derived taste enhancer in process flavors. *J. Agric. Food. Chem.* 67 (20): 5857–5865. https://doi.org/10.1021/acs.jafc.9b01832.

**26** Duyvis, M.G., Hilhorst, R., Laane, C. et al. (2002). Role of riboflavin in beer flavor instability: determination of levels of riboflavin and its origin in beer by fluorometric apoprotein titration. *J. Agric. Food. Chem.* 50 (6): 1548–1552. https://doi.org/10.1021/jf010511k.

**27** Running, C.A. and Hayes, J.E. (2016). Chapter 10 – Individual differences in multisensory flavor perception. In: *Woodhead Publishing Series in Food Science, Technology and Nutrition, Multisensory Flavor Perception* (ed. B. Piqueras-Fiszman and C. Spence), 185–210, ISBN 9780081003503. Woodhead Publishing https://doi.org/10.1016/B978-0-08-100350-3.00010-9.

**28** Sabikun, N., Allah Bakhsh, M., Rahman, S. et al. (2021). Volatile and nonvolatile taste compounds and their correlation with umami and flavor characteristics of chicken nuggets added with milkfat and potato mash. In: *Food Chemistry*, vol. 343, 128499, ISSN 0308-8146, https://doi.org/10.1016/j.foodchem.2020.128499.

**29** Zhang, P. and Zhou, Z. (2019). Postharvest ethephon degreening improves fruit color, flavor quality and increases antioxidant capacity in 'Eureka' lemon (Citrus limon (L.) Burm. f.). *Sci. Horticul.* 248: 70–80, , ISSN 0304-4238. https://doi.org/10.1016/j.scienta.2019.01.008.

**30** Escribano-Viana, R., González-Arenzana, L., Portu, J. et al. (2018). Wine aroma evolution throughout alcoholic fermentation sequentially inoculated with non-Saccharomyces/Saccharomyces yeasts. *Food Res. Int.* 112: 17–24, ISSN 0963-9969, https://doi.org/10.1016/j.foodres.2018.06.018.

**31** van Wyk, N., Grossmann, M., Wendland, J. et al. (2019). The whiff of wine yeast innovation: strategies for enhancing aroma production by yeast during wine fermentation. *J. Agric. Food. Chem.* 67 (49): 13496–13505. https://doi.org/10.1021/acs.jafc.9b06191.

**32** Qin, Z., Petersen, M.A., and Bredie, W.L.P. (2018). Flavor profiling of apple ciders from the UK and Scandinavian region. *Food Res. Int.* 105: 713–723, , ISSN 0963-9969. https://doi.org/10.1016/j.foodres.2017.12.003.

**33** Zhao, X., Procopio, S., and Becker, T. (2015). Flavor impacts of glycerol in the processing of yeast fermented beverages: a review. *J. Food Sci. Technol.* 52 (12): 7588–7598. https://doi.org/10.1007/s13197-015-1977-y.

**34** Ramalingam, V., Song, Z., and Hwang, I. (2019). The potential role of secondary metabolites in modulating the flavor and taste of the meat. *Food Res. Int.* 122: 174–182, , ISSN 0963-9969. https://doi.org/10.1016/j.foodres.2019.04.007.

**35** Sánchez-Rodríguez, L., Ali, N.S., Cano-Lamadrid, M. et al. (2019). Chapter 18 – Flavors and aromas. In: *Postharvest Physiology and Biochemistry of Fruits and Vegetables* (ed. E.M. Yahia), 385–404, ISBN 9780128132784. Woodhead Publishing https://doi.org/10.1016/B978-0-12-813278-4.00019-1.

**36** Hu, K., Jin, G.J., Mei, W.C. et al. (2018 Jan). Increase of medium-chain fatty acid ethyl ester content in mixed *H. uvarum/S. cerevisiae* fermentation leads to wine fruity aroma enhancement. *Food Chem.* 15 (239): 495–501. https://doi.org/10.1016/j.foodchem.2017.06.151. PMID: 28873596.

**37** Wink, M. (2016). Secondary Metabolites, the Role in Plant Diversification of. In: *Encyclopedia of Evolutionary Biology* (ed. R.M. Kliman), 1–9, ISBN 9780128004265. Academic Press https://doi.org/10.1016/B978-0-12-800049-6.00263-8.

**38** Dey, P., Kundu, A., Kumar, A. et al. (2020). Analysis of alkaloids (indole alkaloids, isoquinoline alkaloids, tropane alkaloids). *Recent Adv. Nat. Prod. Anal.* 505–567: https://doi.org/10.1016/B978-0-12-816455-6.00015-9.

**39** Bassoli, A., Borgonovo, G., and Busnelli, G. (2007). Alkaloids and the bitter taste. In: *Modern Alkaloids* (ed. E. Fattorusso and O. Taglialatela-Scafati). https://doi.org/10.1002/9783527621071.ch3.

**40** Allison, A.-M.A. and Chambers, E. (2000). The importance of caffeine as a flavor component in beverages. *J. Sens. Stud.* 15: 449–457. https://doi.org/10.1111/j.1745-459X.2000.tb00282.x.

**41** Gorgani, L., Mohammadi, M., Najafpour, G.D., and Nikzad, M. (2017). Piperine—the bioactive compound of black pepper: from isolation to medicinal formulations. *Compr. Rev. Food Sci. Food Saf.* 16: 124–140. https://doi.org/10.1111/1541-4337.12246.

**42** Mikaili, P., Maadirad, S., Moloudizargari, M. et al. (2013). Therapeutic uses and pharmacological properties of garlic, shallot, and their biologically active compounds. *Iran. J. Basic Med. Sci.* 16 (10): 1031–1048.

**43** Bischoff, K.L. (2016). Chapter 40 – Glucosinolates. In: (ed. R.C. Gupta), 551–554, ISBN 9780128021477. Nutraceuticals, Academic Press https://doi.org/10.1016/B978-0-12-802147-7.00040-1.

**44** de Carvalho, C.C.C.R. and da Fonseca, M.M.R. (2006). Biotransformation of terpenes. *Biotechnol. Adv.* 24 (2): 134–142, ISSN 0734-9750, https://doi.org/10.1016/j.biotechadv..2005.08.004.

**45** Vezzaro, A., Krause, S.T., Nonis, A. et al. (2012). Isolation and characterization of terpene synthases potentially involved in flavor development of ripening olive (Olea europaea) fruits. *J. Plant Physiol.* 169 (9): 908–914, , ISSN 0176-1617. https://doi.org/10.1016/j.jplph.2012.01.021.

**46** Caputi, L. and Aprea, E. (2011). Use of terpenoids as natural flavouring compounds in food industry. *Recent Pat. Food Nutr. Agri.* 3 (1): https://doi.org/10.2174/2212798411103010009.

**47** Chen, W., Vermaak, I., and Viljoen, A. (2013). Camphor--a fumigant during the Black Death and a coveted fragrant wood in ancient Egypt and Babylon--a review. *Molecules* 18 (5): 5434–5454. https://doi.org/10.3390/molecules18055434.

**48** Buckle, J. (2015). Chapter 3 – Basic plant taxonomy, basic essential oil chemistry, extraction, biosynthesis, and analysis. In: *Clinical Aromatherapy*, 3e (ed. J. Buckle), 37–72. Churchill Livingstone, ISBN 9780702054402, https://doi.org/10.1016/B978-0-7020-5440-2.00003-6.

**49** Ren, Y., Wang, C., Xu, J., and Wang, S. (2019). Cafestol and kahweol: a review on their bioactivities and pharmacological properties. *Int. J. Mol. Sci.* 20 (17): 4238. https://doi.org/10.3390/ijms20174238.

**50** Allouche, Y., Jiménez, A., Uceda, M. et al. (2009). Triterpenic content and chemometric analysis of virgin olive oils from forty olive cultivars. *J. Agric. Food. Chem.* 57 (9): 3604–3610. https://doi.org/10.1021/jf803237z. PMID: 19326867.

**51** Becerra, M.O., Contreras, L.M., Lo, M.H. et al. (2020). Lutein as a functional food ingredient: Stability and bioavailability. *J. Funct. Foods* 66: 103771, ISSN 1756-4646, https://doi.org/10.1016/j.jff.2019.103771.

**52** Yahia, E.M. and Gutierrez-Orozco, F. (2011). Chapter 17 – Sapodilla (*Manilkara achras* (Mill) Fosb., syn *Achras sapota* L.). In: *Woodhead Publishing Series in Food*

*Science, Technology and Nutrition, Postharvest Biology and Technology of Tropical and Subtropical Fruits* (ed. E.M. Yahia), 351–363e, ISBN 9780857090904. Woodhead Publishing https://doi.org/10.1533/9780857092618.351.

**53** Praet, T., Van Opstaele, F., Jaskula-Goiris, B. et al. (2012). Biotransformations of hop-derived aroma compounds by Saccharomyces cerevisiae upon fermentation. *Cerevisia* 36 (4): 125–132, , ISSN 1373-7163. https://doi.org/10.1016/j.cervis.2011.12.005.

**54** Lehner, T.B. and Siegmund, B. (2020). The impact of ventilation during postharvest ripening on the development of flavour compounds and sensory quality of mangoes (*Mangifera indica* L.) cv. Kent. *Food Chem.* 320: 126608, ISSN 0308-8146, https://doi.org/10.1016/j.foodchem.2020.126608.

**55** Swanson, B.G. (2003). Tannins and polyphenols. In: *Encyclopedia of Food Sciences and Nutrition*, 2e (ed. B. Caballero), 5729–5733, ISBN 9780122270550. Academic Press https://doi.org/10.1016/B0-12-227055-X/01178-0.

**56** Martillanes, S., Rocha-Pimienta, J., Cabrera-Bañegil, M. et al. (2017). Application of phenolic compounds for food preservation: food additive and active packaging, phenolic compounds – biological activity. In: (ed. M. Soto-Hernandez, M. Palma-Tenango and M. del Rosario Garcia-Mateos). IntechOpen https://doi.org/10.5772/66885.

**57** Luna-Guevara, M.L., Luna-Guevara, J.J., Hernández-Carranza, P. et al. (2018). Chapter 3 – Phenolic compounds: a good choice against chronic degenerative diseases. In: *Studies in Natural Products Chemistry*, vol. 59 (ed. Atta-ur-Rahman), 79–108. Elsevier, ISSN 1572-5995, ISBN 9780444641793, https://doi.org/10.1016/B978-0-444-64179-3.00003-7.

**58** Zhao, H. (2015). Chapter 64 – Effects of processing stages on the profile of phenolic compounds in beer. In: *Processing and Impact on Active Components in Food* (ed. V. Preedy), 533–539, ISBN 9780124046993. Academic Press https://doi.org/10.1016/B978-0-12-404699-3.00064-0.

**59** He, F., Liang, N.N., Mu, L. et al. (2012). Anthocyanins and their variation in red wines I. monomeric anthocyanins and their color expression. *Molecules* 17 (2): 1571–1601. https://doi.org/10.3390/molecules17021571.

**60** Kandylis, P. and Kokkinomagoulos, E. (2020). Food applications and potential health benefits of pomegranate and its derivatives. *Foods* 9 (2): 122. https://doi.org/10.3390/foods9020122.

**61** Coumarin (2016). *Meyler's Side Effects of Drugs*, 16e (ed. J.K. Aronson), 701, ISBN 9780444537164. Elsevier https://doi.org/10.1016/B978-0-444-53717-1.00555-2.

**62** Zhang, Y., Smuts, J.P., Dodbiba, E. et al. (2012). Degradation study of carnosic acid, carnosol, rosmarinic acid, and rosemary extract (*Rosmarinus officinalis* L.) assessed using HPLC. *J. Agric. Food. Chem.* 60 (36): 9305–9314. https://doi.org/10.1021/jf302179c.

**63** Kiokias, S., Proestos, C., and Oreopoulou, V. (2020). Phenolic acids of plant origin-a review on their antioxidant activity in vitro (O/W emulsion systems) along with their in vivo health biochemical properties. *Foods* 9 (4): 534. https://doi.org/10.3390/foods9040534.

**64** Jones, M.G. (2008). Chapter 5 – Formation of vegetable flavour. In: *Woodhead Publishing Series in Food Science, Technology and Nutrition, Fruit and Vegetable Flavour* (ed. B. Brückner and S.G. Wyllie), 71–102, ISBN 9781845691837. Woodhead Publishing https://doi.org/10.1533/9781845694296.2.71.

**65** Ardö, Y. (2006). Flavour formation by amino acid catabolism. *Biotechnol. Adv.* 24 (2): 238–242. https://doi.org/10.1016/j.biotechadv.2005.11.005., PMID: 16406465.

**66** Bettenhausen, H.M., Barr, L., Broeckling, C.D. et al. (2018). Influence of malt source on beer chemistry, flavor, and flavor stability. *Food Res. Int.* 113: 487–504. ISSN 0963-9969. https://doi.org/10.1016/j.foodres.2018.07.024.

**67** Almarie, A.A.A. (2020). Roles of terpenoids in essential oils and its potential as natural weed killers: recent developments. In: *Essential Oils – Bioactive Compounds, New Perspectives and Applications* (ed. M.S. de Oliveira, W.A. da Costa and S.G. Silva). IntechOpen https://doi.org/10.5772/intechopen.91322. https://www.intechopen.com/chapters/72424.

**68** Pott, D.M., Sonia, O., and Vallarino José, G. (2019). From central to specialized metabolism: an overview of some secondary compounds derived from the primary metabolism for their role in conferring nutritional and organoleptic characteristics to fruit. *Front. Plant Sci.* 10: 835. https://www.frontiersin.org/article/10.3389/fpls.2019.00835.

**69** Schwab, W., Davidovich-Rikanati, R., and Lewinsohn, E. (2008). Biosynthesis of plant-derived flavor compounds. *Plant J.* 54: 712–732. https://doi.org/10.1111/j.1365-313X.2008.03446.x.

**70** Zhang, X., Wei, J., Zhao, S. et al. (2021). Flavor differences between commercial and traditional soybean paste. *LWT* 142: 111052, ISSN 0023-6438, https://doi.org/10.1016/j.lwt.2021.111052.

**71** Kumari, A. (2018). Chapter 17 – Purine structures. In: *Sweet Biochemistry* (ed. A. Kumari), 89–91, ISBN 9780128144534. Academic Press https://doi.org/10.1016/B978-0-12-814453-4.00017-0.

**72** O'Connor, S.E. (2010). Chapter 1.25 – Alkaloids. In: *Comprehensive Natural Products II* (ed. H.-W. (Ben) Liu and L. Mander), 977–1007, ISBN 9780080453828. Elsevier https://doi.org/10.1016/B978-008045382-8.00013-7.

**73** Kotani, K., Toyota, T., and Yasumatsu, K. (1992). 5′-Ribonucleotides as flavor enhancers. In: *Developments in Food Science*, vol. 28 (ed. G. Charalambous), 523–546, ISSN 0167-4501, ISBN 9780444885586. Elsevier https://doi.org/10.1016/B978-0-444-88558-6.50026-X.

**74** Howitt, C.A. and Pogson, B.J. (2006). Carotenoid accumulation and function in seeds and non-green tissues. *Plant Cell Environ.* 29: 435–445. https://doi.org/10.1111/j.1365-3040.2005.01492.x.

**75** Ianni, A., Bennato, F., Martino, C. et al. (2020). Volatile flavor compounds in cheese as affected by ruminant diet. *Molecules* 25 (3): 461. https://doi.org/10.3390/molecules25030461.

**76** Cole, A.S. and Eastoe, J.E. (1988). Chapter 17 – Carbohydrate metabolism. In: *Biochemistry and Oral Biology*, 2e (ed. A.S. Cole and J.E. Eastoe), 224–248, ISBN 9780723617518. Butterworth-Heinemann https://doi.org/10.1016/B978-0-7236-1751-8.50024-7.

**77** Doelle, H.W. (1975). Chapter 5 – Carbohydrate metabolism. In: *Bacterial Metabolism*, 2e (ed. H.W. Doelle), 208–311, ISBN 9780122193521. Academic Press https://doi.org/10.1016/B978-0-12-219352-1.50009-8.

**78** Siegmund, B. (2014). Chapter 7 – Biogenesis of aroma compounds: Flavor formation in fruits and vegetables. In: *Flavor Development, Analysis and Perception in Food and Beverages*, Woodhead Publishing Series in Food Science, Technology and Nutrition (ed. J.K. Parker, S. Elmore and L. Methven). Elsevier, ISBN 1782421114, 9781782421115.

**79** Kulkarni, R., Chidley, H., Deshpande, A. et al. (2013). An oxidoreductase from 'Alphonso' mango catalyzing biosynthesis of furaneol and reduction of reactive carbonyls. *Springerplus* (2): 494. https://doi.org/10.1186/2193-1801-2-494, PMID: 24133645; PMCID: PMC3797322.

**80** French, D. (2017). Chapter 5 – Advances in clinical mass spectrometry. In: *Advances in Clinical Chemistry*, vol. 79 (ed. G.S. Makowski), 153–198, ISSN 0065-2423, ISBN 9780128120767. Elsevier https://doi.org/10.1016/bs.acc.2016.09.003.

**81** Huang, Y., Rasco, B.A., and Cavinato, A.G. (2009). Chapter 13 – Fruit juices. In: *Infrared Spectroscopy for Food Quality Analysis and Control* (ed. D.-W. Sun), 355–375, ISBN 9780123741363. Academic Press https://doi.org/10.1016/B978-0-12-374136-3.00013-4.

**82** Papagianni, M. (2011). Chapter 1.09 – Organic acids. In: *Comprehensive Biotechnology*, 2e (ed. M. Moo-Young), 109–120. Academic Press, ISBN 9780080885049, https://doi.org/10.1016/B978-0-08-088504-9.00011-8.

# 3

# A Sense of Design: Pathway Unravelling and Rational Metabolic Flow Switching for the Production of Novel Flavor Materials

*Nimisha P. Sukumaran[1], Joby Jacob[1], Sabu Thomas[2], and Sreeraj Gopi[1]*

[1] *Aurea Biolabs (P) Ltd, R&D Centre, Kolenchery, Cochin, Kerala, 682311, India*

[2] *International and Inter University Center for Nanoscience and Nanotechnology, Kottayam, Kerala, 686560, India*

## 3.1 Introduction

As globalization progresses, consumers are exposed to many unique food products from various cultures, and a need to satisfy consumers' growing demands also increases. Moreover, one of the biggest challenges in creating a new food product is predicting how it will be accepted by consumers. Individual liking is critical for product acceptance and hence for product quality, thereby making flavor a critical feature [1]. Thus, flavors play an important role in consumer preferences for many products, and flavor researchers are constantly pushed toward further research strategies. The strategies include breeding efforts to focus on genetic improvement of a smaller set of important volatiles that would benefit much toward more efficient and cheaper ways of stabilizing and crafting delicate and enhanced flavor ingredients. Flavors encountered in everyday life are almost always the result of multicomponent mixtures of volatiles [2] that are evolved from primary metabolism through specific and specialized metabolic pathways. In particular, secondary metabolites thus formed from these pathways serve to protect the plant against biotic and abiotic stresses. Consequently, several organoleptic characteristics such as aroma, color, and fruit nutritional value rely upon the secondary metabolite content of a plant [3]. Thus, a strategy to modulate or control biosynthetic pathways endogenously has to be devised. This endogenous modulation by fine-tuning the expression of pathway enzymes affects target production levels of multiple commercially important flavor metabolites. This diminishes the unwanted intermediates and increases the product yield, which always represented a challenge.

In general terms, targeted and untargeted primary and secondary metabolome fingerprinting can be used to explore and delineate informative patterns and to provide insights into the key flavor molecules and its precursors [4]. This is because the unique molecules found after targeted and untargeted metabolic profiling often

*Natural Flavours, Fragrances, and Perfumes: Chemistry, Production, and Sensory Approach*, First Edition. Edited by Sreeraj Gopi, Nimisha Pulikkal Sukumaran, Joby Jacob, and Sabu Thomas.

perform as excellent candidates for artisanal and industrial metabolic engineering of flavor-active compounds. The substantial interest in this includes both the biosynthesis and large-scale production of flavor-active compounds. Figure 3.1 depicts examples of interesting pathways leading to unique flavors. Briefly, metabolic profiling is a method for the identification and quantification of as many pre-defined small molecule metabolites as possible occurring within a system and generally associated with a specific metabolic pathway. For instance, once natural genetic variations responsible for volatile biosynthesis within heirloom tomato populations were exploited, DNA marker-assisted breeding techniques were utilized to introgress desirable alleles with modern genetic backgrounds into genes regulating biosynthesis of volatiles. Thus, the metabolomics field can be used to catalog new compounds on the basis of their functions, and roles beyond the metabolism can be elucidated. Similarly, another promising approach is metabolic fingerprinting, which is a high-throughput screening tool for samples with a different biological status or origin [5].

In another way, metabolic engineering is the intentional modification of cellular metabolism for the production of desired compounds. Thus, with recombinant DNA technology, the metabolic pathways of host organisms can be manipulated. Host organisms include bacteria, fungi, plant, and animal cells, and their selection is based on the properties of their interesting metabolic pathways and intermediates. Important industrial products include amino acids, biofuels, secondary metabolites, recombinant proteins, and polymers [6]. Despite being used chiefly for fermenting sugars of grape to alcohol, wine yeasts (most prominently *Saccharomyces cerevisiae*) play a pivotal role in the final aroma profiles of wines. For instance, strain selection, intentionally incorporating non-*Saccharomyces* yeast in mixed culture fermentation, and genetic modifications of *S. cerevisiae* enhance the chemical composition and sensory profile of wines [7]. Similarly, acetic acid, butanoic acid, dimethyl trisulfide, methional, hexanal, (*E*)-2-nonenal, acetoin, 1-octen-3-one, δ-dodecalactone, furaneol, hexanoic acid, heptanal, and ethyl caproate are the main compounds in cheddar cheese, and their content differ with ripening time. Thus, a study on metabolism and its manipulation could provide information for researching and developing cheddar cheese-related products [8]. Likewise, a metabolite analysis on peach aroma suggested a crucial association between peach-like volatiles and "ethylene production and modulation of fatty acid levels." Briefly, along with fatty acid metabolism and lipoxygenase/β-oxidation pathway, the ethylene signal transduction through transduction genes like PpaSAMS1/2, PpaACS1/2, PpaACO1, PpaETR1/2, PpaERS1, PpaEIN4, and PpaCTR1 has molecular cross-talks for the characteristic peach-like aroma [9].

Genome sequencing provides a list of genes involved in the biosynthesis of flavor compounds and evidence for lineage-specific expansions of genes associated with flavor biosynthesis. It also helps to reveal the variations in metabolites and gene expression among tissues and different plant species [10]. Continued efforts to identify more functional genes and enzymes that control secondary metabolite production, and a multidisciplinary approach with the integration of information from genomics, proteomics, metabolomics, and synthetic biology are essential for rapid progress and successful and economically viable biotechnological production of secondary metabolites. For instance, fruit ripening is now considered a genetically programmed event involving the regulated expression of specific genes, where ethylene has a role in controlling fruit respiration, softening, and color changes [11]. The postharvest ripening process increases flavor quality due to changes in enzymatic ripening reactions, which change the key taste compounds. The general trends for the taste compounds are an increase in sweetness due to accumulation of glucose and fructose (reflected as an increase in total soluble solids) and a decrease of sourness due to degradation of organic acids (reflected as a decrease in titratable acidity) [12]. Similarly, the degradation of terpenes leads to reduced formation of reaction products from the lipoxygenase pathway, for instance, less pronounced fruitiness and flavor in mango. Thus, the degradation of terpenes occurs when the fruit is arranged densely after harvest [13]. On the other hand, the trends for volatile compounds differ according to their chemical families: aldehydes and alcohols decrease while esters increase during postharvest [12].

Systems metabolic engineering, which integrates traditional metabolic engineering with systems biology, synthetic biology, and evolutionary engineering, has recently emerged and been used to facilitate the construction of high-performance strains. Moreover, many tools and strategies of systems metabolic engineering are developed for construction of cell factories and chemical biosynthesis [14]. Various engineering strategies include rational metabolic engineering, evolutionary engineering, inverse metabolic engineering, engineering of secretory pathways, and process optimization. Consequently, integration of transcriptomic and metabolomic analyses and identifying the *in vivo* functions of protein-encoding genes can be implemented to increase productivity. Generally, metabolic engineering applications are driven by a desire to decrease costs, increase yields, increase efficiency, increase consumer acceptability, and use renewable feedstocks to provide a more environmentally friendly process than traditional chemical approaches [6]. Genome-scale bioinformatics approaches are also highly useful as the first step toward predicting gene or enzyme function, which subsequently can be validated via the use of reverse genetic or heterologous expression followed by biochemical characterization of recombinant proteins.

In particular, with plants producing miscellaneous series of low-molecular-weight compounds, these secondary metabolites confer metabolic flexibility. Briefly, Figure 3.1 represents the pictorial representation of commercially important metabolic pathways and the exceptional flavors generated from them. For all these reasons, metabolic engineering and biotechnological approaches require an understanding of the regulation of the secondary metabolite pathways involved and the identification of enzymes and genes, which have to be explored and exploited. The chapter highlights the great potential for research strategies to be developed without major collateral metabolic changes and with the generation of more sensobolome data from multiple sources. Moreover, general strategies of systems metabolite engineering with functional expansion and application for metabolic model construction, design, and construction of cell factories and biological systems are discussed.

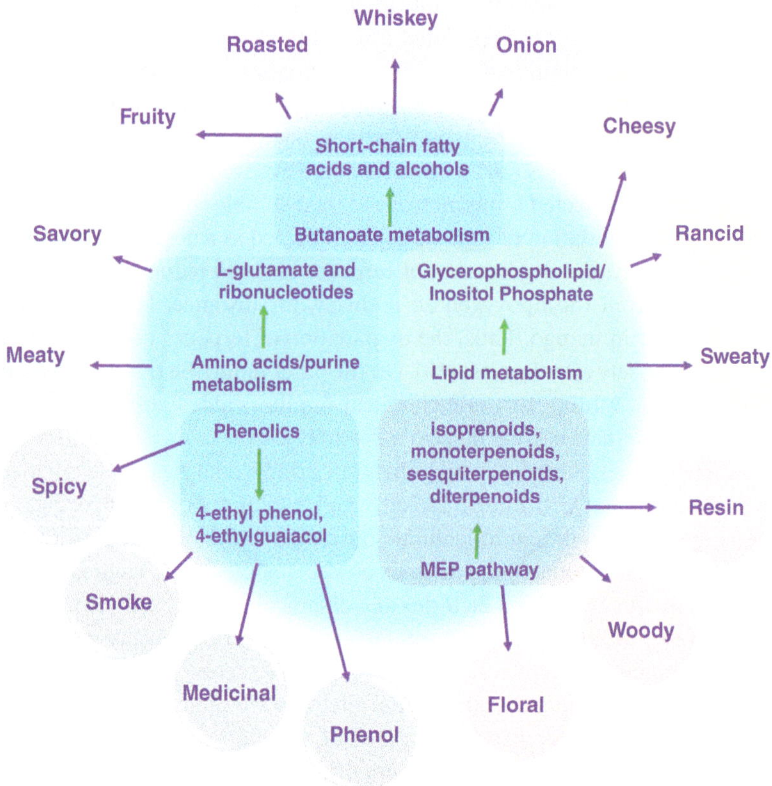


**Figure 3.1** Commercially important metabolic pathways and examples of flavors generated.

## 3.2 Elicitation of Plants

Elicitation is one of the approaches used for enhanced commercial production of secondary metabolites from the plant cell culture system. Elicitors are biological, chemical, and physical external factors able to switch on enzymatic responses by inducing specific responses to boost the production of secondary metabolites in plant cultures; elicitors are categorized according to their origin and molecular structure as biotic or abiotic [15]. For the most part, the elicitation response induced to enhance the production of secondary metabolites depends on the complex interactions between the elicitor and the plant cell, which in turn is affected by certain factors, including specificity, culture conditions (growth stage, medium composition, light), concentration, and treatment interval [16]. Nevertheless, elicitors that can elicit the production of a particular metabolite in certain plants can be inactive in other species; moreover, different plant species can be responsive to the same elicitor [17]. In detail, the molecular mechanism of action of elicitors on secondary metabolism is through the intracellular signal transduction systems, chiefly, plant elicitor receptors, GTP binding proteins, PLC/IP3-DAG/PKC and adenylyl cyclase/cAMP/PKA pathways, $Ca^{2+}$ messenger system, and PI3K-III type (yeast Vps34-like). Furthermore, the recognition of elicitation stimulus leads to the activation of specific genes through mitogen-activated protein kinases (MAPKs): SIMK, SAMK, SIPK, and WIPK [16].

### 3.2.1 Biotic Elicitors

The elicitors of biological origin derived from the pathogen or from the plant itself are termed to be endogenous or biotic elicitors [16]. For decades, the use of biological preparations as elicitors has become one of the most effective strategies to induce phenylpropanoid/flavonoid biosynthetic pathways in plant cells [18]. For instance, during the fermentation of cocoa beans, a positive interaction between fructophilic lactic acid bacteria and yeast is a microbial consortium that could improve sugar metabolism and aroma formation. This occurs due to the notable genomic features of fructophilic lactic acid bacteria that carry the alcohol/acetaldehyde dehydrogenase gene, which results in an improved formation of primary (ethanol, lactic acid, and acetic acid) and secondary (2-methyl-1-butanol, isoamyl acetate, and ethyl acetate) metabolites during fermentation [19]. Thus, fructophilic lactic acid bacteria act as a biotic elicitor that causes important reactions of color development and formation of key flavor molecules in cocoa. Similarly, significant elicitation in the activity of peroxidase, a key enzyme in the phenyl-propanoid (PP) pathway which catalyzes the conversion of PP compounds to vanillin/vanillic acid, was achieved by the addition of the dry cell powder of *Candida versatalis* or glutathione [20]. Biotic elicitors (having biological origin) such as methyl jasmonate and salicylic acid have been reported to increase the production of secondary metabolites, namely isoflavonoid, betalain, saponin, and glucosinolates, through various mechanisms; these metabolites are natural plant growth regulator phytohormones that play key roles in the signal interplay during biotic and abiotic stresses [21].

### 3.2.2 Abiotic Elicitors

Different from biotic elicitors, the abiotic elicitors are the ones that do not have a biological origin, for instance, physical factors or chemical compounds [16]. Abiotic elicitors are produced by factors responsible for environmental stress, for example, environmental factors like temperature, humidity, light intensity, and supply of water, minerals, and $CO_2$ influence the growth of a plant and its secondary metabolite production [22]. These factors can be of chemical (inorganic salts, metal ions, and other compounds that disturb membrane integrity) and physical origin (UV irradiation, wound, saline stress, ozone, etc.) [23]. Multiple postharvest stresses are exerted on oolong tea during the manufacturing process, which induce the formation and accumulation of many important aroma compounds such as indole – a key floral aroma contributor of oolong tea. Briefly, under stress conditions, epigenetic modifications like changes in the DNA methylation levels of promoter sequences can regulate gene expression and subsequently regulate the mechanisms of aroma compound formation [24] For instance, exposure of 3-old-day sprouts of alfalfa, broccoli, and radish to high light intensity or chilling temperature resulted in higher total phenolic content [23]. Similarly, the application of polysaccharide based edible coatings with phenylalanine, an elicitor of fruit defense response, on avocado fruits reduced the natural decay of these fruits. Interestingly, the coated fruit had a better flavor than the control fruit, with a minor transcript upregulation of several genes such as lipoxygenase, heat-shock protein, and several transcripts in the phenylpropanoid pathway [25].

## 3.3 Transformation Within Cells

Advances in plant biotechnology provide various means to improve crop productivity and greatly contribute to sustainable agriculture. A significant advance in plant biotechnology is the cloning of secondary metabolite pathway genes or biotransformation within cells [26]. Biotransformed volatile compounds derived from primary metabolites have relevance for improving value-added flavor and health benefits from short-chain fatty acids along with potential probiotic benefits. For example, in horse gram sprouts, the prime contributors to the sensory attributes of fermented samples are fermentation-mediated volatile metabolites, namely acetic acid, eugenol, benzyl alcohol (benzene methanol), acetoin, 2,3-butanediol, and ethyl palmitate. Of which, eugenol was detected for the first time in fermented horse gram, which thus holds a new bioprocessing strategy for the production of health-beneficial aroma compounds [27]. Another example is lactone flavors with fruity, milky, coconut, and other aromas that are widely used in food and fragrance industries, which can be produced by biotransformation of plant-sourced hydroxyl fatty acids using yeast cell factories. The engineered strains could produce γ-dodecalactone from oleic acid and δ-decalactone from linoleic acid, as shown in Figure 3.2 [28].

Considerable progress has been made in the development of the process for producing 2-phenylethanol (2-PE) from L-phenylalanine through the Ehrlich pathway [29]. Similarly, the biotransformation of terpenes is of great interest because it allows the production of enantiomerically pure flavors and fragrances under mild reaction conditions [30]. Advances in biotechnology make it possible to synthesize natural flavors economically and successfully at the commercial scale. Also entire cells of microorganisms are very promising in biotransformation because the microorganisms can be easily generated and used in the fermenters. Moreover, the use of biotransformation systems allows biotechnology products to be labeled as natural [31]. There is still a typical concern about the use of engineered organisms for food production due to a negative consumer perception of genetically modified organisms (GMOs). However, one cannot neglect the fact that *S. cerevisiae*, the yeast used widely for the production of beer, bread, cider, and wine is the most resourceful eukaryotic model used for genetic engineering [32].

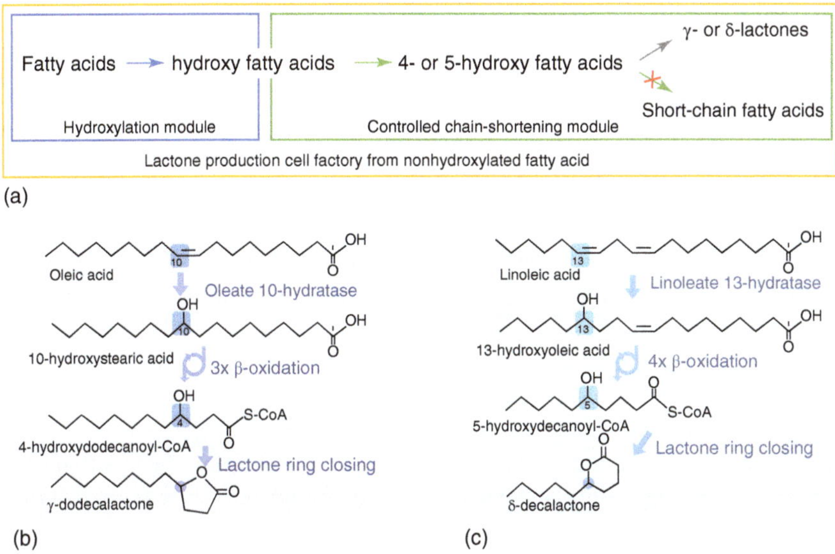


**Figure 3.2 Lactone production from nonhydroxylated fatty acids.** a) Free fatty acid is hydroxylated at positions that generate 4- or 5-hydroxy fatty acid after several β-oxidation cycles, which spontaneously converts into γ- or δ-lactone, respectively. Further degradation of these hydroxy fatty acids is prevented by replacing the native acyl-CoA oxidases with long chain-specific oxidases. Transformation of (b) oleic acid into γ-dodecalactone and (c) linoleic acid into δ-lactone. Source: Ref. [28]/Elsevier/ CC BY 4.0.

## 3.4 Metabolic Engineering

Systems metabolic engineering, which integrates traditional metabolic engineering with systems biology, synthetic biology, and evolutionary engineering, is enabling the development of microbial cell factories capable of efficiently producing a myriad of flavor and fragrance-related chemicals and materials [33]. Precisely, genetic engineering or recombinant DNA technology involves cutting and pasting desired DNA fragments that encode data for the expression of industrially important enzymes. Moreover, the production of natural flavors and fragrances through the microbial transformation process is an environmentally friendly process, and the products are considered "natural." For instance, 2-PE is important aromatic alcohol with a rose-like fragrance. It has been widely applied in cosmetic, perfume, and food industries and is mainly produced by chemical synthesis [29]. Consequently, strategies focusing on developments in metabolic model construction, design and construction of novel biological parts/systems, metabolic pathway reconstruction and optimization, and evolutionary strategies to improve cell factories remain the main challenge and perspective.

### 3.4.1 Upregulating Pathways with Transcription Factors

Transcription factors play many different roles, which vary according to the organism. Transcription factors are proteins possessing domains that bind to the DNA of promoter or enhancer regions of specific genes, which also possess a domain that interacts with RNA polymerase II or other transcription factors and consequently regulates the amount of messenger RNA (mRNA) produced by the gene [34]. Moreover, transcription factors are specific for certain types of cells and stages of development. Specific transcription factors are often very important in initiating patterns of gene expression that result in major developmental changes [34]. Thus, understanding how the sequential deployment of transcription factors controls differentiation and development is a vibrant current area of research, and it is important to understand how transcription factors drive development. Accordingly, the context of their function within the cell and in cellular adaptation, differentiation, and development is crucial [35].

For instance, the ripening process is regulated by transcription factors that activate or inhibit the expression of downstream ripening-related target genes by binding to their promoter regions and regulate the fruit ripening process. The transcription factors in the regulation of fruit ripening, including the MADS-box (MCM1, AGAMOUS, DEFICIENS, SRF-box) family and MYB (v-myb avian myeloblastosis viral oncogene homolog) family [36]. In peach fruits, alcohol acyltransferase PpAAT1 contributes to ester formation. Since volatile esters are major contributors to fruity flavor notes in many species, a study on the regulation of volatile synthesis pathways identified the transcription factor PpNAC1 as an activator of PpAAT1 expression and ester production [37]. Similarly, several hub genes for flavor compound metabolism were identified from apricot, and a regulatory transcriptional network was constructed. The co-expression network of transcription factors and structural genes potentially involved in the metabolism of sugars, organic acids, and the main aroma volatiles are depicted in Figure 3.3 [38]. Likewise, a tomato fruit ripening-specific transcription factor, SlLOB1, predominantly influences fruit cell wall-related gene regulation and textural changes during fruit maturation and thus induces softening of fruits [30].

### 3.4.2 Redirecting with Tailored Enzymes

The research on uncovering new enzymes or their new applications has become a focus of the food and fragrance industry. Moreover, microbial production of enzymes can lead to a greater availability as compared to other resources, which makes enzymes a competitive alternative [39]. By and large, tailored enzyme mixtures are in great demand and have extensive application in flavors, perfumery, cosmetics, luxury products, food and beverage industries, and pharmaceutical industry. Recent advances in developing tailored enzymes include reshaping enzyme specificities and mechanisms, and engineering biocatalysts through molecular assembly [40]. For instance, the study of the effect of yeast enzymes on hop glycosides indicates their high potential as precursors of highly flavor-active aroma compounds that contribute to the hoppy aroma of beer [39]. Hence, knowledge-based studies along with the combination of rational and combinatorial methods open up new vistas in

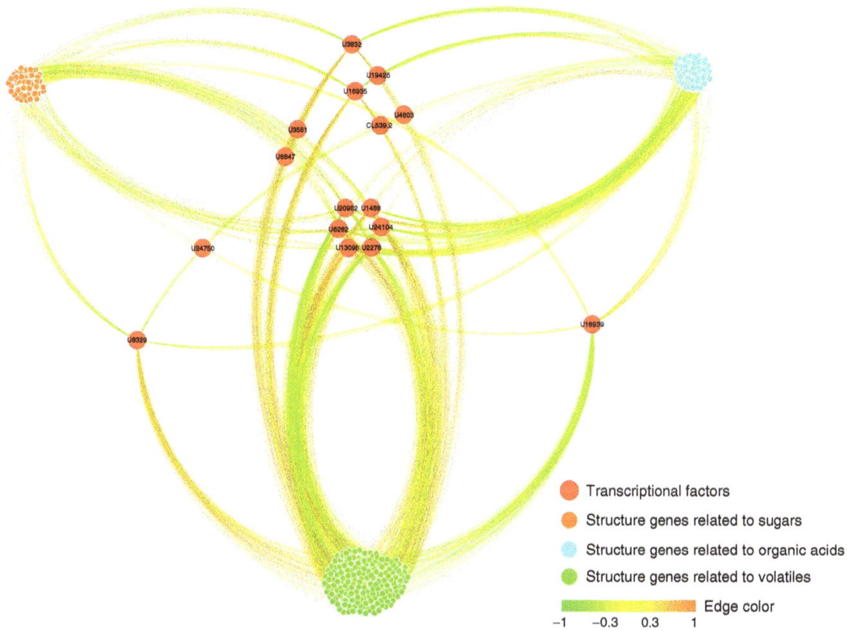


**Figure 3.3 A network of transcription factors related to flavor compound metabolism.** Red dots indicate transcription factors, orange dots indicate sugar structural genes, cyan dots indicate organic acid structural genes, and green dots indicate volatile structural genes. Source: Ref. [38]/Springer Nature/CC BY 4.0.

the design of stable and efficient enzymes. For instance, a wide range of common modifications are made to basic skeletons while tailoring enzymes in flavonoid biosynthesis, including the frequently occurring glycosylation, methylation, hydroxylation, acylation, oxidation, phosphorylation, and prenylation as well as rarer chemical modifications such as the addition of shikimate or quinol (Figure 3.4) [41].

### 3.4.3 Downregulating Pathways Using Knockout of the Gene/Enzyme

Gene knockouts are an important source of genetic variation that can be defined as alleles that disrupt or break a gene's molecular function. Since gene content varies considerably between and within crop species, discoveries in the functional molecular genomics promise its importance in crop evolution. Thus, breaking the right genes can have a positive impact on numerous traits, including yield [42]. Similarly, the protein-knockout system offers a powerful and versatile proteomic tool to dissect diverse functional properties of cellular proteins in cells. Thus, this approach involves specific engineering of the cellular ubiquitination machinery to directly remove specific proteins through accelerated proteolysis [43]. For instance, the enzyme alcohol acyltransferase is critical for ester biosynthesis, which is regulated by ethylene. For that reason, the emission of ethylene can reflect the production of volatile aroma compounds of apple to some extent; it has been demonstrated that

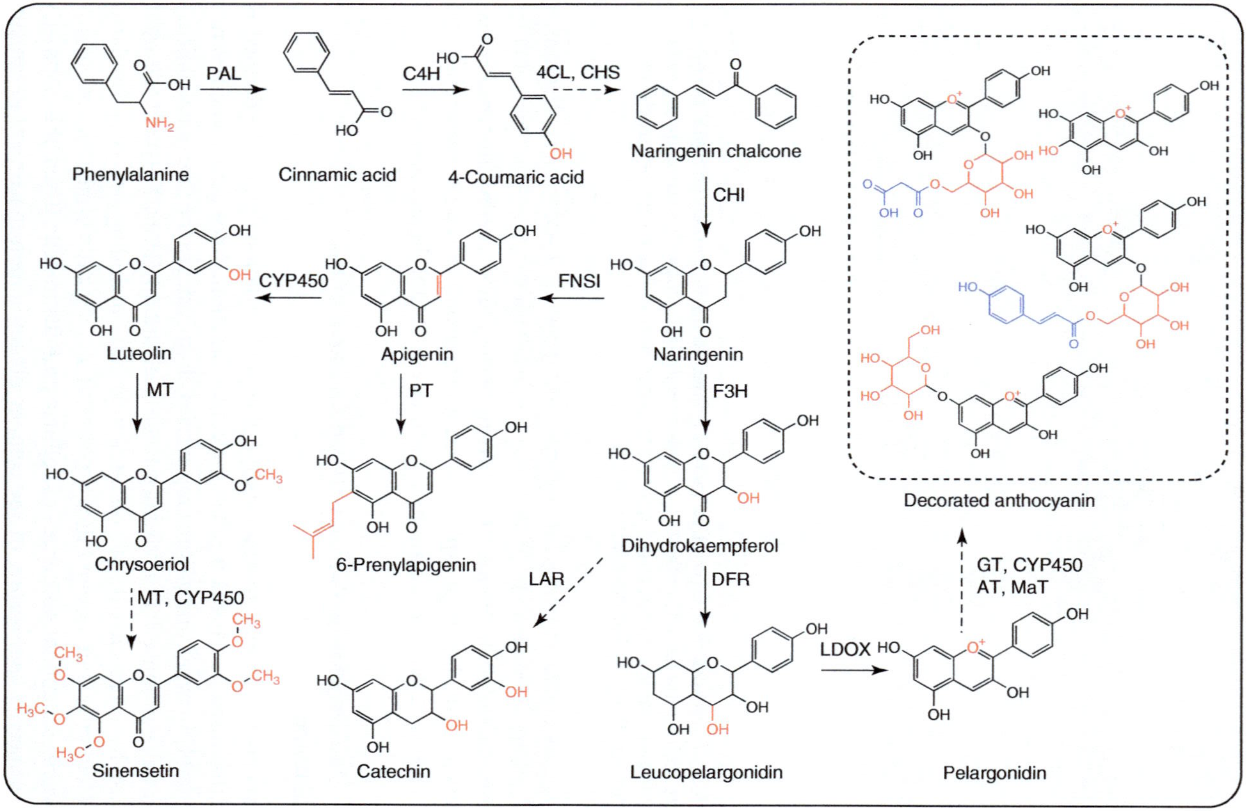


**Figure 3.4** Major tailoring enzymes and reactions in flavonoid biosynthesis. Source: Ref. [41]/Elsevier.

ester production in transgenic apple fruits, which was suppressed under ethylene biosynthesis because of carrying an antisense ACC oxidase gene, was significantly lower than that of ordinary apples [44]. A number of approaches that aim to downregulate the gene/enzyme have been developed over the years, which include RNA interference (RNAi), antisense deoxyoligonucleotides, ribozyme, and triple-helix DNA.

*RNA Interference*: RNAi has emerged as an endogenous cellular mechanism for controlling gene expression by a highly precise mechanism of sequence-directed gene silencing at the stage of translation [45]. Research on the use of RNAi for industrial applications has gained considerable momentum. This is mainly because the manipulations in gene expression for quality traits can now be easily achieved by RNAi. It can be employed by identifying the target gene(s) and developing vectors as an RNAi construct, transforming the plant, and finally screening and evaluating the traits [46]. RNAi technology has been used successfully for the modification of fatty acid composition in oils. Nutritional enhancement of cotton seed oil with high oleic and stearic contents was achieved by RNAi with two key fatty acid desaturase genes [47]. Similarly, the flavor and stability of soybean oil were improved using hairpin RNA-mediated RNAi strategy (glycinin promoter was used for seed-specific silencing) to downregulate omega-3 fatty acid desaturase (an enzyme converting linoleic acid to alpha-linolenic acid) [48].

## 3.5 Plant Tissue Culture

The production of high-value secondary metabolites, including flavors and food additives, can be achieved through plant cell cultures, shoot cultures, root cultures, and transgenic roots obtained through biotechnological means. Thus, plant cell and transgenic hairy root cultures serve as a promising potential alternative source for the production of high-value secondary metabolites of industrial importance [49]. For instance, the pungent food additive capsaicin, the natural color anthocyanin, and the natural flavor vanillin are produced on a large scale through plant tissue culture technology. Thus, bioengineering provides promising technical options for increasing the productivity of bioflavors, such as different fermentation strategies (batch, fed-batch, and continuous fermentation); gas-phase or two-phase reactions; specific reactor constructions such as membrane, solid-state, or closed loop reactors; optimization and modeling approaches of the bioprocess; and in situ recovery of the product [50]. Moreover, plant cell culture technology has now reached the point where a variety of culture types can be critically assessed as potential sources of existing novel flavors and pigments [51]. Yet, consistent production and high yields are the important factors in the commercial development of secondary metabolites. Hence, traditional and metabolic engineering approaches have been utilized to enhance the construction and yields of secondary metabolites [52].

## 3.6 Transgenic (Genetically Modified Organisms) Organisms

Latest developments in transgenic research have opened up the prospect of metabolic engineering of synthetic pathways to harvest commercially important secondary metabolites. Agricultural plants are one of the most often cited illustrations of GMOs [53]. Precisely, to achieve a transgenic organism, specific DNA sequences are characteristically introduced into the genome of an organism and expressed under the control of a specific promoter sequence to chaperon the cellular, temporal, and spatial localization of transgene expression [54]. It typically is a shortcut to the traditional genetic breeding and selection experiments that have led to varieties of food animals and plant crops for human consumption [55]. Moreover, the flavor quality of many fruits has significantly declined over recent decades, which is linked to the selection of certain traits such as firmness and postharvest shelf life that run counter to good flavor [56]. By addressing these challenges involved in traditional strategies, exceptional flavors and aromas from these transgenic organisms could come to dominate the natural product industry.

## References

1 Brückner, B. (2008). Chapter 2 – Consumer acceptance of fruit and vegetables: the role of flavour and other quality attributes. In: *Woodhead Publishing Series in Food Science, Technology and Nutrition, Fruit and Vegetable Flavour* (ed. B. Brückner and S.G. Wyllie), 11–17. Woodhead Publishing, ISBN 9781845691837, https://doi.org/10.1533/9781845694296.1.11.

2 Johnson, A.J., Hjelmeland, A.K., Heymann, H. et al. (2019). GC-Recomposition-Olfactometry (GC-R) and multivariate study of three terpenoid compounds in the aroma profile of Angostura bitters. *Sci. Rep.* 9: 7633. https://doi.org/10.1038/s41598-019-44064-y.

3 Pott, D.M., Sonia, O., and Vallarino José, G. (2019). From central to specialized metabolism: an overview of some secondary compounds derived from the primary metabolism for their role in conferring nutritional and organoleptic characteristics to fruit. *Front. Plant Sci.* 10: 835. https://www.frontiersin.org/article/10.3389/fpls.2019.00835.

4 Cialiè Rosso, M., Stilo, F., Bicchi, C. et al. (2021). Combined untargeted and targeted fingerprinting by comprehensive two-dimensional gas chromatography to track compositional changes on hazelnut primary metabolome during roasting. *Appl. Sci.* 11 (2): 525. https://doi.org/10.3390/app11020525.

5 Lang, C.P., Merkt, N., Klaiber, I. et al. (2019). Different forms of nitrogen application affect metabolite patterns in grapevine leaves and the sensory of wine. *Plant Physiol. Biochem.* 143: 308–319, ISSN 0981-9428, https://doi.org/10.1016/j.plaphy.2019.09.009.

6 Wuest, D.M., Hou, S., and Lee, K.H. (2011). Chapter 3.52 – Metabolic engineering. In: *Comprehensive Biotechnology*, 2e (ed. M. Moo-Young), 617–628. Academic Press, ISBN 9780080885049, https://doi.org/10.1016/B978-0-08-088504-9.00229-4.

7 van Wyk, N., Grossmann, M., Wendland, J. et al. (2019). The whiff of wine yeast innovation: strategies for enhancing aroma production by yeast during wine fermentation. *J. Agric. Food. Chem.* 67 (49): 13496–13505. https://doi.org/10.1021/acs.jafc.9b06191.

8 Wang, J., Yang, Z.J., Xu, L.Y. et al. (2021). Key aroma compounds identified in Cheddar cheese with different ripening times by aroma extract dilution analysis, odor activity value, aroma recombination, and omission. *J. Dairy Sci.* 104 (2): 1576–1590, ISSN 0022-0302, https://doi.org/10.3168/jds.2020-18757.

9 Cai, H., Han, S., Jiang, L. et al. (2019). 1-MCP treatment affects peach fruit aroma metabolism as revealed by transcriptomics and metabolite analyses. *Food Res. Int.* 122: 573–584, ISSN 0963-9969, https://doi.org/10.1016/j.foodres.2019.01.026.

10 Yu, X., Xiao, J., Chen, S. et al. (2020). Metabolite signatures of diverse *Camellia sinensis* tea populations. *Nat. Commun.* 11: 5586. https://doi.org/10.1038/s41467-020-19441-1.

11 Zhu, H.L., Zhu, B.Z., Fu, D.Q. et al. (2005). Role of ethylene in the biosynthetic pathways of aroma volatiles in ripening fruit. *Russ. J. Plant Physiol.* 52: 691–695. https://doi.org/10.1007/s11183-005-0103-9.

12 Sánchez-Rodríguez, L., Ali, N.S., Cano-Lamadrid, M. et al. (2019). Chapter 18 – Flavors and aromas. In: *Postharvest Physiology and Biochemistry of Fruits and Vegetables* (ed. E.M. Yahia), 385–404. Woodhead Publishing, ISBN 9780128132784, https://doi.org/10.1016/B978-0-12-813278-4.00019-1.

13 Lehner, T.B. and Siegmund, B. (2020). The impact of ventilation during postharvest ripening on the development of flavour compounds and sensory quality of mangoes (*Mangifera indica* L.) cv. Kent. *Food Chem.* 320: 126608, ISSN 0308-8146, https://doi.org/10.1016/j.foodchem.2020.126608.

14 Teng, M., Chen, S., Yin, X., and Zhang, G. (2020). Chapter 6 – System metabolic engineering strategies for cell factories construction. In: *Systems and Synthetic Metabolic Engineering* (ed. L. Liu, G. Du and Y. Liu), 125–151. Academic Press, ISBN 9780128217535, https://doi.org/10.1016/B978-0-12-821753-5.00006-X.

15 Lucini, L., Baccolo, G., Rouphael, Y. et al. (2018). Chitosan treatment elicited defence mechanisms, pentacyclic triterpenoids and stilbene accumulation in grape (*Vitis vinifera* L.) bunches. *Phytochemistry* 156: 1–8, ISSN 0031-9422, https://doi.org/10.1016/j.phytochem.2018.08.011.

16 Vasconsuelo, A. and Boland, R. (2007). Molecular aspects of the early stages of elicitation of secondary metabolites in plants. *Plant Sci.* 172 (5): 861–875, ISSN 0168-9452, https://doi.org/10.1016/j.plantsci.2007.01.006.

17 Héloir, M.C., Adrian, M., Brulé, D. et al. (2019). Recognition of elicitors in grapevine: from MAMP and DAMP perception to induced resistance. *Front. Plant Sci.* 10: 1117. https://doi.org/10.3389/fpls.2019.01117.

18 Naik, P.M. and Al–Khayri, J.M. (2016). *Abiotic and Biotic Elicitors–Role in Secondary Metabolites Production through in vitro Culture of Medicinal Plants, Abiotic and Biotic Stress in Plants - Recent Advances and Future Perspectives*

(ed. A.K. Shanker and C. Shanker). IntechOpen https://doi.org/10.5772/61442. https://www.intechopen.com/chapters/49289.

**19** Viesser, J.A., de Melo Pereira, G.V., de Carvalho Neto, D.P. et al. (2021). Co-culturing fructophilic lactic acid bacteria and yeast enhanced sugar metabolism and aroma formation during cocoa beans fermentation. *Int. J. Food Microbiol.* 339: 109015, ISSN 0168-1605, https://doi.org/10.1016/j.ijfoodmicro.2020.109015.

**20** Sreedhar, R.V., Roohie, K., Maya, P. et al. (2009). Biotic elicitors enhance flavour compounds during accelerated curing of vanilla beans. *Food Chem.* 112 (2): 461–468, , ISSN 0308-8146. https://doi.org/10.1016/j.foodchem.2008.05.108.

**21** Kumari, S., Dahuja, A., Vinutha, T. et al. (2015). Changes in the levels of off-flavor generation in soybean through biotic elicitor treatments. *J. Agric. Food. Chem.* 63 (2): 700–706. https://doi.org/10.1021/jf505199a, PMID: 25552290.

**22** Ramakrishna, A. and Ravishankar, G.A. (2011). Influence of abiotic stress signals on secondary metabolites in plants. *Plant Signaling Behav.* 6 (11): 1720–1731. https://doi.org/10.4161/psb.6.11.17613.

**23** Baenas, N., García-Viguera, C., and Moreno, D.A. (2014). Elicitation: a tool for enriching the bioactive composition of foods. *Molecules* 19 (9): 13541–13563. https://doi.org/10.3390/molecules190913541.

**24** Yang, J., Zhou, X., Wu, S. et al. (2021). Involvement of DNA methylation in regulating the accumulation of the aroma compound indole in tea (Camellia sinensis) leaves during postharvest processing. *Food Res. Int.* 142: 110183, ISSN 0963-9969, https://doi.org/10.1016/j.foodres.2021.110183.

**25** Saidi, L., Duanis-Assaf, D., Galsarker, O. et al. (2021). Elicitation of fruit defense response by active edible coatings embedded with phenylalanine to improve quality and storability of avocado fruit. *Postharvest Biol. Technol.* 174: 111442, ISSN 0925-5214, https://doi.org/10.1016/j.postharvbio.2020.111442.

**26** Ali, S. and Kim, W.C. (2019). A fruitful decade using synthetic promoters in the improvement of transgenic plants. *Front. Plant Sci.* (10): 1433. https://doi.org/10.3389/fpls.2019.01433.

**27** Goswami, R.P., Jayaprakasha, G.K., Shetty, K., and Patil, B.S. (2018). *Lactobacillus plantarum* and natural fermentation-mediated biotransformation of flavor and aromatic compounds in horse gram sprouts. *Process Biochem.* 66: 7–18, ISSN 1359-5113. https://doi.org/10.1016/j.procbio.2018.01.001.

**28** Marella, E.R., Dahlin, J., Dam, M.I. et al. (2020). A single-host fermentation process for the production of flavor lactones from non-hydroxylated fatty acids. *Metab. Eng.* 61: 427–436, ISSN 1096-7176, https://doi.org/10.1016/j.ymben.2019.08.009.

**29** Hua, D. and Xu, P. (2011). Recent advances in biotechnological production of 2-phenylethanol. *Biotechnol. Adv.* 29 (6): 654–660, ISSN 0734-9750. https://doi.org/10.1016/j.biotechadv.2011.05.001.

**30** Shi, Y., Vrebalov, J., Zheng, H. et al. (2021). A tomato lateral organ boundaries transcription factor, SlLOB1, predominantly regulates cell wall and softening components of ripening. *Proc. Natl. Acad. Sci. U.S.A.* 118 (33): e2102486118; https://doi.org/10.1073/pnas.2102486118.

**31** Verma, D.K., Al-Sahlany, S.T.G., Niamah, A.K. et al. (2021). Recent trends in microbial flavour compounds: a review on chemistry, synthesis mechanism and their application in food. *Saudi J. Biol. Sci.*, ISSN 1319-562X, https://doi.org/10.1016/j.sjbs.2021.11.010.

**32** Carrau, F., Gaggero, C., and Aguilar, P.S. (2015). Yeast diversity and native vigor for flavor phenotypes. *Trends Biotechnol.* 33 (3): 148–154, ISSN 0167-7799. https://doi.org/10.1016/j.tibtech.2014.12.009.

**33** Ko, Y.-S., Kim, J.W., Lee, J.A. et al. (2020). Tools and strategies of systems metabolic engineering for the development of microbial cell factories for chemical production. *Chem. Soc. Rev.* 49: 4615–4636. https://doi.org/10.1039/D0CS00155D.

**34** (2014). *Transcription Factors, Reference Module in Biomedical Sciences.* Elsevier, ISBN 9780128012383, https://doi.org/10.1016/B978-0-12-801238-3.05466-0.

**35** Minchin, S.D. and Busby, S.J.W. (2013). Transcription factors. In: *Brenner's Encyclopedia of Genetics*, 2e (ed. S. Maloy and K. Hughes), 93–96, ISBN 9780080961569. Academic Press https://doi.org/10.1016/B978-0-12-374984-0.01552-7.

**36** Liu, G.-S., Li, H.-L., Grierson, D., and Fu, D.-Q. (2022). NAC transcription factor family regulation of fruit ripening and quality: a review. *Cells* 11 (3): 525. https://doi.org/10.3390/cells11030525.

**37** Cao, X., Wei, C., Duan, W. et al. (2021). Transcriptional and epigenetic analysis reveals that NAC transcription factors regulate fruit flavor ester biosynthesis. *Plant J.* 106: 785–800. https://doi.org/10.1111/tpj.15200.

**38** Zhang, Q., Feng, C., Li, W. et al. (2019). Transcriptional regulatory networks controlling taste and aroma quality of apricot (*Prunus armeniaca* L.) fruit during ripening. *BMC Genom.* 20: 45. https://doi.org/10.1186/s12864-019-5424-8.

**39** Praet, T., Van Opstaele, F., Jaskula-Goiris, B. et al. (2012). Biotransformations of hop-derived aroma compounds by *Saccharomyces cerevisiae* upon fermentation. *Cerevisia* 36 (4): 125–132, , ISSN 1373-7163. https://doi.org/10.1016/j.cervis.2011.12.005.

**40** Cedrone, F., Ménez, A., and Quéméneur, E. (2000 Aug). Tailoring new enzyme functions by rational redesign. *Curr. Opin. Struct. Biol.* 10 (4): 405–410. https://doi.org/10.1016/s0959-440x(00)00106-8. , PMID: 10981626.

**41** Wang, S., Alseekh, S., Fernie, A.R., and Luo, J. (2019). The structure and function of major plant metabolite modifications. *Mol. Plant* 12 (7): 899–919, , ISSN 1674-2052. https://doi.org/10.1016/j.molp.2019.06.001.

**42** Flores, T., Karpova, O., Su, X. et al. (2008). Silencing of GmFAD3 gene by siRNA leads to low alpha-linolenic acids (18:3) of fad3-mutant phenotype in soybean [*Glycine max* (Merr.)]. *Trans. Res.* 17: 839–850. https://doi.org/10.1007/s11248-008-9167-6.

**43** Monroe, J.G., Arciniegas, J.P., Moreno, J.L. et al. (2020). The lowest hanging fruit: beneficial gene knockouts in past, present, and future crop evolution. *Curr. Plant Biol.* 24: 100185, ISSN 2214-6628, https://doi.org/10.1016/j.cpb.2020.100185.

**44** Zhang, J., Zheng, N., and Zhou, P. (2003). Exploring the functional complexity of cellular proteins by protein knockout. *Biological Sciences* 100 (24): 14127–14132. https://doi.org/10.1073/pnas.2233012100.

**45** Qi, W., Wang, H., Zhou, Z. et al. (2020). Ethylene emission as a potential indicator of fuji apple flavor quality evaluation under low temperature. *Horticul. Plant J.* 6 (4): 231–239, ISSN 2468-0141. https://doi.org/10.1016/j.hpj.2020.03.007.

**46** Skorecki, K. and Galun, E. (2012). Chapter 43 – Cell and gene therapy. In: *Goldman's Cecil Medicine*, 24e (ed. L. Goldman and A.I. Schafer), 203–211. W.B. Saunders, ISBN 9781437716047, https://doi.org/10.1016/B978-1-4377-1604-7.00043-9.

**47** Saurabh, S., Vidyarthi, A.S., and Prasad, D. (2014). RNA interference: concept to reality in crop improvement. *Planta* 239: 543–564. https://doi.org/10.1007/s00425-013-2019-5.

**48** Liu, Q., Singh, P.S., and Green, A.G. (2002). High-stearic and high-oleic cotton seed oils produced by hairpin RNA-mediated post-transcriptional gene silencing. *Plant Physiol.* 129: 1732–1743. https://doi.org/10.1104/pp.001933.

**49** Rao, S.R. and Ravishankar, G.A. (2002). Plant cell cultures: chemical factories of secondary metabolites. *Biotechnol. Adv.* 20 (2): 101–153, ISSN 0734-9750, https://doi.org/10.1016/S0734-9750(02)00007-1.

**50** Hosoglu, M.I., Guneser, O., and Yuceer, Y.K. (2018). Chapter 2 – Different bioengineering approaches on production of bioflavor compounds. In: *Handbook of Food Bioengineering, Role of Materials Science in Food Bioengineering* (ed. A.M. Grumezescu and A.M. Holban), 37–71, ISBN 9780128114483. Academic Press https://doi.org/10.1016/B978-0-12-811448-3.00002-4.

**51** Dörnenburg, H. and Knorr, D. (ed.) (1996). Generation of colors and flavors in plant cell and tissue cultures. *Crit. Rev. Plant Sci.* 15 (2): 141–168. https://doi.org/10.1080/07352689.1996.10393184.

**52** Chandran, H., Meena, M., Barupal, T., and Sharma, K. (2020). Plant tissue culture as a perpetual source for production of industrially important bioactive compounds. *Biotechnol. Rep.* 26: e00450. https://doi.org/10.1016/j.btre.2020.e00450.

**53** Phillips, T. (2008). Genetically modified organisms (GMOs): transgenic crops and recombinant DNA technology. *Nat. Edu.* 1 (1): 213.

**54** Payne, E. and Look, T. (2008). Chapter 8 – Animal models: flies, fish, and yeast. In: *The Molecular Basis of Cancer*, 3e (ed. J. Mendelsohn, P.M. Howley, M.A. Israel, et al.), 115–127, ISBN 9781416037033. W.B. Saunders https://doi.org/10.1016/B978-141603703-3.10008-1.

**55** Bock, J.H. and Norris, D.O. (2016). *Chapter 11 - Summation and a Look to the Future* (ed. J.H. Bock and D.O. Norris), 149–168, ISBN 9780128014752. Forensic Plant Science, Academic Press https://doi.org/10.1016/B978-0-12-801475-2.00011-7.

**56** Klee, H.J. (2010). Improving the flavor of fresh fruits: genomics, biochemistry, and biotechnology. *New Phytol.* 187: 44–56. https://doi.org/10.1111/j.1469-8137.2010.03281.x.

# Part III

# Flavor Technology

# 4

# Flavor Technology and Flavor Delivering Systems

*Anjali Anil*[1,2], *Józef T. Haponiuk*[2], *and Kunnirikka Sumith*[1]

[1] *Tropical flavors (P) Ltd., Kolenchery, Cochin, Kerala, 682311, India*
[2] *Gdansk University of Technology, Department of Polymer Technology, Gdańsk, Poland*

## 4.1 Introduction

Flavor is one of the most important components responsible for the overall sensory properties of taste and smell in any food products. Flavors can be the most expensive ingredients in any food formulation because they are usually delicate and volatile aroma compounds. Among the many organoleptic quality components such as color, rheological properties or packaging, Flavor takes a particular place through stimulating the odor and taste receptors when eating. The addition of flavor into foods can significantly influence the final product quality, cost, and customer satisfaction and influences the further consumption of food [1, 2]. However due to the volatility and delicate properties of volatile flavoring compounds, they are unstable [3].

Flavor is the sensory impression of food or other substance and is determined mainly by the chemical senses of taste and smell. The trigeminal senses, which detect chemical irritants in the mouth and throat as well as temperature and texture, are also very important to the overall gestalt of flavor perception. Flavor is also explained as a mixture of taste and odor sensations. It characterizes our eating experience activate physiological and psychological responses and eventually functions as quality control. The odor molecules must have certain volatility to reach nose epithelium, and it includes mainly aldehydes, esters, alcohols, ethers, pyrazines, sulfides, and furans. Taste is generally associated with flavors to provide the overall sensory experience during eating food. The flavor of food, as such, can be altered with natural or artificial flavorants, that affect these senses. Flavorant is defined as substance that gives another substance taste by altering the characteristics of the solute, causing it to become sweet, sour, tangy, bitter, salty, etc.

*Natural Flavours, Fragrances, and Perfumes: Chemistry, Production, and Sensory Approach,*
First Edition. Edited by Sreeraj Gopi, Nimisha Pulikkal Sukumaran, Joby Jacob, and Sabu Thomas.

At the beginning of 1900s, a growing number of food and beverage companies created more demand for commercial flavors. The Flavor and Extract Manufacturers Association (FEMA) is a food industry trade group founded in 1909 by several flavor firms in response to the passage of Pure Food and Drug Act of 1906 that plays an important role in creating program to assess the safety and GRAS status of flavor ingredients. EU regulation (EC) No 1334/2008 was adopted on 16 December 2008 and entered into force on 20 January 2009 that lays down general requirement for safe use of flavorings and provide definitions for different types of flavorings.

## 4.2 Flavor Delivery Systems

Flavor stability is an important factor in quality and acceptability of food. The physiochemical properties, concentration, and interactions of volatile aromatic molecules with food components or external factors cause degradation or loss of flavors during processing and storage [4]. A delivery system consists of a chemical or physical or physiochemical barrier between the active compounds and the external environment. Encapsulation technology is a process that physically entraps the sensitive material in a protective coating matrix. The entrapped material is called core or active material, while the coating material is called shell, carrier, wall material, or encapsulant. The encapsulation process involves the emulsification of sensitive core material with a dense carrier material, followed by drying or cooling of emulsion [1]. The most relevant encapsulation process involved in flavor delivery systems are micro encapsulation and nano encapsulation.

The role of delivery system includes the protection of flavors from oxidation, evaporation, and physical damage, avoid undesirable interaction between matrix and flavor compounds, to reduce interaction between different aromatic compounds, control and prolong release of flavors at specific time and site [5]. The selection of suitable wall material and encapsulation technique depend upon the final product application and processing condition for the manufacture of the food product. The flavor is released by means of diffusion or degradation, swelling or melting of wall material in food matrix [1].

### 4.2.1 Microencapsulation

Microencapsulation is the methodology of entrapping very minute particles of compounds that exist as liquids, solids or gases, using an inert shell or thin film, so as to protect the bioactive compounds from factors such as light, temperature, oxygen, and humidity. Microencapsulation act as a physical barrier to separate the core compound of the product from the physical factors mentioned above. The method also helps to convert liquid to solids, thereby altering the colloidal and surface properties and offering protection to products from environmental factors and control release characteristics. Microencapsulation technique is highly relevant in food industry as the sensitive food flavors are protected from hostile process environment. The organoleptic properties such as color, flavor, and taste are effectively masked, the core materials are released in a controlled manner, and easy handling is facilitated [6].

The tiny particles obtained through the process can be described as microspheres, microcapsules or micro particles and their size range from a few nanometers to a few millimeters. Based on the particle size, researchers categorized the particles into nano (less than 1 μm), micro (1–5000 μm) and macro (more than 5000 μm). The encapsulated food additive is described as the core material, active agent, fill, internal phase, or payload phase. On the other hand, the encapsulating substance is called coating, membrane, shell, or carrier material. The core material may consist of one or many additives such as colors, flavors, volatile oils, essential oils, and antioxidants. The coating may be made of one or several substances such as methyl cellulose, maltodextrins, alginate, gelatin, cyclodextrins, chitosan etc. The methodology of encapsulation includes spray drying, solvent evaporation, air suspension, coacervation, phase separation, freeze drying, extrusion, emulsification, fluidized bed coating, inclusion complexation, and liposome entrapment [6].

### 4.2.2 Nanoencapsulation

Food products under production and storage conditions undergo changes in flavor, loss of aroma because of reactions with other chemical components. Factors such as physicochemical properties, reactions between volatile aromatic compounds, food ingredients also affect the overall quality of product. Flavor stability increases the quality, acceptability, and consumption of the product. This is achieved with the help of nanotechnology by delivering encapsulated bioactive compounds to food thereby preventing the deterioration of flavor during processing and storage, as bioactive compounds play a crucial role in the overall acceptance of a product [7].

Methods for flavor nanoencapsulation [7].

1) Emulsification
2) High-pressure homogenization
3) Microfluidization
4) Phase inversion emulsification
5) Spray drying
6) Spray chilling
7) Molecular inclusion
8) Freeze-drying
9) Electrospraying/electrospinning

## 4.3 Encapsulation Techniques

The encapsulation techniques for flavors are categorized into chemical methods and mechanical methods. Chemical methods include coacervation and molecular inversion, while mechanical methods include spray drying, spray chilling, extrusion, and fluidized bed.

### 4.3.1 Coacervation

Coacervation is a chemical method for producing polymer droplets in suspension, based on the separation of two liquid phases into concentrated colloidal phase called coacervate and another highly dilute colloidal phase [8]. Coacervation involves the separation of a liquid phase of a carrier material from a polymeric solution and enveloping of that phase as a uniform layer around suspended flavor compounds. Coacervation is possible when surface energies of core and carrier material are adjusted with varying system temperature, pH, or composition. The coating material is then dried by means of heat, cross-linking, or solvent removal techniques [9]. The microcapsules are then collected by filtration or centrifugation, washed with a suitable solvent, and dried by standard techniques such as spray drying or fluidized bed drying to yield free-flowing capsules. Simple coacervation deals with system containing only one solute (only gelatin) while complex coacervation deals with more than one solute (gelatin and gum arabica) [10]. The microcapsules formed by coacervation may contain pay load up to 85–90%, and their final application includes chewing gum, paste, and baked foods [1].

### 4.3.2 Molecular Inclusion

In the molecular inclusion method, β-cyclodextrin has been used as a medium for flavor encapsulation in food industry. The β-cyclodextrin molecule is a water-soluble cyclic polymer consists of seven glucopyranose unit linked in a (1→4) position. The β-cyclodextrin molecule has a hollow torus-shaped structure with a hydrophilic outer surface and a hydrophobic hollow center. Due to this structure, the β-cyclodextrin molecule forms the inclusion complex with flavor compounds that can fit dimensionally within the cavity as guest molecules. The resulting complex is pretty stable, and it precipitates out of solution. This precipitate can be dried to conventional methods to yield microcapsules [1]. The resulting microcapsules may contain pay load about 6–15% and their final application includes extruded snacks, instant drinks, and confectionery [1].

### 4.3.3 Spray Drying

Spray drying is the most common used flavor encapsulation method in the food industry. Spray drying involves the preparation of dispersion or emulsion by dispersing flavoring compounds in the solution of carrier material such as modified starch, gum, protein, or in combination of these, which it is immiscible. The typical ratio of carrier to core is 4 : 1. The dispersed mixture is then homogenized to create small droplets of flavoring compounds in the carrier solution. Then the mixture is fed into a spray dryer where it is transformed into controlled size droplets by a nozzle spray/atomizer, and it is sprayed into a drying chamber where the atomized particle treated with hot air, evaporate the water ,and create free flowing discrete particles/capsules fall through the gaseous medium to the bottom of the dryer. The functional properties of optimum carrier material for spray drying includes high water solubility, low viscosity at high concentration, good emulsification and film forming ability, and good drying ability [11]. The final application of spray-dried products includes confectionery, instant dessert, and instant beverages [1].

### 4.3.4 Spray Chilling

Spray chilling or cooling technique involves the emulsification of flavor compounds to the molten lipid carrier and resulting mixture is fed through an atomized nozzle. When the nebulized material comes in contact with the environment which is cooled below the melting point of matrix, due to the heat exchange between the molten material and cool air, the droplets adhere on the flavor compounds and solidify forming a coat film [12]. This type of encapsulation is done to improve heat stability of capsule and to delay release in wet environment [13]. Spray chilled product has application in bakery products, soup premix, and fatty foods [1].

### 4.3.5 Extrusion

The encapsulation by extrusion process is classified into simple and centrifugal extrusion. The simple extrusion involves the dispersion of flavor compounds in molten carbohydrate mass subsequently forcing of dispersed mixture through a die into dehydrating liquid such as isopropyl alcohol, which hardens the coating material to trap the flavoring compounds; ultimately, the strands/filaments of hardened material are broken into small pieces, separated, and dried [14]. The centrifugal extrusion method utilizes the nozzle located on the outside of rotating cylinder of the extruder. The flavoring compounds are pumped through the inner orifice and carrier material through the outer orifice, leading to the formation of a co-extruded rod of flavoring compounds enveloped by the wall material. As device rotates, the extruded rods break and form capsules [15]. Extruded products have application in instant beverages, tea and confectionery [1].

### 4.3.6 Fluidized Bed Coating

Fluidized bed coating is also called air suspension coating which involves the flavoring compounds to be encapsulated are fluidized in the coating chamber that consists of hot atmosphere, and the carrier material is sprayed through a nozzle onto the flavoring compounds in fluidized bed, spread, and coalesce that initiates film formation. The evaporation of the solvent in mixture by hot air leads to the adherence of coating material on the flavoring compound and form capsules [16]. In this method, the carrier material used in the flavoring system voluntarily dissolves and forms strong interparticle bridges on drying [17]. The final application of fluidized bed coating includes confectionery and prepared dishes [1] (Table 4.1).

**Table 4.1** Different methods of flavor encapsulation [18].

| Encapsulation method | Flavor | Encapsulating agents | Particle size | Encapsulated form |
|---|---|---|---|---|
| Coacervation | Lavender oil<br>Vanilla oil | Soy protein<br>Pectin<br>Gelatin<br>Gum Arabic<br>Chitosan | 5–200 μm | Paste<br>Powder<br>Capsule<br>Coreshell<br>Multiwall |
| Molecular inclusion | Olive oil<br>Orange oil<br>Onion oil<br>Black pepper oil | B cyclodextrin | 5–50 μm | Powder |
| Spray drying | Peach<br>Strawberry<br>Rosemary oil<br>Mandarin oil<br>Lime oil<br>Mint oil<br>Ginger oil | Modified starch<br>Maltodextrin<br>Gum Arabic<br>Guar gum<br>Inulin<br>Chitosan | 1–50 μm | Powder<br>Matrix type |
| Spray chilling | Limonene<br>Nicotine<br>Methyl salicylate<br>Cinnamic aldehyde | Erythritol | 20–200 μm | Powder<br>Matrix type |
| Extrusion | Ethyl vanillin | Alginate | 20–2000 μm | Powder<br>Granule<br>Matrix<br>Coreshell<br>Multiwall<br>Multicore |
| Fluidized bed | Menthol | Gelatin<br>Modified starch | >100 μm | Coreshell<br>Multiwall<br>Multicore |

*Source:* Adapted from Estevinho and Rocha [18].

## 4.4 Future Perspectives

The global flavor industry can be marked as highly technical, specialized, innovative, highly competitive, and concentrated as compared to other markets of food and beverage industry. The global flavor market reached around US$ 11 billion in 2013 and is expected to expand at an average annual growth rate of 13.70% rising to US$ 19.35 billion by 2025 [19]. Approximately 80% of global sales are concentrated in Asia-Pacific, North America, and Western Europe. Flavors have a wide scope in food industry applications such as beverage sectors, dairy industry, and savory and convenient foods. The main advantages of micro and nano encapsulation process in flavor technology includes physical protection of flavors, reduction of flavor reactivity with internal and external factors (flavor–flavor interactions, light-induced reaction and oxidation), decreasing transfer rate from core to outside, controlling the release of flavor and promote easier handling. Depending on the physicochemical properties of the core, type of encapsulating agent, and the microencapsulation technique used, different types of particles can be obtained such as a simple sphere surrounded by a coating of uniform thickness (single core), particles containing an irregularly shaped core, several core particles embedded in a continuous matrix of an encapsulating agent (multicore), several distinct cores within the same capsule, and multilayer microcapsules [18]. These different forms of encapsulated particles are now employed in various food applications such as beverages, dairy, and bakery and confectionary and can be utilized in other wide range of food applications in the future.

## References


1 Madene, A., Jacquot, M., Scher, J., and Desobry, S. (2006). Flavour encapsulation and controlled release–a review. *Int. J. Food Sci. Technol.* 41 (1): 1–21.

2 Zeller, B.L. and Saleeb, F.Z. (1996). Production of microporous sugars for adsorption of volatile flavors. *J. Food Sci.* 61 (4): 749–752.

3 Tan, S.P., Tuyen, C.K., Parks, S.E. et al. (2015). Effects of the spray-drying temperatures on the physiochemical properties of an encapsulated bitter melon aqueous extract powder. *Powder Technol.* 281: 65–75.

4 Landy, P., Druaux, C., and Voilley, A. (1995). Retention of aroma compounds by proteins in aqueous solution. *Food Chem.* 54 (4): 387–392.

5 Gutiérrez, T.J. (2018). Processing nano-and microcapsules for industrial applications. In: *Handbook of Nanomaterials for Industrial Applications* (ed. C.M. Hussain), 989–1011. Elsevier.

6 Klinjapo, R. and Krasaekoopt, W. (2018). Chapter 14 – Microencapsulation of color and flavor in confectionery products. In: *Handbook of Food Bioengineering, Natural and Artificial Flavoring Agents and Food Dyes* (ed. A.M. Grumezescu and A.M. Holban), 457–494. Academic Press, ISBN 9780128115183, https://doi.org/10.1016/B978-0-12-811518-3.00014-4.

7 Mohebbi, M. (2020). Chapter 11 – Nanoencapsulation of flavors for beverage manufacturing. In: *Micro and Nano Technologies*Nanotechnology in the Beverage Industry (ed. A. Amrane, S. Rajendran, T.A. Nguyen, et al.), 317–336, ISBN 9780128199411. Elsevier https://doi.org/10.1016/B978-0-12-819941-1.00011-0.

8 De Kruif, C.G., Weinbreck, F., and de Vries, R. (2004). Complex coacervation of proteins and anionic polysaccharides. *Curr. Opin. Colloid Interface Sci.* 9 (5): 340–349.

9 Bakan, J.A. (1975). Microcapsule drug delivery systems. In: *Polymers in Medicine and Surgery* (ed. R.L. Kronenthal, Z. Oser and E. Martin), 213–235. Boston, MA: Springer.

**10** Luzzi, L.A. and Gerraughty, R.J. (1964). Effects of selected variables on the extractability of oils from coacervate capsules. *J. Pharm. Sci.* 53 (4): 429–431.

**11** Reineccius, G.A. (1988). Spray-drying of food flavors. *ACS Symp. Ser.* 370.

**12** Okuro, P.K., de Matos Junior, F.E., and Favaro-Trindade, C.S. (2013). Technological challenges for spray chilling encapsulation of functional food ingredients. *Food Technol. Biotechnol.* 51 (2): 171.

**13** Gouin, S. (2004). Microencapsulation: industrial appraisal of existing technologies and trends. *Trends Food Sci. Technol.* 15 (7–8): 330–347.

**14** Risch, S.J. (1995). Encapsulation: overview of uses and techniques. In: *Encapsulation and Controlled Release of Food Ingredients*, ACS Symposium Series, vol. 590 (ed. S.J. Risch and G.A. Reineccius), 2–7. ACS Publications.

**15** Schlameus, W. (1995). Centrifugal extrusion encapsulation. In: *Encapsulation and Controlled Release of Food Ingredients*, ACS Symposium Series, vol. 590, 96–103. ACS Publications.

**16** Jacquot, M. and Pernetti, M. (2004). Spray coating and drying processes. In: *Fundamentals of Cell Immobilisation Biotechnology*, 343–356. Dordrecht: Springer.

**17** Buffo, R.A., Probst, K., Zehentbauer, G. et al. (2002). Effects of agglomeration on the properties of spray-dried encapsulated flavours. *Flavour Fragrance J.* 17 (4): 292–299.

**18** Estevinho, B.N. and Rocha, F. (2017). A key for the future of the flavors in food industry: nanoencapsulation and microencapsulation. In: *Nanotechnology Applications in Food* (ed. A.E. Oprea and A.M. Grumezescu), 1–19. Academic Press.

**19** Yang, M., Liang, Z., Wang, L. et al. (2020). Microencapsulation delivery system in food industry—challenge and the way forward. *Adv. Polym. Tech.* 2020.

# 5

# Flavor Signatures of Beverages and Confectionaries

*Neha N. Areekal[1], Sneha George[1], Irene M. Peter[1], Roshin Thankachan[2, 3], Józef T. Haponiuk[2], and Sreeraj Gopi[4]*

[1] *Karunya Institute of Technology and Sciences, Department of Food Processing Engineering, Coimbatore, Tamil Nadu, 641114, India*
[2] *Gdansk University of Technology, Department of Polymer Technology, G. Narutowicza 11/12, 80-233, Gdańsk, Poland*
[3] *R&D Centre, Tropical flavors (P) Ltd., Kolenchery, Cochin, Kerala, 682311, India*
[4] *R&D Centre, Aurea Biolabs (P) Ltd, Kolenchery, Cochin, Kerala, 682311, India*

## 5.1 Introduction

In the modern era, beverage and confectionary industries have been improvising on several sensory attributes such as sight, smell, temperature, touch, and taste so that there is an increase in the acceptability of a product by consumers [1]. Unless the food we consume is of great flavor and tastes good, consumption of any food, be it a beverage or a confectionary, would not be satisfactory [2]. Moreover, flavor signature plays an important role in most of the products available in the market, especially for beverages like carbonated drinks, energy drinks, coffee, and tea and confectionaries like candies, gum, and jelly. Indeed, flavor can give many characteristic properties to a tasteless product, which can reduce unpleasant aroma that might evolve during heating, fermentation, or smell of fish of the products [3]. Presumably, during new product development, mainly functional foods or nutritional foods, there are tendencies to generate an undesirable flavor that gives a low quality to the product in the market. In the food industry, there is a lack of effective modern strategies to mask an off flavor or bad taste. Moreover, the addition of flavor can create an aroma in the product, which is incorporated along with the pre-existing bad taste and off flavor, thus compromising the quality of the product and increasing the preference of this product by consumers by diverting attention from undesirable taste and flavor [4].

A multisensory experience that involves all the five senses, namely auditory, olfactory, visual, gustatory, and tactile, is felt on consumption of a food or beverage.

*Natural Flavours, Fragrances, and Perfumes: Chemistry, Production, and Sensory Approach*, First Edition. Edited by Sreeraj Gopi, Nimisha Pulikkal Sukumaran, Joby Jacob, and Sabu Thomas.

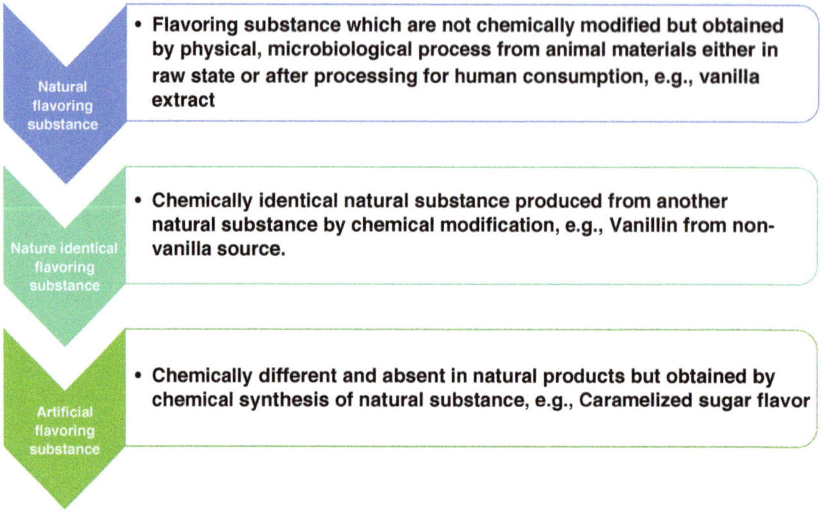


**Figure 5.1** Three main categories of flavor substances used in industries with examples.

There are two categories of senses that determine the flavor experience altogether, namely interceptive or consummatory senses and exteroceptive or anticipatory senses. While interoceptive senses are stimulated after the food reaches our mouth, exteroceptive senses are responsible for giving us expectation before a food or beverage reaches our mouth [5]. Accordingly, the use of flavors is pertinent wherever the product interacts with the oral system [6]. Flavor is also influenced by other factors like food texture, and color. Color influences not only the perception of taste but also the quality and preference of food items [7]. For instance, yellow is associated with increasing energy, while white such as milk have a calming effect; a clear beverage such as water represents a refreshing experience. This is why beverage manufacturers consider color with the same importance as much as they deliberate flavor and nutrition platforms when developing new products [8].

Flavor compounds, which is the sensory impression of a food, can either be present naturally in foods or can be added during its manufacturing process; they can be classified into three main groups, viz. natural flavoring substance, nature identical flavoring substance, and artificial flavoring substance [9]. The main three categories of flavoring substances used in food industry are depicted in Figure 5.1 with examples. According to the US Food and Drug Administration, flavoring agents are defined as substances added to impart or help impart a taste or an aroma in foods, while flavor enhancers are substances used to supplement, enhance, or modify the original taste or aroma [10]. However, with growing concerns about food safety and potential health risk, increasing attention is being currently given to food additives [11]. Flavoring substances and their ingredients are standardized and tested in accordance with the Food Additive Amendment (1958) under the federal Food Drug and Cosmetic Act (FDCA) by the Flavor and Extract Manufacturers Association (FEMA). The ingredients must acquire Generally Recognized AS Safe (GRAS) status to be fit for consumption. Flavoring ingredients can be either complex mixtures (NFC) or pure chemically defined compounds [12].

Moreover, according to the market size, US flavors and fragrances were valued US$ 20.75 billion in 2018 and were expected to expand at a compound annual growth rate (CAGR) of 4.7% over the forecast period. Increasing demand for industries such as food and beverages are the main reason to drive the growth [13]. Not only in food industry but also in other sectors, flavor and fragrance play a major role, for instance, right from the toothpaste we use to the coffee and tea we drink, and from the deodorants and tissues we select to fine wine and perfumes we choose, they play a sufficient role. Notably, beverage and confectionaries are a big and constantly evolving market, with 30% of the entire food flavors being used in the beverage industry [14]. In food industry, especially in beverages and confectionaries, "congruency" is one of the most important characteristics of a product. It involves a combination or pairing of food product, for example, while adding a sweet flavor, a sweet-related aroma should be coming from the product. On the other hand, if the next pair of aroma we are adding is not congruent to the first one, it can suppress the taste intensity [15]. Hence, the pairing of flavors is a challenging assignment that needs keen attention. Consider red wine and sea food as an example; both are very delicious when consumed individually, but when combined, they create an unpleasant sensation, metallic taste, an unpleasant fishy odor, and in some cases, it can form even bitterness in the mouth [16]. Therefore, while pairing, immense considerations should be given to the presence of shared flavor molecules like aromatic volatiles and similarity and contrast between them [17].

There are variety of sources from which flavoring materials can be extracted, one of which is plants, wherein every part of a plant, including the flower, leaf, stem, or bark, contains flavoring materials. The materials are usually extracted from the plant material to be used in food products. Various methods, including solvent extraction (often ethanol), steam distillation, and supercritical fluid extraction, may be used for isolation [18]. In the beginning, the ingredients used in flavor and fragrance industry were of natural origin. The main strategies used to prepare natural extracts were mainly steam distillation, cold-press methods, and organic solvent extraction. The resulting free-flowing oil produced as a result of steam distillation or cold-press method is referred to as an "essential oil," and the others are commonly called as "extracts" [19]. Similarly, the residual waxy mass obtained after the process, which are left after evaporation of extraction solvents is called "concrete," and those obtained from other sources is called resinoid (oleoresin) [6].

## 5.2 Classification of Flavor Compounds

### 5.2.1 Based on Type of Flavor Compounds

There are three principal types of flavor compounds used in foods under definitions agreed in the EU and Australia, namely natural flavoring substances, nature-identical flavoring substances, and artificial flavoring substances [10]. Of these,

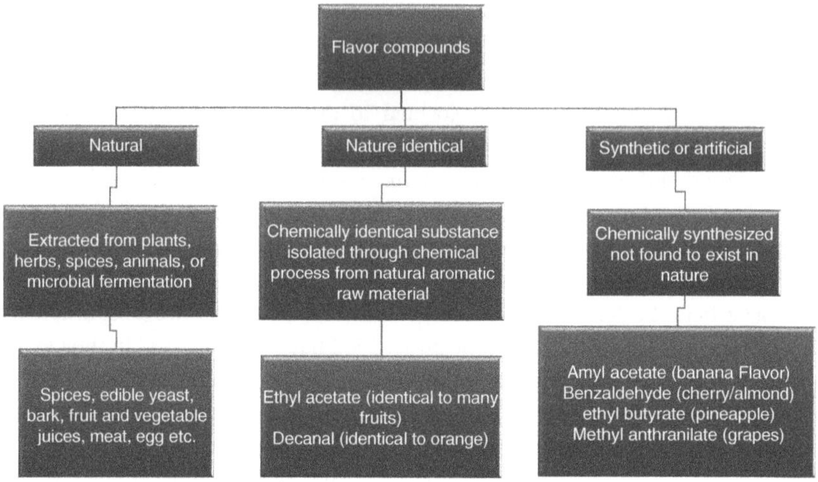


**Figure 5.2** Classification based on flavoring compounds.

natural flavors are all about the very best of nature; they are isolated from a natural aromatic raw material by physical methods. They are extracted mainly from herbs, which are a large group of aromatic plants whose leaves, stems, or flowers are used in dry or fresh forms. Spices are a large group of aromatic plants whose barks, roots, seeds, buds, or berries can be used in dry, whole, or ground forms. The best example of this is citral obtained by fractionation from oil of lemongrass. Natural-identical flavor substances-obtained by synthesis or isolated through chemical processes from a natural raw material-have the same chemical composition as that of substances present in natural products; they are intended for human consumption in the processed form. The best example of this is vanillin from lignin and citral obtained by chemical synthesis or from oil of lemongrass through its bisulfate derivative. Artificial or synthetic compounds are simply chemical mixtures of synthetic flavors that taste and smell like natural flavors. They are chemically different and absent in natural products. The best example of this is the artificial flavor of vanilla, the chemical compound that gives taste and flavor similar to vanilla and is created in laboratory and diluted with alcohol [20]. Thus, the flavoring compounds mentioned above help to increase and restore the flavors in a food product. Figure 5.2 shows a flow chart of classification based on flavoring compounds.

### 5.2.2 Based on Flavor Generation

Volatile and nonvolatile components present in a product bring out the flavor. Nonvolatile compounds influence taste sensation, while volatile compounds influence both taste and aroma, which plays a major role in the overall quality of product [21]. Depending upon the generation, flavors can be divided as plant-derived flavor, processed flavor, and biotechnological flavor as shown in Figure 5.3.

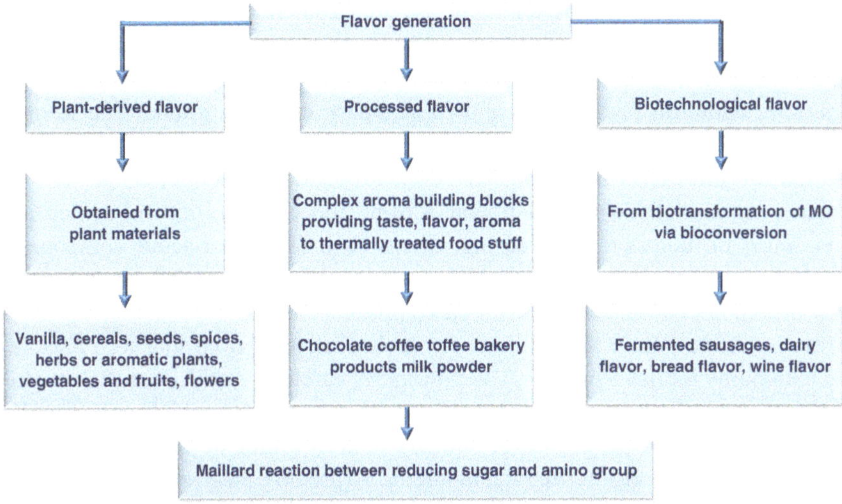


**Figure 5.3** Classification based on flavor generation.

As the name implies, the processed flavor is produced by decomposition during the processing of food items. It can also be produced by the combination of other compounds that are present in food items. The flavor that is produced while cooking, heating, toasting, baking, roasting, and fermentation of products also comes under the processed flavor. For example, while cutting onion and garlic, a type of odor can be sensed, which is a type of processed flavor due to the presence of diallyl disulfide [2]. On the other hand, biotechnological advances in the field of flavor are expanding, and their approach implies additional advantages like increasing the quality and marketing of food products. Biotechnological approaches include biotransformation, which is the transformation or conversion of compounds by using microorganisms like fungus, enzymes, and plant cells. The best example of this is vanillin, where raw materials like lignin and eugenol are found to match vanillin for transformation [22]. In contrast, in the case of plant-derived flavors, the plant parts such as roots, leaves, and flowers have the capability to synthesize volatile compounds that can produce both aromas and flavors by interacting with human receptors. Spices, cereals, herbs, fruits, and flowers used in food items come under plant-derived flavors [23]. Thus, this classification provides insights into understanding various compounds and their properties and how flavors influence the products.

## 5.3 Plant Parts as Flavoring Compounds

A culinary herb lavender, which is used in food and beverages, is quite a popular herb for the flavoring of savory food and is a component in "Herbes de Provence" [24]. Another plant of the Lamiaceae family, called the Clary Sage - historically has been used as a culinary herb -, is used in wine as a replacement of hops in brewing and/ or as an addition to its flavoring, especially those wines that originate from the Rhine region of Germany [12]. In addition to Lavender, there are various flowers from other botanical plants that are used as a part of flavoring food. The scented flower from the bitter orange tree is thoroughly distilled to collect its clear and essential oil, known as neroli oil. However, to yield an absolute, it can also be extracted sequentially with a nonpolar solvent and ethanol, both of which are important contributors to flavored ingredients [25]. The flavor-active compounds like phenolic compounds present in olive and olive oils play a major role in the flavor of beer, hops, cocoa, tea, and fruit-based products [26].

Likewise, flowers from Michelia Alba trees that originate from southern Asia are used commonly for flavoring teas. The flowers are carefully distilled for their essential oil and used as a flavoring agent [27]. Blackcurrant which is widely grown in the colder regions of Europe, is used in liqueurs such as Crème de Cassis and for the preparation of juice concentrates and in purees. To produce an absolute that imparts a blackcurrant-like flavor, the flower buds of blackcurrant are extracted. The buds are used in the flavoring of food and alcoholic and nonalcoholic beverages [12]. Yet other important flavor is carotenoid, which is found in ketchups, fruits, and vegetables like tomato and watermelon. Briefly, tomatoes contain high levels of several carotenoids, including lycopene and β-carotene, which impart the characteristic tomato flavor [28]. Tea, one of the common and popular beverages in the world, is the most extensively studied plant because of its many nutritional and healthy functions. These functions are due to the presence of catechins, which is a main compound of green tea having properties like antioxidant and antibacterial effects, which prevent cell damage, reduce the formation of free radicals in the body, and protect cells and molecules from damage. Apart from catechins, tea contains glutamic acid, caffine, amino acid, etc.; these compounds present in tea contributes to various sensory attributes such as bitterness, astringency, umami, and sweet, which also have many health benefits [29]. Cinnamon is another example that is used for flavoring snack foods, tea, sweets, and savory dishes [30].

## 5.4 Flavor Signatures

Flavors are used widely in the food industry to strengthen the natural flavors or to add new flavors to food products. For instance, products like ice cream, chocolate chewing gums, and various beverages and confectionaries like candies, cakes, and cookies are flavored. Food flavors are also used in other industries like toothpastes and tobacco to impart flavors [31]. Some examples of flavoring agents and their application are given in Table 5.1. Food materials in restaurants and bars, like sauces, seasonings, and dressings, are also flavored, which help in keeping the aroma quality consistent in industrial manufacturing. Despite this, flavors are often used to mask an unpleasant flavor in a product, for example, cough syrup for children and barium swallow used to obtain an X-ray of the esophagus [32].

**Table 5.1** Flavoring compounds and their application in confectionaries and beverages.

| Flavoring compound | Application | References |
|---|---|---|
| Methyleugenol | Gelatin and puddings, hard candy | [32] |
| Benzoin | Soft candy, chewing gum | [33] |
| Beta-asarone | Cake | [34] |
| Coumarins | Granola bar, Muffins | [35] |
| Hypericins | Herbal tea | [36] |
| Pulegone | Mint-flavored confectionaries and beverages | [37] |
| Safrole | Chewing gum, cinnamon roll | [38] |
| Thujone | Alcoholic beverages | [39] |
| Isoamyl acetate | Wine, grape-derived alcoholic beverage | [40] |
| Cinnamaldehyde | Fruit juices, chewing gum, baked goods | [41] |

### 5.4.1 Effect of Maillard Reaction

Food flavors are unstable, and the decline in their quality (change in aroma profiles caused by the degradation process and low boiling compound) is a matter of major concern to food industries. Similarly, penetration of oxygen into the package, change in storage temperatures, and loss of volatiles through diffusion result in rancidity. Moisture can also contribute to the spoilage process. Many Maillard reaction products are considered to play a role in the prevention of off-flavors produced as a result of lipid oxidation due to the antioxidant properties of the same [42]. The Maillard reaction is a chemical reaction between amino acids and reducing sugars that gives browned food its distinctive flavor. Seared steaks, fried dumplings, cookies, and other kinds of biscuits, breads, toasted marshmallows, and many other foods undergo this reaction [43].

For instance, furans and pyrazines, produced as a result of Maillard reaction, possess cooked or roasted flavors, which in turn enhances the palatability of beverages. Notably, pyrazines are the largest heterocyclic compounds formed as a result of Maillard reaction, and their presence in heat-treated beverages and foods has been well studied. Coffee also requires various heterocyclic compounds formed by the Maillard reaction. Yet, some of the beverages do not gain from the formation of heterocyclic compounds as such compounds affect their freshness. Recently, other heterocyclic compounds such as furfurals and methoxypyrazines, which are naturally occurring, are also used in flavoring beverages. In addition to their flavors, their antioxidant properties have captured the attention of many researchers [44]. However, a slow Maillard reaction that occurs at room temperature during storage is very injurious to flavor quality. The perfect example for this is the browning and flavor reduction in citric fruit juices. Thus, to avoid color and flavor reduction along with the detectable change in color and a flavor that is unacceptable, the product must be maintained at low temperature [42].

### 5.4.2 Effect of Baking

Baking is a method of preparing food that uses dry heat, typically in an oven or done in hot ashes or on hot stones, where heat is gradually transferred from the surface of cakes, cookies, and breads to their center. As the heat travels through, it transforms batters and doughs into baked goods and more with a firm dry crust and a softer center, with unique, specific flavors [45]. Thus, the process of baking induces many changes: structural enhancement, development of the desired texture, and improved digestibility, but the major effect is the transformation of the sensory attributes, specifically aroma formation [46]. Baking promotes thermal reactions and other interactions within the matrix, which are thought to be the main precursors of the "characterizing" volatile aroma compounds associated with baked goods. Fat and sugar have been identified as the most important contributors to the overall acceptability of sweet baked confectionaries with both contributing to texture, mouthfeel, volume, color, and flavor [47].

Similarly, flavor stability being important for the quality of tea beverages, baking plays an important role in tea processing. Briefly, baking is a typical processing technology to improve the flavor of tea leaves, where it lowers the amount of water present in tea leaves, thereby enhancing the ripe taste and quality. In the Oolong tea leaves processing, for example, wuyi rock tea and tie guanyine, baking plays a crucial role as it heightens the quality of aroma and flavor of tea leaves. But, in green tea processing, baking is used to regulate the flavor of green tea leaves after drying and fixation. Notably, as the baking deepens, the scent transforms from faint to floral, sweet, and fruity flavor, but on further deep baking, it develops a baking flavor, with a notable increment in the proportion of pyrazine furan and pyrole [48].

### 5.4.3 Enhancement by Addition of Flavorings

Flavor enhancement is the use of substances to augment the original flavors of foods and beverages. Moreover, flavor enhancement ingredients are components that may be used singularly or in combination to create flavorful mixtures for the benefit of beverages, foods, products, recipes, and/or meals [2]. Thus, the addition of flavorings could contribute to enhancement of flavor or reduce the presence of off-flavors. For instance, hydrocolloids – which can alter the rate and strength of flavor release by diffusion, gel-effect caging, micro-region trapping, molecular interactions, and molecular inclusion – are among the most commonly used additives in industrial beverage foods [18]. Similarly, though hops are only added in relatively small amounts at particular stages of the brewing process, they have a high impact on the final beer flavor characteristics [49].

#### 5.4.3.1 Flavor-Active Esters

In wines and beer, ethyl esters play an important role in contributing to the positive sensory attributes, as they are responsible for the fruity characteristics of fermented beverages, and they constitute an important group of aromatic compounds in beer [50]. For beer, it is indicated by its freshness; however, that decreases with ageing.

The fruity/estery and floral aroma compounds are the positive flavor attributes of beer, which may lead to a decrease in its intensity during the process of ageing. In contrast, ethyl esters help in contributing the overall balance in flavor when it comes to cream liqueurs. When at high concentrations, they have the chances of developing a very fruity flavor which may then turn out to be an overpowering flavor in the liqueur and thus increase in concentration as the cream liqueurs age [51].

#### 5.4.3.2 Xyloligosaccharides

Xylooligosaccharides (XOS) are considered food ingredients that have functional properties that include excellent acid and heat stability, and the ability to offer lower available energy. Moreover, it has excellent potential as candidate prebiotic compounds in maintaining and promoting normal microbiota balance [52]. Xyloligosaccharides on addition into strawberry flavored whey beverages has proven to be a feasible alternative as it turned out to have better functional characteristics, volatile profile, rheological properties (increased viscosity) and most importantly sensory acceptance, which indicate flavor and aroma. Xylose on thermal treatment releases flower-like flavor compounds that can enhance the fresh flavor of various food products [53].

#### 5.4.3.3 Flax Seeds

Flax seed is an important functional food ingredient because of its rich content of alpha-linolenic acid (ALA, omega-3 fatty acid), lignans, and fiber [54]. Bakery products containing flax seed with ALA may provide health benefits to consumers who are seeking functional foods that contain omega-3 fatty acids [55]. The addition of milled flax seeds in the development of bakery products has potential health benefits due to its high content of lignin, omega-3 fatty acid, alpha-linoleic acid, and antioxidants. However, on the addition of an excess amount of flax seeds (for clinical trials), its flavor characteristics gets affected. Omega-3 fatty acids are highly essential for the body and flax seeds are amongst the leading products containing omega-3 fatty acids [56]. The high content of ALA in flax seeds increases its susceptibility for lipid oxidation. It affects the aroma and flavor. However, flax seed addition in a small amount is not a problem, and the antioxidants are a solution to flavor improvement. It was found that a musty smell, which may be an indication of rancidity, was significantly reduced after 4 weeks of storage in room temperature for flax bread [50].

#### 5.4.3.4 1,2-Dicarbonyl Compounds

Maltol, ethyl maltol, and diacetyl are representatives of the 1,2-dicarbonyl class of chemicals [57]. Maltol (3-hydroxy-2-methyl-4H-pyran-4-one) can be used as a potential flavor enhancer in beverages and foods. Baked cereals, beverages, coffee, bread, caramelized food, starch, etc. have undergone thermal breakdown or pyrolysis are the sources of maltol [49]. Maltol, ethyl maltol, and diacetyl are weak mutagens, where maltol is formed when glucose has a 4-O glycoside linkage, like in disaccharides maltose and lactose, and its formation from monosaccharides is negligible [57]. Maltol can modify or enhance the flavor of food products and beverages in high concentration. Therefore, it has been used as a flavor enhancer in various food products such as cakes, breads, beer, malt beverages, and chocolate milk. However, Food and Agricultural Organisation (FAO) and World Health Organization (WHO) have allowed a maximum of 2 mg $kg^{-1}$ of body weight per day as acceptable for consumption by humans. Studies also concluded that consumption of more than the prescribed level could cause nausea, neurotic effects, headache, and vomiting [58]. Ethyl maltol, 3-hydroxy-2-ethyl-pyran-4-one, which is an analogue of maltol, has a sweet caramel-like odor and fruity taste, and it is lot more powerful than maltol. This material is almost the same in its odor and flavor profile as Maltol, except that it is about five times stronger in effect and is included on the GRAS list of approved flavoring agents by the FDA [59].

## 5.5 Role of Flavor Compounds in Sensory Attributes

The first and foremost benefit of flavor in foods such as confectionery and beverage is that it attracts consumers. Without flavors, be it natural or synthetic, every food item or beverage will taste the same. So the most important benefit is that flavors make food and beverages acceptable among consumers. The use of flavors and flavor enhancement play a big role in intensifying the smell and taste of confectionaries, beverages etc., which may also compensate in the diet of old people [60]. Interestingly, it was observed that as a person grows older, his or her taste, smell may decline, which results in low intake of food and beverages such as fruit and vegetable juices which are very necessary for their immune system. Moreover, their chemosensory decline may change their food choices and violate their medical conditions that may lead to weight loss and nutrient deficiency disorders. This is where flavor enhancing comes into application. On addition of flavor enhancers, food products and beverages became much more acceptable by compensating nutrition losses that would have taken place if there was no flavor added to food as flavor enhancers can increase their craving for food, improve palatability of food, increase the salivary flow, and reduce their disinterest in food. As a result, there was a rapid increase in the level of acceptability of various products [2]. Beyond that, some studies have shown the role of volatile flavor compounds in human health, which include antioxidant, anti-obesity activities, anti-microbial agents, etc. Because of these various properties of flavor, studies on the pharmacological acts of flavors and volatile compounds have been increased.

Addition of flavor to food and other consumable product contain certain unique features such as:

- It gives a different flavor characteristic for a product arising from a basic ingredient, thereby modifying the existing flavor profile
- Masking and blocking are two common functional flavor modulators for beverages. These are designated to eliminate one or more undesirable attributes. Masking can make the final taste of a product stay neutral. Sweeteners, acid, and

salts are examples for this. Blocking involves the changing of taste attributes, where an ingredient, which acts as a blocker interacts with bitter or off taste and hence gets blocked.

- Undesirable flavor attributes that come from a product, due to several chemical reactions and processing, can be covered by the addition of flavoring, which in turn increases the quality, preference ability, and attraction of the product in the market.
- They can boost weak intrinsic flavors or replace flavor during processing.
- Flavoring can replace basic ingredient or expensive ingredient, which in turn can reduce the price of production and can make the product economically cheap in the market, thereby overcoming seasonal variability.
- When the natural products possess toxic or other hazards, these artificial flavors can replace them (Table 5.2).

Dairy products, seafood, spice blends, fruits, nuts, and vegetables, especially confectionery products, have flavor applications in foods and beverages, as existence without flavorings is challenging. It should not, however, be added to foods to conceal or deceive users about the product's quality and nutritional value. In addition, flavoring agents are added not only to preserve appearance and taste of the product for a long shelf-life; they also result in better acceptability and demand of various products [10]. The manufacturers apply taste and flavor to create tasty food and beverage products. Basic tastes of flavor include bitter, salt, sour, sweet, and umami and sensations such as cool and heat. To design or improve a flavor for a particular product, a detail analysis of the finished product as well as the raw material is required for its manufacture. Analysis involves concentration and extraction of flavoring materials, instrumental analysis of resulting extract, and interpretation of the results obtained [61].

As stated, various segments of food industry have been incorporating flavors. For instance, dairy segment, the application of flavors is used in main products like ice creams, flavored milk, and yoghurt; bakery products and confectionaries products like biscuits, cakes, cookies, chocolate, chewing gums, and bubble gum use flavorings for the taste and smell of the products [2]. In the manufacturing of canned soups, the liquid flavoring is very benefitable, as it gets readily dispersed throughout the bulk and can create a uniform flavor throughout the mixture. In ice cream and frozen goods, flavoring plays a major role, but these products should be eaten before they freeze-out, otherwise fading of flavor occurs, which reduces the taste of the product. Further, flavorings used in soft drinks like carbonated beverages and in noncarbonated drinks such as squash and cordials can make the product free from spoilage organisms, can give correct physical appearance, and can make them stable towards light, heat, and preservatives. The addition of flavors to a product can be determined by the effect of temperature, time, and flavor balance; type of preservatives; and moisture content of the end product. The techniques ensure stability of the flavors during processing and storage, thereby increasing shelf life of products [10].

**Table 5.2** Basic tastes and methods to reduce, complement, and heighten them.

| | Reduce/mask | Complement | Heighten |
|---|---|---|---|
| Acidity | 1) Add fat such as butter or cream sauce<br>2) Counter acidity with a touch of salt<br>3) Dilute – using a vegetable stalk or water<br>4) Increase sweetness to negate sourness (sweet carrots in marinara sauce)<br>5) Intensify savoriness (using soy sauce) | 1) Fat – "rounds out" acidity of vinegar<br>2) Salt – balances tartness and suppresses bitterness<br>3) Sugars – softens and neutralize acidity | 1) Add a hint of bitterness/saltiness<br>2) Deepen – by adding bitter ingredients<br>3) Combine citrus fruits with other sour ingredients<br>4) Cook acidic food with other acidic food (onions or tomatoes in lemon juice or vinegar) |
| Bitterness | 1) Acid – lessens the perception of bitterness, adds freshness<br>2) Blanching – dilutes peppery taste<br>3) Cooking – lessens mustard like bitterness<br>4) Dilute – on dilution with water or addition of fat<br>5) Remove the source of bitterness<br>6) Salt – extracts bitter juice<br>7) Umami – enhances the inherent sweetness | 1) Add – sweet extracts<br>2) Balance – offset<br>3) Greener/unripeolives and olive oils with peppery flavor<br>4) Salt – sprinkle sea salt on grilled or steamed vegetables (broccoli)<br>5) Sugar – roll chocolate truffles in powdered sugar | 1) Condense – reduce amount of bitter ingredients<br>2) Deep fry- submerge bitter ingredients in oil<br>3) Oven dry – dehydrate bitter ingredients (citrus zests or herb sprigs) |
| Salt | 1) Unpeeled potato- absorbs some of the salt and then discard the potato before consumption<br>2) Addition of small amount of brown sugar, lemon juice, sour cream, vinegar, yogurt may be useful to bring some sweetness to balance the salt | 1) Pairing salt and bitter. Salt may suppress the bitterness and sourness or sweetness will surface<br>2) Blending salty or sugary taste (salted nuts mixed with chocolate or honey glaze) | 1) Adding umami rich ingredients – the salt enhances umami rich ingredient which rounds out salt<br>2) Boost the other four basic tastes; acidity, bitterness, sweetness, and umami (barbecue sauces on protein foods such as fish, meat and poultry) |
| Sweetness | 1) Combine sweetness and bitterness<br>2) Use caramel and coffee along with sweetness as both contains acidic and bitter tastes<br>3) Adding little salt to acidic, bitter and sweet mix will give a product containing all basic tastes and may activate many taste receptors | 1) Sweetness and sourness complement each other<br>2) Cloying degree of sweetness can be curtailed by using citrus fruits, juice or even zest | 1) Layering sweet aromas and different sweeteners serve to amplify sweetness to some point<br>2) Layering fattiness with sweetness can intensify the experience as fats can carry the sweetness throughout the oral cavity (cookies, pastries, pies) |

## 5.6 Conclusion

Flavor plays a major role in consumer satisfaction and in creating influence of further consumption of foods. Hence, flavor has a large scope of application in the food industry, especially in beverages and confectionaries. In the food and beverage industry, flavor is an important part, and optimization of flavor in food is important for manufactures in adding value and quality to their products by the consumers. Therefore, flavor, as an ever-evolving art, plays a major role in food industry and has opportunities for innovation and creativity. Flavors are known not only for taste but also for many other functions like replacing other ingredients and boosting least intrinsic flavors. Absence of flavor can in turn develop food products with unsatisfactory taste. Flavor can also give a characteristic to a tasteless product. Flavor is known to be a vast developing field in the food industry. Research shows that the role of flavors will be increased in the upcoming era, as long as taste and aroma play an important role.

## References


1. Considine, J.A. and Frankish, E. (2014). Chapter 2 – Flavors and aromas in foods and beverages. In: *A Complete Guide to Quality in Small-Scale Wine Making* (ed. J.A. Considine and E. Frankish), 11–21, ISBN 9780124080812. Academic Press https://doi.org/10.1016/B978-0-12-408081-2.00002-0.
2. Marcus, J.B. (2019). Chapter 6 – Flavor enhancement ingredients. In: *Aging, Nutrition and Taste* (ed. J.B. Marcus), 173–206, ISBN 9780128135273. Academic Press https://doi.org/10.1016/B978-0-12-813527-3.00006-5.
3. Schiffman, S.S. (2003). Taste enhancers. In: *Encyclopedia of Food Sciences and Nutrition*, 2e (ed. B. Caballero), 5734–5737, ISBN 9780122270550. Academic Press https://doi.org/10.1016/B0-12-227055-X/01180-9.
4. Gonçalves, B., Oliveira, I., Bacelar, E. et al. Chapter 2 – Aromas and flavours of fruits. In: *Generation of Aromas and Flavours* (ed. A. Vilela). https://doi.org/10.5772/intechopen.72489.
5. Kpossa, M.R. and Lick, E. (2020). Visual merchandising of pastries in foodscapes: The influence of plate colours on consumers' flavour expectations and perceptions. *J. Retail. Consum. Serv.* 52: 101684, ISSN 0969-6989, https://doi.org/10.1016/j.jretconser.2018.10.001.
6. Ishikawa, M., Warita, Y., Takahisa, E., and Ohkubo, Y. (2010). Chapter 4.15 – Human–environment interactions (1): flavor and fragrance. In: *Comprehensive Natural Products II* (ed. H.-W. (Ben) Liu and L. Mander), 595–629, ISBN 9780080453828. Elsevier https://doi.org/10.1016/B978-008045382-8.00107-6.
7. Barrett, D.M., Beaulieu, J.C., and Shewfelt, R. (2010). Color, flavor, texture, and nutritional quality of fresh-cut fruits and vegetables: desirable levels, instrumental and sensory measurement, and the effects of processing. *Crit. Rev. Food Sci. Nutr.* 50 (5): 369–389. https://doi.org/10.1080/10408391003626322.

**8** Roger Harker, F., Anne Gunson, F., and Jaeger, S.R. (2003). The case for fruit quality: an interpretive review of consumer attitudes, and preferences for apples. *Postharvest Biol. Technol.* 28 (3): 333–347, ISSN 0925-5214, https://doi.org/10.1016/S0925-5214(02)00215-6.

**9** Xu, J., He, Z., Zeng, M. et al. (2017). Effect of xanthan gum on the release of strawberry flavor in formulated soy beverage. *Food Chem.* 228: 595–601, ISSN 0308-8146, https://doi.org/10.1016/j.foodchem.2017.02.040.

**10** Klinjapo, R. and Krasaekoopt, W. (2018). Chapter 14 – Microencapsulation of color and flavor in confectionery products. In: *Handbook of Food Bioengineering, Natural and Artificial Flavoring Agents and Food Dyes* (ed. A.M. Grumezescu and A.M. Holban), 457–494. Academic Press, ISBN 9780128115183,https://doi.org/10.1016/B978-0-12-811518-3.00014-4.

**11** Amchova, P., Kotolova, H., and Ruda-Kucerova, J. (2015). Health safety issues of synthetic food colorants. *Regul. Toxicol. Pharm.* 73 (3): 914–922, ISSN 0273-2300, https://doi.org/10.1016/j.yrtph.2015.09.026.

**12** Fukushima, S., Cohen, S.M., Eisenbrand, G. et al. (2020). FEMA GRAS assessment of natural flavor complexes: Lavender, Guaiac Coriander-derived and related flavoring ingredients. *Food Chem. Toxicol.* 145: 111584, ISSN 0278-6915, https://doi.org/10.1016/j.fct.2020.111584.

**13** Flavours and fragrances market share, size & trends analysis report by product (natural, aroma) by application (flavours, fragrances), by region, and segment forecasts. (2019–2025). https://www.researchandmarkets.com/reports/4761193/flavors-and-fragrances-market-size-share-and (accessed 09 November 2021).

**14** Food flavor market-segmented by type, application, and geography (2020–2025). https://www.industryarc.com/Report/7460/food-flavors-market.html (accessed 09 November 2021).

**15** Li, S., Zeng, Y., and Zhou, S. (2020). The congruence effect of food shape and name typeface on consumers' food preferences. *Food Qual. Preference* 86: 104017, ISSN 0950-3293,https://doi.org/10.1016/j.foodqual.2020.104017.

**16** Tamura, T., Taniguchi, K., Suzuki, Y. et al. (2009). Iron is an essential cause of fishy aftertaste formation in wine and seafood pairing. *J. Agric. Food. Chem.* 57 (18): 8550–8556. https://doi.org/10.1021/jf901656k.

**17** Spence, C. (2020). Food and beverage flavour pairing: A critical review of the literature. *Food Res. Int.* 133: 109124, ISSN 0963-9969, https://doi.org/10.1016/j.foodres.2020.109124.

**18** Risch, S.J. (2009). Chapter 7 – Flavors and colors for microwave foods. In: *Development of Packaging and Products for Use in Microwave Ovens* (ed. M.W. Lorence and P.S. Pesheck), 176–191. Woodhead Publishing, ISBN 9781845694203,https://doi.org/10.1533/9781845696573.1.176.

**19** Romm, A., Ganora, L., Hoffmann, D. et al. (2010). Chapter 3 – Fundamental principles of herbal medicine. In: *Botanical Medicine for Women's Health* (ed. A. Romm, M.L. Hardy and S. Mills), 24–74. Churchill Livingstone, ISBN 9780443072772, https://doi.org/10.1016/B978-0-443-07277-2.00003-9.

**20** Potter, N.N. and Hotchkiss, J.H. (1995). *Food Science*, 5, 5e. Springer US https://doi.org/10.1007/978-1-4615-4985-7.

**21** Vilela, A. (2018). Introductory chapter: generation of aromas and flavours. In: *Generation of Aromas and Flavours* (ed. A. Vilela). IntechOpen https://doi.org/10.5772/intechopen.81630.

**22** Krings, U. and Berger, R. (1998). Biotechnological production of flavours and fragrances. *Appl. Microbiol. Biotechnol.* 49: 1–8. https://doi.org/10.1007/s002530051129.

**23** Schwab, W., Davidovich-Rikanati, R., and Lewinsohn, E. (2008). Biosynthesis of plant-derived flavor compounds. *Plant J.* 54: 712–732. https://doi.org/10.1111/j.1365-313X.2008.03446.x.

**24** Laget, F. (2005). From its birthplace in Egypt to Marseilles, an ancient trade: 'drugs and spices.'. *Diogenes* 52 (3): 131–139. https://doi.org/10.1177/0392192105055941.

**25** Bonaccorsi, I., Sciarrone, D., Schipilliti, L. et al. (2011). Composition of Egyptian nerolì oil. *Nat. Prod. Commun.* 6 (7): 1009–1014, PMID: 21834247.

**26** de Lourdes, M. and Giada, R. (2013). Food phenolic compounds: main classes, sources and their antioxidant power. In: *Oxidative Stress and Chronic Degenerative Diseases – A Role for Antioxidants* (ed. J.A. Morales-González). IntechOpen https://doi.org/10.5772/51687.

**27** Koomhin, P., Sattayakhom, A., Chandharakool, S. et al. (2020). Michelia essential oil inhalation increases fast alpha wave activity. *Sci. Pharm.* 88: 23. https://doi.org/10.3390/scipharm88020023.

**28** Vogel, J.T., Tieman, D.M., Sims, C.A. et al. (2010). Carotenoid content impacts flavor acceptability in tomato (*Solanum lycopersicum*). *J. Sci. Food Agric.* 90 (13): 2233–2240. https://doi.org/10.1002/jsfa.4076., PMID: 20661902.

**29** Xia, E.H., Zhang, H.B., Sheng, J. et al. (2017). The tea tree genome provides insights into tea flavor and independent evolution of caffeine biosynthesis. *Mol. Plant* 10 (6): 866–877. https://doi.org/10.1016/j.molp.2017.04.002.Epub, PMID: 28473262.

**30** Gooderham, N.J., Cohen, S.M., Eisenbrand, G. et al. (2020). FEMA GRAS assessment of natural flavor complexes: Clove, cinnamon leaf and West Indian bay leaf-derived flavoring ingredients. *Food Chem. Toxicol.* 145: 111585. https://doi.org/10.1016/j.fct.2020.111585.Epub , PMID: 32702506.

**31** Reineccius, G. (1995). The flavor industry. In: *Source Book of Flavors* (ed. G. Reineccius). Boston, MA: Springer https://doi.org/10.1007/978-1-4615-7889-5_1.

**32** IARC Working Group on the Evaluation of Carcinogenic Risks to Humans (2013). Some Chemicals Present in Industrial and Consumer Products, Food and Drinking-Water. Lyon (FR): International Agency for Research on Cancer (IARC Monographs on the Evaluation of Carcinogenic Risks to Humans, No. 101.) METHYLEUGENOL. https://www.ncbi.nlm.nih.gov/books/NBK373178/ (accessed 11 November 2021).

**33** Bedigian, D. (2003). Monograph on Benzoin (Balsamic resin from Styrax species). *Econ. Bot.* 57 (3): 427–428. https://doi.org/10.1663/0013-0001(2003)057[0427:BREDFA]2.0.CO;2.

**34** European Commission Health & Consumer Protection Directorate-General (2002). Opinion of the Scientific Committee on Food on the presence of -asarone in flavourings and other food ingredients with flavouring properties.

**35** Lončar, M., Jakovljević, M., Šubarić, D. et al. (2020). Coumarins in food and methods of their determination. *Foods* 9: 645. https://doi.org/10.3390/foods 9050645.

**36** Tawaha, K., Gharaibeh, M., El-Elimat, T., and Alali, F.Q. (2010). Determination of hypericin and hyperforin content in selected Jordanian Hypericum species. *Ind. Crops Prod.* 32 (3): 241–245, ISSN 0926-6690,https://doi.org/10.1016/j.indcrop .2010.04.017.

**37** (2005). European Commission Health & Consumer Protection Directorate-General Opinion of the Scientific Panel on Food Additives, Flavourings, Processing Aids and Materials in contact with Foods on a request from the Commission on Pulegone and Menthofuran in flavourings and other food ingredients with flavouring properties. *EFSA J.* 298: 1–32.

**38** European Commission Health & Consumer Protection Directorate-General (2002). Opinion of the Scientific Committee on Food on the safety of the presence of safrole (1-allyl-3,4- methylene dioxy benzene) in flavourings and other food ingredients with flavouring properties, SCF/CS/FLAV/FLAVOUR/6 ADD3 Final.

**39** European Commission Health & Consumer Protection Directorate-General 2003. Opinion of the Scientific Committee on Food on Thujone, SCF/CS/FLAV/ FLAVOUR/23 ADD2 Final.

**40** Lopes, D.B., Madeira, J.V., Júnior, L.V. et al. (2017). Chapter 1 – Microbial production of added-value ingredients: state of the art. In: *Handbook of Food Bioengineering, Microbial Production of Food Ingredients and Additives* (ed. A.M. Holban and A.M. Grumezescu), 1–32. Academic Press, ISBN 9780128115206, https://doi.org/10.1016/B978-0-12-811520-6.00001-5.

**41** Gowder, S.J.T. (2014). *Safety Assessment of Food Flavour – Cinnamaldehyde*, Gowder, Biosafety, 3:1. https://doi.org/10.4172/2167-0331.1000e147.

**42** Arnoldi, A. (2003). Chapter 7 – The Maillard reaction as a source of off-flavours. In: *Woodhead Publishing Series in Food Science, Technology and Nutrition, Taints and Off-Flavours in Foods* (ed. B. Baigrie), 162–175. Woodhead Publishing, ISBN 9781855734494, https://doi.org/10.1533/9781855736979.162.

**43** Murata, M. (2021). Browning and pigmentation in food through the Maillard reaction. *Glycoconjugate J.* 38 (3): 283–292. https://doi.org/10.1007/s10719-020-09943-x. , PMID: 32910400.

**44** Shibamoto, T. (2019). Formation of selected heterocyclic flavor chemicals in beverages. In: *Encyclopedia of Food Chemistry* (ed. L. Melton, F. Shahidi and P. Varelis), 363–373. Academic Press, ISBN 9780128140451,https://doi.org/10.1016/ B978-0-08-100596-5.21671-5.

**45** Figoni, P.I. (2010). Chapter 2 – Heat transfer. In: *How Baking Works: Exploring the Fundamentals of Baking Science*, 3e, ISBN: 978-0-470-39267-6 (ed. P.I. Figoni), 19.

**46** Mohsen, S.M., Fadel, H.H.M., Bekhit, M.A. et al. (2009). Effect of substitution of soy protein isolate on aroma volatiles, chemical composition and sensory quality of wheat cookies. *Int. J. Food Sci. Technol.* 44: 1705–1712. https://doi.org/10.1111/j .1365-2621.2009.01978.x.

**47** Garvey, E.C., O'Sullivan, M.G., Kerry, J.P., and Kilcawley, K.N. (2020). Factors influencing the sensory perception of reformulated baked confectionary products.

*Crit. Rev. Food Sci. Nutr.* 60 (7): 1160–1188. https://doi.org/10.1080/10408398.2018.1562419. PMID: 30668147.

**48** Yan-Qing, F., Wang, J.-Q., Chen, J.-X. et al. (2020). Effect of baking on the flavor stability of green tea beverages. *Food Chem.* 331: 127258, ISSN 0308-8146, https://doi.org/10.1016/j.foodchem.2020.127258.

**49** Altunay, N., Gürkan, R., and Orhan, U. (2017). Indirect determination of the flavor enhancer maltol in foods and beverages through flame atomic absorption spectrometry after ultrasound assisted-cloud point extraction. *Food Chem.* 235: 308–317, ISSN 0308-8146, https://doi.org/10.1016/j.foodchem.2017.05.066.

**50** Verstrepen, K.J., Derdelinckx, G., Dufour, J.-P. et al. (2003). Flavor-active esters: adding fruitiness to beer. *J. Biosci. Bioeng.* 96 (2): 110–118. ISSN 1389-1723, https://doi.org/10.1016/S1389-1723(03)90112-5.

**51** O'Sullivan, M.G. (2017). Chapter 6 – Shelf life and sensory quality of foods and beverages. In: *A Handbook for Sensory and Consumer-Driven New Product Development* (ed. M.G. O'Sullivan) Woodhead Publishing Series in Food Science, Technology and Nutrition,, 103–123. Woodhead Publishing, ISBN 9780081003527, https://doi.org/10.1016/B978-0-08-100352-7.00006-3.

**52** Tungland, B. (2018). Chapter 7 – Overview of prebiotics: membership, physiological effects and their health attributes. In: *Human Microbiota in Health and Disease* (ed. B. Tungland), 289–348. Academic Press, ISBN 9780128146491, https://doi.org/10.1016/B978-0-12-814649-1.00007-7.

**53** Souza, F.P., Balthazar, C.F., Guimarães, J.T. et al. (2019). The addition of xyloligoosaccharide in strawberry-flavored whey beverage. *LWT* 109: 118–122, ISSN 0023-6438, https://doi.org/10.1016/j.lwt.2019.03.093.

**54** Goyal, A., Sharma, V., Upadhyay, N. et al. (2014). Flax and flaxseed oil: an ancient medicine & modern functional food. *J. Food Sci. Technol.* 51 (9): 1633–1653. https://doi.org/10.1007/s13197-013-1247-9.

**55** Aliani, M., Ryland, D., and Pierce, G.N. (2012). Effect of flax addition on the flavor profile and acceptability of bagels. *J. Food Sci.* 77 (1): S62–S70. https://doi.org/10.1111/j.1750-3841.2011.02509.x. , PMID: 22260132.

**56** Shahidi, F. and Ambigaipalan, P. (2018). Omega-3 polyunsaturated fatty acids and their health benefits. *Annu. Rev. Food Sci. Technol.* 25 (9): 345–381. https://doi.org/10.1146/annurev-food-111317-095850.PMID: 29350557.

**57** Shibamoto, T. and Bjeldanes, L.F. (1993). Chapter 5 – Natural toxins in plant foodstuffs. In: *Food Science and Technology, Introduction to Food Toxicology* (ed. T. Shibamoto and L.F. Bjeldanes), 67–96. Academic Press, ISBN 9780080925776, https://doi.org/10.1016/B978-0-08-092577-6.50010-1.

**58** Praet, T., Van Opstaele, F., Jaskula-Goiris, B. et al. (2012). Biotransformations of hop-derived aroma compounds by *Saccharomyces cerevisiae* upon fermentation. *Cerevisia* 36 (4): 125–132. ISSN 1373-7163, https://doi.org/10.1016/j.cervis.2011.12.005.

**59** Gascon, M. (2007). Chapter 13 – Masking agents for use in foods. In: *Woodhead Publishing Series in Food Science, Technology and Nutrition, Modifying Flavour in Food* (ed. A. Taylor and J. Hort), 232–242. Woodhead Publishing, ISBN 9781845690748, https://doi.org/10.1533/9781845693367.232.

**60** Lopes, D.B., Speranza, P., and Macedo, G.A. (2016). Chapter 18 – A new approach for flavor and aroma encapsulation. In: *Nanotechnology in the Agri-Food Industry, Novel Approaches of Nanotechnology in Food* (ed. A.M. Grumezescu), 623–661. Academic Press, ISBN 9780128043080, https://doi.org/10.1016/B978-0-12-804308-0.00018-2.

**61** Saraiva, A., Carrascosa, C., Raheem, D. et al. (2020). Maltitol: analytical determination methods, applications in the food industry, metabolism and health impacts. *Int. J. Environ. Res. Public Health* 17 (14): 5227. https://doi.org/10.3390/ijerph17145227. PMID: 32698373; PMCID: PMC7400077.

# 6

# Flavor Biochemistry of Fermented Alcoholic Beverages

*Maurício B.M. de Castilhos[1], Ana P.G. de Queiroga[1], Lia L. Sabino[1], Jorge R. dos Santos Júnior[2], Jorge A. Santiago-Urbina[3], Hipócrates Nolasco-Cancino[4,5], Francisco Ruíz-Terán[6], and Vanildo L. Del Bianchi[2]*

[1]Minas Gerais State University, Department of Agricultural Sciences and Biology, Escócia Avenue 1001, Universitário, 38200-000, Frutal, Minas Gerais, Brazil
[2]São Paulo State Univeristy, Department of Food Technology and Engineering, Cristóvão Colombo street 2265, Jardim Nazareth, São José do Rio Preto, São Paulo, 15054-000, Brazil
[3]Universidad Tecnológica de los Valles Centrales de Oaxaca, Division of Career Management of Sustainable and Protected Agriculture, Universidad Avenue S/N, San Pablo Huixtepec, Zimatlán, Oaxaca, 71265, Mexico
[4]Universidad Autónoma Benito Juárez de Oaxaca, Chemistry Science Faculty, Universidad Avenue S/N, Ex-Hacienda 5 Señores, 68120, Oaxaca, Mexico
[5]Consejo Regulador del Mezcal, Cofre de Perote 325, Col. Volcanes, 68020, Oaxaca, Mexico
[6]Universidad Nacional Autónoma de México, Food and Biotechnology Department, Faculty of Chemistry, Universidad Avenue 3000, Coyoacán, 04510, Ciudad de México, México

## 6.1 Introduction

The flavor is one of the sensory attributes that are evaluated by panelists in sensory acceptance and descriptive assessments, and due to its complex behavior, sensory research studies have accounted for several discussions regarding the definition of the flavor concept. On the one side, the authors describe the flavor as the sum of perceptions resulting from the stimulation of the senses by the food in the digestive and respiratory tracts. In this context, some researchers support the idea that aromatic compounds influence the flavor (volatiles perceived by the olfactory system in the mouth), taste (soluble compounds perceived by the gustatory system in the mouth), and stimuli provided by chemical compounds (stimulation of nerve ends in the mouth and nasal cavities, such as astringency) [1]. On the other side, other authors describe the flavor as a result of the combination of two factors: taste and mouthfeel. Taste is directly related to specialized receptor cells, which generate basic perceptions such as sweet, sour, salty, bitter, and umami (savory). Mouthfeel is described as perceptions of astringency, touch, dryness, viscosity, burning, heat, coolness, body, prickling, and pain scattered throughout the oral cavity and generated by the free nerve endings [2].

*Natural Flavours, Fragrances, and Perfumes: Chemistry, Production, and Sensory Approach,*
First Edition. Edited by Sreeraj Gopi, Nimisha Pulikkal Sukumaran, Joby Jacob, and Sabu Thomas.

Despite all the discussion about the subject of flavor regarding its definition and its interaction with the other human sense organs, it is a fact that flavor is unquestionable, i.e., all sensory changes are based on chemical changes. Among the studies that have already been published worldwide, the relationship between the sensory features of food and beverage matrices and chemical properties is undeniable, and there is a current trend in applying some valuable statistical tools to make this relationship possible.

Some research studies have reported the importance of sensory drivers for fermented beverages, i.e., chemical properties that can drive the acceptance or the description of a beverage sensory profile. De Castilhos et al. [3] studied the sensory acceptance drivers of pre-fermentation dehydration and submerged cap winemaking in red wines produced from *Vitis labrusca* hybrid grapes, and they reported that a variation in wine technology could guide a relevant sensory change. The use of grape dehydration to avoid the chaptalization process improves wine flavor since heat can produce different volatile compounds in grapes, transferring them to wine during maceration. In this context, pre-drying of grapes can produce red wines with different flavors, which draws the attention of specific wine consumers.

Based on the above-mentioned information, if chemical changes can explain all the sensory changes, it is crucial to analyze these chemical changes by using the biochemical point of view, i.e., when it comes to sophisticated fermented alcoholic beverages, the biochemistry phenomena are essential to explain some positive sensory effects or some sensory off-flavors. The biochemistry of the fermentation processes is quite complicated since it is difficult (almost impossible) to regulate the fermentation process considering a unique microorganism. It is well known that fruits and vegetables, in general, have in their composition a bunch of microorganisms naturally present in their skins, which are known as autochthonous microorganisms. They also participate in the fermentation process and guide the sensory features of the produced fermented beverage. Some studies have reported the possible use of autochthonous microorganisms in fermented beverages intending to provide a distinctive sensory profile. Essential information about the potential of producing fermented beverages with their sensory profile was assessed, highlighting their sensory uniqueness and singularity, due to the use of these wild microorganisms in the alcohol fermentation process [4–6].

Based on this context, this chapter will provide some essential information regarding the flavor biochemistry of fermented alcoholic beverages. Among the several different types of fermented beverages, this chapter provides essential information about wine, due to its high complexity, and mezcal, a valuable beverage that is fermented and distilled. The synthesis of all the flavor compounds is explained based on the biochemical phenomena that have already been studied, and these reactions will be used to explain the sensory effects of fermented beverages.

## 6.2 General Aspects of Alcohol Fermentation

In general, alcohol fermentation is a biochemical reaction that transforms sugars, mainly glucose and fructose, into ethanol and carbon dioxide by using yeasts (overall reaction, Figure 6.1). *Saccharomyces cerevisiae* is considered the primary

**Figure 6.1** Overall reaction of alcohol fermentation. Source: Adapted from Jackson [7].

$$\underset{\text{Hexoses}}{C_6H_{12}O_6} \longrightarrow \underset{\text{Ethanol}}{2\,C_2H_5OH} + \underset{\text{Carbon dioxide}}{2\,CO_2} + \underset{\text{Energy}}{2\,ATP}$$

yeast responsible for alcohol fermentation that produces fermented beverages since it shows high osmotolerance in comparison to other yeasts [7]. Other yeasts such as *Hanseniaspora uvarum*, *Torulopsis bacillaris*, and *Kloeckera apiculata* are responsible for alcohol fermentation in some punctual steps of the fermentation process since some of them lose their activity due to ethanol toxicity. Usually, the genera *Kloeckera*, *Hanseniaspora*, and *Candida* participate in the beginning stages of alcohol fermentation. The genus *Pichia* then prevails in the middle stages, directing the alcohol fermentation process to attain the prevalence of *S. cerevisiae* due to its excellent resistance to high ethanol content. Some other yeasts can also be observed during the alcohol fermentation process, such as *Kluyveromyces*, *Schizosaccharomyces*, *Zygosaccharomyces*, and *Brettanomyces*, and some of them can cause off-flavors in the fermented beverages. In addition to yeasts, some bacteria such as *Zymomonas mobilis* can also perform this bioprocess [8].

In some cases, the presence of these yeasts can provide positive results for the fermented beverage produced; in other cases, they can provide off-flavors and severe defects. The case of *Brettanomyces* is a typical example since this yeast is considered a wine spoilage organism; however, there is a controversy among researchers regarding this fact. One group reported that *Brettanomyces* in wines is responsible for the Brett character, i.e. strong wine off-flavors (medicinal, horse sweat, and barnyard) caused by the formation of volatile phenols, mainly 4-ethylphenol (4-EP) and 4-ethyl guaiacol (4-EG) [9]. The other group argued that these compounds mentioned above in specific concentrations caused this off-flavor effect, and *Brettanomyces* can provide a fruity aroma for fruity wines due to its potential in synthesizing relevant esters as follows: ethyl acetate, ethyl decanoate, ethyl caprylate, and ethyl lactate [10, 11].

Alcohol fermentation can be considered a complex biochemical reaction due to the formation of many secondary compounds, which are primarily synthesized by yeast metabolisms, such as higher alcohols, esters, glycerol, and acids. During the overall reaction of alcohol fermentation, secondary reactions (chemical, biochemical, and physicochemical) take place and produce a bunch of compounds responsible for the flavor of fermented beverages [12].

At the beginning of alcohol fermentation, the latency phase (lag phase), the yeasts start to metabolize the substrates (fermentable sugars) and other nutrients to obtain energy and increase the number of viable cells (Figure 6.2). During this initial phase, the yeast species do not increase their population due to the necessary adaptation of the viable cells to the external environment conditions. Depending on the occasion, the yeast cells may present a different population at the beginning of alcohol fermentation. In some fermented beverages, yeasts are inoculated at the beginning of alcohol fermentation by using this as a standardization procedure and increasing the initial yeast population in the fermentation environment. After the adaptation of yeasts to the new environment, they start to grow and multiply following an exponential pattern [7].

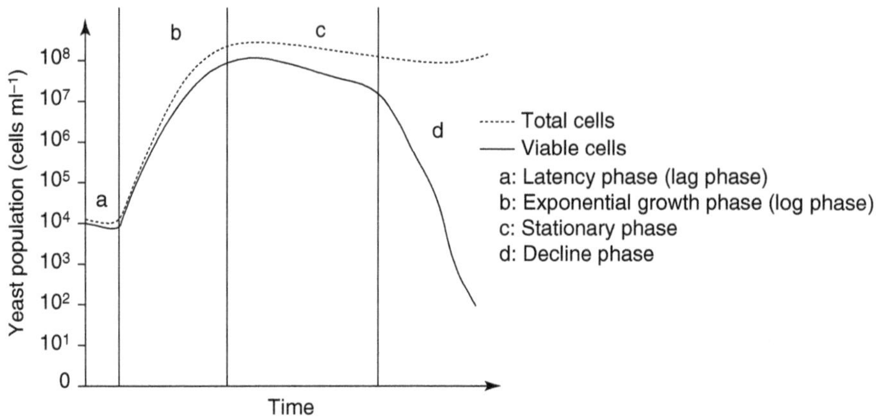


**Figure 6.2** Yeast growth behavior during alcohol fermentation. Source: Adapted from Del Nobile et al. [13].

This fermentation stage is commonly known as the exponential growth phase or log phase, and several factors influence its development, such as temperature and nutrient concentrations such as ammonia, amino acids, and oxygen [7]. At this stage, yeasts reach a population ranging from $10^7$ to $10^9$ viable cells per ml and can survive around three to six days depending on the alcoholic beverage. After this time, the yeast population rate decreases due to the lack of crucial nutrients and the toxicity of ethanol. Also, in the final period of this phase, the yeasts are unable to synthesize essential sterols and long-chain unsaturated fatty acids; simultaneously, there is an accumulation of toxic, mid-size, and carboxylic by-products resulting from yeast metabolism. After the log phase, the yeasts begin a new phase known as the stationary phase, which is characterized by the stabilization of the yeast population and lasts for approximately 2–10 days. In the stationary phase, there is a change in the enzyme profile, i.e. the synthesis of stress-related proteins and the accumulation of trehalose and glycerol. This sugar maintains cell fluidity and avoids protein denaturation, and this is a possible explanation for the cell viability duration until this phase [7, 14].

After the stationary phase, the decline stage begins, and the yeast population gradually decreases until their disappearance. In this stage, the lack of nutrients and the toxic effects of ethanol and other compounds produced during the alcohol fermentation process are the reason for yeast death. Also, there is an additional explanation for the relative decrease of yeast activity in this phase, which is directly related to the membrane disruption of yeasts resulting from the combined toxicity effect of ethanol and mid-chain fatty acid [7, 15, 16].

## 6.3 General Aspects of Flavor

As previously mentioned, the meaning of the term flavor designates all the sensory properties that are indirectly perceptible by the retronasal olfactory, i.e. the perceptions provided by the olfactory system when tasting. The flavor term denotes an interaction between the olfactory and gustatory properties perceived by taste and can be directly influenced by tactile, thermal, painful, and kinesthetic effects [17]. In the same way, researchers in the field of sensory science have used some essential definitions for the description of taste and flavor. There are five basic tastes: salty, sweet, sour, bitter, and umami (savory), and all these basic tastes are felt by the consumers or by a sensory panel directly in the mouth by a specific group of cavities within taste buds. Free nerve endings generate mouthfeel perceptions such as astringency, viscosity, dryness, burning, coolness, and prickling, and these nerves are scattered along the oral cavity. The combination of these sensations with those produced by the olfactory system (orthonasal or odor) compose the flavor perception [2].

The gustatory sensations are detected by epidermal cells called taste buds, and each of them is associated with neuroepithelial cells. Cranial nerve fibers are linked to the base of the taste buds with synapses with one or more receptor cells. Almost 70% of the total taste buds are located on the tongue, and the other ones are distributed on the pharynx, larynx, and upper portion of the esophagus. Taste buds are usually associated with four classes of papillae: fungiform papillae, circumvallate papillae, foliate papillae, and filiform papillae. Fungiform papillae are essential for the palate due to their density that is directly correlated with taste acuity. Circumvallate papillae are located mainly on the backside of the tongue, and foliate papillae are distributed along the margins of the tongue. Filiform papillae are the most common form and are distributed throughout the entire tongue surface; however, they have no taste buds. In the middle portion of the tongue, it is not possible to observe the presence of taste buds, which indicates that this portion does not have relevant participation in taste phenomena (Figure 6.3) [2].

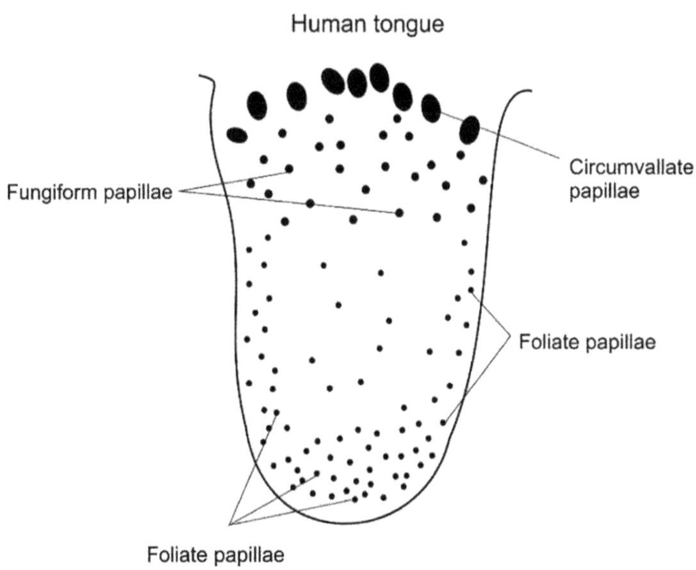


**Figure 6.3** Indication of the major types of papillae on the human tongue. Source: Adapted from [2].

A discussion regarding the taste areas of the tongue has been performed over the years. Before the neurophysiological studies on the taste phenomena, the tongue was divided into areas that were responsible for the sensation of punctual and different basic tastes. According to this thesis, salty and sweet basic tastes were felt at the tip of the tongue, the basic sour taste was felt at the tongue borders, the bitter taste at the back of the tongue, and umami was felt at the center region of the tongue [18]. This thesis fell apart with the new sensory studies that showed the distribution of different taste buds along the tongue, and all of them are responsible for the sensation provided by the basic tastes and all the mouthfeel sensations as a result of the interaction of all forms of taste buds. The standard distribution of cranial nerves in the tongue partially reflects the different stimuli in the separate areas of the tongue; however, these differences are small, and the differentiation among the tongue areas regarding the taste sensations is unpractical [2].

Three cranial nerves are responsible for the enervation of taste buds: the geniculate ganglion of the facial nerve is responsible for the taste buds of fungiform papillae at the back of the tongue. The petrous ganglion of the glossopharyngeal nerve is linked to the taste buds of the foliate and circumvallate papillae. The nodose ganglion of the vagus nerve is responsible for the taste buds of the epiglottis, larynx, and upper sides of the esophagus [19]. These cranial nerves are responsible for recognizing the five basic tastes mentioned above. The umami basic taste is a taste response to some L-amino acids, mainly glutamate and aspartate, as well as 5′-ribonucleotides, mainly inosinate and guanylate [20]. The high or low sensitivity to several tastes seems to be related to specific protein receptors or the combination of them [21]. A specific receptor cell can produce only one or a few pairs of protein receptors responsible for generating impulses for that specific taste. However, individual taste buds may respond to most taste sensations. A group of genes known as TAS, the taste genes, is responsible for taste sensations. The basic tastes of sweetness, umami, and bitterness are associated with a group of approximately 30 genes. In contrast, sour and salty sensations are associated with a group of genes that present no relationship with themselves. Studies have also reported that, receptors that present the same sensitivity are grouped, and an individual receptor may react in a different form by the interaction with the same compound or stimuli [2].

In another study, the authors reported that sweetness, umami, and bitterness receptors are expressed not only by the taste buds present on the tongue but also by digestive organs, kidneys, and brain, i.e. these organs can influence the sensitivity of these basic tastes mentioned above together with the receptors present on the tongue surface [22]. This context can explain the complexity of the flavor as a sensory attribute since it can be explained by the interaction of several factors and by the influence of organs instead of human sense organs. This complexity explains the reason why sensory researchers have difficulties regarding the sensory analysis of food and beverages. Despite the sensory expertise of the panelists, in most sensory sessions, the panelists did not provide results with objectivity and reproducibility, and research studies have been performed to incorporate technology in this field using proper equipment with high precision and sensitivity such as the electronic tongue and electronic nose [23].

Both taste and aroma play an essential role in the flavor concept. To reach the olfactory receptors, the odorant compounds need to be volatile and do not need to have a specific odor since the interaction between the volatile odorant and the mouth environment can modify all the perceptions of the panelist. A total of 12000 volatile compounds have already been identified in food and beverages, of which only 5% play a significant role in aroma perception [24]. All these compounds can only be perceived if the concentration exceeds their respective thresholds, i.e. the minimum concentration that allows the sensory perception of the compound into the matrix (food or beverage). Depending on the volatile compound, the threshold can assume values ranging from $\mu g\,l^{-1}$ to $mg\,l^{-1}$, and the identification of these compounds with significantly lower thresholds depends on the separation and identification technique employed. Techniques such as gas chromatography coupled with mass spectrometry (GC-MS) present higher accuracy and sensitivity to identify and quantitate these compounds showing minimum concentrations [25].

According to the same authors, three types of approach can be used for the analysis of flavor compounds in foods and beverages:

(i) Perform a sensory analysis, mainly descriptive sensory assessment;
(ii) Analyze the volatile/flavor compounds to obtain the volatile profile into the matrix using multivariate statistical techniques aiming at comparing different matrices and different volatile profiles;
(iii) Use a target analysis to know the quality of the matrix and collect applicable information about the technology, processing, and storage. This approach generally uses more accurate and precise instrumental techniques such as nuclear magnetic resonance (NMR).

In terms of alcoholic beverages, the flavor sensory analysis and the instrumental data can provide essential and feasible information concerning the specific volatile compounds responsible for specific aroma sensations, characterizing an essential aroma or pivotal odorant study. This type of approach is commonly used by several authors worldwide [26–29], especially for wines produced in countries considered more fabulous worldwide wine producers to confirm their geographical indication or their controlled denomination of origin.

Based on the information regarding the general flavor aspects, the complexity of this subject in terms of sensory and instrumental analysis, and the interaction between the sensory perceptions with the human sense organs, the analysis of flavor biochemistry becomes necessary and useful. The study of the flavor biochemistry aspects is feasible to comprehend the mechanisms that guide taste, mouthfeel, and aroma and all the interactions between them, with an aim to understand the taste and all the mechanisms around this sensory attribute.

## 6.4 Flavor Biochemistry in Fermented Beverages

### 6.4.1 Wines

The alcohol fermentation process consists of metabolizing glucose and fructose into ethanol primarily via glycolysis, also known as the Embden-Meyerhof pathway. Additional metabolites are generated by the secondary yeast metabolism, and they are responsible for the most relevant aroma compounds in fermented beverages. The changes occurring during alcohol fermentation, mainly during the log phase, are responsible for significant modifications in yeast metabolism. Thus, the primary and intermediary yeast cell metabolism is responsible for the aroma and flavor of the fermented beverage [8].

Winemaking is considered a complicated procedure that involves several steps, that transform grape must into wine, mainly by alcohol fermentation. The transformation of the grape must shows high acidity, sweetness, and deep flavor in wine with unique sensory features, and the complexity is rendered by the yeasts responsible for alcohol fermentation. In general, the alcohol fermentation process consists of metabolizing the sugars into ethanol and carbon dioxide via biochemical pathways. Also, the glycolytic and associated pathways form nonvolatile and volatile metabolites that contribute to the wine flavor. Several factors such as the yeast species, the fermentation conditions, and the nutrient content of the must modulate the production of these compounds responsible for the wine flavor complexity [30, 31]. The interaction between the yeasts and the grape must during the alcohol fermentation process contributes to wine's appearance, aroma, flavor, and body. When wood is used during the fermentation process, some wood-derived substances can also modify the sensory attributes and helps the wineries to enhance the distinctive wine varietal character.

It is well known that yeasts are one of the leading agents that contribute to sensory wine complexity. Few wineries around the world use autochthonous yeasts naturally present in the grape pomace as fermentation agents to provide a varietal character for wine as a result of the synthesis of key odorant and key flavor compounds. The majority of the wineries worldwide produce wines using selected yeast strains such as *S. cerevisiae* to avoid the formation of off-flavors and guarantee the wine quality [32].

Generally, wine production begins only when the yeasts have direct contact with their respective substrates, i.e. fermentable sugars, mainly glucose and fructose. When the grape clusters are crushed and the grape juice is released, the wild yeasts begin the alcohol fermentation process. The addition of specific and isolated *S. cerevisiae* yeasts enables standardization of the fermentation process. Sulfur dioxide is also added as an antioxidant agent. After the alcohol fermentation process, the wine is dejuiced and racked three times after bottling. Between the second and third racking, the wine is submitted to a second fermentation known as malolactic fermentation, which comprises the decarboxylation of malic acid into lactic acid, while generating a small amount of carbon dioxide. This fermentation is performed using the lactic acid bacteria *Oenococcus oeni*. After malolactic fermentation, the wines are placed in a refrigerated ambient environment to allow tartrate stabilization; thus, the wines are stabilized for a specific time and then bottled [33–35].

The grape crushing step at the beginning of winemaking releases volatile and nonvolatile compounds associated with the grape berry, and the addition of pectinolytic enzymes in an exogenous way during the maceration step facilitates the release of flavor and precursor flavor compounds associated with the grape skin and seeds [31]. Specific grape compounds synthesize some punctual compounds belonging to a specific yeast metabolism pathway, and each metabolite will be responsible for a wine sensory feature (Figure 6.4).

#### 6.4.1.1 Flavor Precursors

The alcohol fermentation process is a biochemical reaction that provides the formation of a set of volatile compounds that influence the wine flavor. The synthesis and accumulation of these compounds depend on several factors such as the yeast strain, must composition (chemical, nutritional, and physical), and fermentation conditions (aeration and temperature). Punctual biochemical pathways that the yeasts choose according to the conditions imposed by the environment produce many of these compounds. In this section, we present some essential information about the biochemistry of these compounds that influence the wine flavor.

#### 6.4.1.2 Esters

Esters derived from the alcohol fermentation process are responsible for wine fruitiness and play an essential role in the sensory profile of both red and white wines. Studies revealed that these volatile compounds are present in young wines and remain in wines that were oak aged for years. Therefore, the esters derived from alcohol fermentation are essential compounds for the flavor of both young and aged wines [36]. Two groups of esters provide the fruity character for wines, namely acetate esters such as ethyl acetate, isobutyl acetate, and isoamyl acetate and ethyl fatty acid esters such

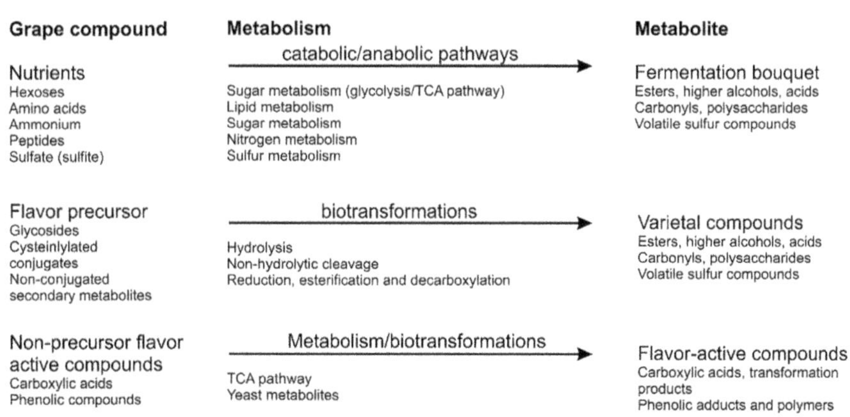


**Figure 6.4** Interaction between grape compounds and yeast resulting in different metabolites as a result of the different metabolic pathways. Source: Adapted from [31]. Note: TCA pathway, tricarboxylic acid cycle pathway.

as ethyl hexanoate, ethyl octanoate, and ethyl decanoate. In a recent study by Sánchez-Palomo et al. [28], the authors studied the aroma profile of La Mancha Chelva wines (*Vitis vinifera*). They reported the relevant concentration of some ethyl fatty acid esters such as ethyl hexanoate, ethyl octanoate, and ethyl decanoate, which provide an intensely fruity aroma and flavor for these wines. De Castilhos et al. [27] studied the volatile aroma profile of red wines produced from *V. labrusca* grapes known as BRS Rúbea and BRS Cor. They reported the higher concentration of some esters such as isoamyl acetate, ethyl hexanoate, ethyl octanoate, and 2-phenylethyl acetate, the latter providing a rose/floral aroma with a honey note. In the same study, the authors reported that these volatile compounds were responsible for driving the sensory acceptance of the red wines mentioned above, considering them crucial sensory acceptance drivers.

Acetate esters are synthesized by a condensation reaction between higher alcohols with acetyl-CoA catalyzed by the alcohol acetyltransferase enzymes [37, 38]. The final concentrations of these compounds depend on the balance between alcohol acetyltransferase enzymes that promote their synthesis and esterase enzymes that promote their hydrolysis [39]. The genes responsible for the synthesis of acetate esters by the yeast *Saccharomyces* that encodes the alcohol acetyltransferase enzymes are ATF1 and ATF2 (Figure 6.5). Both genes, mainly ATF1, contribute to the formation of these volatile compounds, and they are responsible for the formation of fruity wine features. The variations in the expression levels of these genes (deletion and modulation) afford the basis of the production of acetate esters by the different yeast strains and alcohol fermentation conditions.

The biochemical pathways related to the synthesis of fatty acid esters are quite unknown. Studies have reported that the significant esters derived from the fatty acids, regardless of the carbon chain size, are formed enzymatically by the esterification of the respective fatty acids (acyl-CoA) during the beginning of lipid biosynthesis. In a recent study, two enzymes were identified as critical factors for the formation of ethyl esters of medium-chain fatty acids, known as Eht1p and Eeb1p. In general, some research studies on wines have shown that the Eht1p enzyme is directly responsible for the formation of ethyl hexanoate, ethyl octanoate, and ethyl decanoate, which present fresh fruity aroma. There is also a yeast genetic bias that guides the concentration of these esters derived from the fatty acids, i.e. the deletion of one or both of the genes responsible for the synthesis of the enzymes reduces the concentration of these respective esters; however, the overexpression of these genes had limited or minimal effects on the final concentration of fatty acid esters in wines. This biochemical phenomenon is possibly explained by the bifunctional synthetic and hydrolytic activities [31, 40].

Ethyl acetate, one of the primary esters produced in wines, is synthesized by several different types of non-*Saccharomyces* yeasts, and these yeasts generally produce

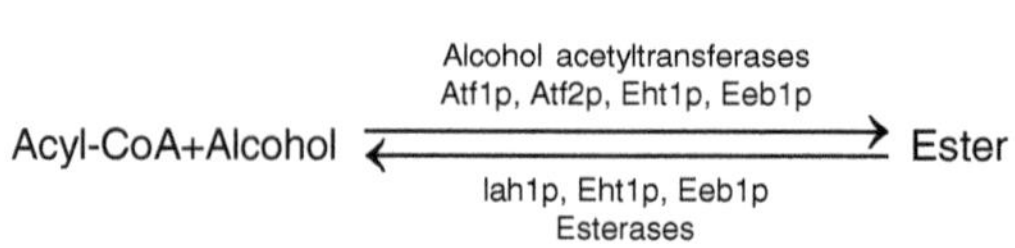


**Figure 6.5** The biochemical pathway for ester synthesis and degradation. Source: Adapted from Ugliano and Henschke [31]/ Springer Nature.

higher amounts of this volatile compound as compared to the genus *Saccharomyces*. These non-*Saccharomyces* yeasts are *H. uvarum*, *K. apiculata*, *Hanseniaspora guilliermondii*, *Candida krusei*, *Pichia anomala*, among others. In most cases, wineries standardize the alcohol fermentation process by inoculating *S. cerevisiae* strains to promote a considerable variability of aromas and flavors. Important factors such as environmental nutrients (variations in sugar content, oxygen, lipids, assimilable nitrogen) and fermentation temperature can also be responsible for the significant variability of wine aroma and flavors. It is well known that low temperature during the alcohol fermentation process preserves the fruity scents in wines and affects the accumulation of acetate and ethyl fatty acid esters. In general, the lower temperature during the fermentation facilitates the synthesis of ethyl esters; however, high temperature also facilitates the synthesis of 2-methyl acetate, 2-phenyl ethanol, and 2-phenyl ethyl acetate synthesis [41].

Despite the environmental conditions, fermentation temperature and fatty acid wine must composition are the main factors for the biochemical formation of esters by the wine yeasts; oxygen availability also plays a vital role in the synthesis of esters. The aeration of the fermentation environment represses ATF1 gene expression, and this causes a decrease in the concentration of acetates [39]. The wine acetate concentration does not have a direct relationship with the concentration of its own higher alcohol, instead of the esters derived from fatty acids, which depend on their respective fatty acid precursor [40].

***Higher Alcohols*** These alcohols, also named fusel alcohols, are considered one of the most important volatile compounds that contribute to wine flavor complexity. The majority of the fusel alcohols are produced from yeasts during the alcohol fermentation process, and their concentration can be decisive for the wine flavor complexity and quality. De Castilhos et al. [27] reported the relevant concentration of 2-pentanol in non-*V. vinifera* red wines produced from alternative winemaking procedures. They reported that this alcohol provided mild, oily, green, and fermented notes for the assessed wines, and they also reported the presence of 2-phenylethanol that was responsible for the floral/fruity typical aroma and flavor of these wines prepared from *V. labrusca* grapes and their hybrids.

According to Ugliano and Henschke [31], higher alcohols are formed by the decarboxylation and posterior reduction of α-keto-acids produced as intermediate compounds of amino acid biosynthesis and degradation (Figure 6.6). Some specific amino acids and specific transport proteins are responsible for higher alcohol formation in wines through Ehrlich and biosynthesis pathways. In this context, some studies have already reported that the amino acids phenylalanine, tyrosine, and tryptophan formed 2-phenylethanol, tyrosol, and tryptophol aromatic alcohols, respectively, while valine, leucine, and isoleucine are responsible for the synthesis of branched-chain aliphatic alcohols 2-methylpropanol, 2-methylbutanol, and 3-methylbutanol, respectively. According to the above-mentioned information, the formation of these specific alcohols as a result of their respective amino acid precursors is mediated by transport proteins as follows: branched-chain amino acid permease Bap2p and Bap3p and the general permease named Gap1p. The amino

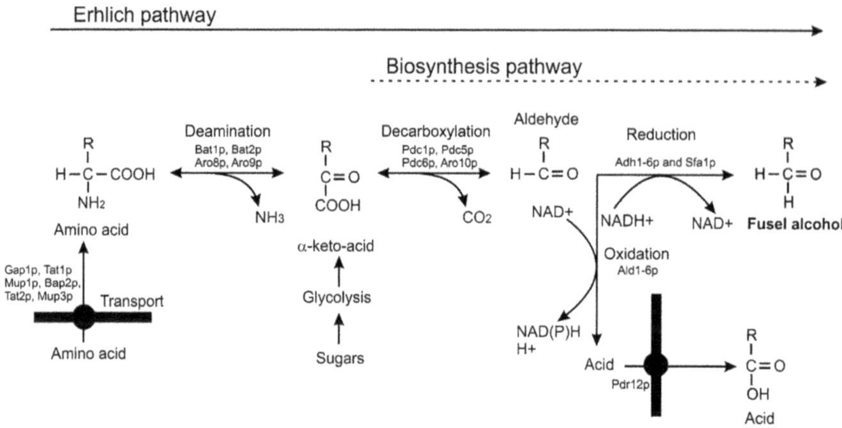


**Figure 6.6** The biochemical Ehrlich pathway for fusel alcohol synthesis. Source: Adapted from Ugliano and Henschke [31]/Springer Nature.

acids with aromatic structures in their specific chain are transported by Tat1p, Tat2p, Gap1p, and Bap2p, and methionine is transported by Mup1p, Mup3p, and Gap1 [42].

The Ehrlich pathway begins with the amino acid transamination/deamination process to form the respective α-keto-acid, catalyzed by branched-chain (Bat1p and Bat2p) and aromatic amino acid transferases (Aro8p and Aro9p). After this initial reaction, the α-keto-acids are transformed into their respective aldehydes as a result of the decarboxylation reaction by the pyruvate decarboxylases (Pdc1p, Pdc5p, and Pdc6p) and then reduced to alcohols by the alcohol dehydrogenase activity (Adh1p and Adh6p). The aldehydes are oxidized to their corresponding acids, which are removed from the cell environment by the action of the organic acid permease (Pdr12p) [43]. All the alterations regarding the genes that modulate the transport proteins or enzymes might provide relevant changes in the wine flavor profile. According to Figure 6.6, two pathways can generate the wine fuse alcohols responsible for wine flavor complexity. The environment presents a deficiency in amino acid; the glycolysis forms the α-keto via yeast sugar metabolism; however, if the environment has a sufficient concentration of amino acid (higher medium nitrogen concentration), these amino acids are transformed into their respective α-keto acids as a result of a deamination reaction. Both pathways are essential for wine flavor since they will be responsible for differences in wine flavor composition and complexity.

Several factors influence the lower or higher concentrations of fusel alcohols as cited in the latter section. Yeast strains, fermentation temperature, sugar concentration, pH, aeration, grape variety, grape must composition, maceration time, and assimilable nitrogen are the main factors that modulate higher alcohol synthesis. The latter factor, the assimilable nitrogen, is one of the most critical factors in this case due to its highlighted participation in the biosynthesis of fusel alcohol. The lower nitrogen composition in the grape must will result in a higher concentration of fusel alcohol since the yeasts will provide the formation of α-keto acids by sugar glycolysis as expected. Considering the existence of a chemical equilibrium between the formation of α-keto acids and their respective amino acids, in this above-mentioned condition, the α-keto acid formed by sugar glycolysis would not be converted to its respective amino acid due to the lack of assimilable nitrogen in the grape must environment. In contrast, if the grape must has higher assimilable nitrogen content, α-keto acids can be converted by their primary form, i.e. their respective amino acid, thereby decreasing fusel alcohol production.

#### 6.4.1.3 Carbonyl Compounds

Among the main carbonyl compounds, acetaldehyde plays a vital role in wine flavor due to its contribution to bruised apple and nutty wine character when it has concentrations above its threshold ($100\,mg\,l^{-1}$ in wine). In general, the aldehydes in wines provide a vegetal scent with grassy and green descriptors as well as fatty, fruity, and pungent flavors. In addition to acetaldehyde, hexanal and benzaldehyde are other notable examples of volatile compounds that contribute to wine flavor; the former provides grass and beefy flavor, while the latter provides an almond and burnt sugar scent for wines. Both the above-mentioned compounds were identified in both the *V. vinifera* and non-*V. vinifera* red wines. Some authors have reported the presence of benzaldehyde and hexanal in wines. Martins et al. [44] identified benzaldehyde in wine samples called *Vinhos de Talha* or amphora wines (*V. vinifera*); Ubeda et al. [29] identified and quantitated a significant concentration for hexanal in sparkling wines produced from the País grape cultivar (*V. vinifera*). De Castilhos et al. [27] identified benzaldehyde in wines produced from hybrid grapes (*V. labrusca*) and reported the lower contribution of this volatile compound to wine flavor.

In addition to aldehydes, ketones and lactones are also considered great contributors to wine flavor complexity and quality. Among the ketones, 2,3-butanedione is one of the most important in wines, and butyrolactone is one of the most important representatives of lactones. Butanedione provides nutty, toasty, and buttery wine flavor, and this formation is associated with wine malolactic fermentation. Butyrolactone and its alternative forms are responsible for rubbery and buttery flavors and can be considered as a sensory descriptor driver for wines produced from non-*V. vinifera* grapes, especially BRS Carmem and BRS Violeta grapes [26]. Geffroy et al. [45] also identified γ-butyrolactone in wines produced from Grenache and Carignan grape varieties (*V. vinifera*).

Acetaldehyde is formed as a result of ethanol production by the fermentative yeast. Primarily, pyruvic acid is decarboxylated to acetaldehyde by pyruvate decarboxylase (Pdcp). The concentration of acetaldehyde in wine depends on the reactions performed during the alcoholic fermentation (Figure 6.7). It is well known that the lack of thiamine can lead to a decrease in acetaldehyde concentration since this amino acid is considered a pyruvate decarboxylase cofactor. The common biochemical pathway for acetaldehyde degradation is its reduction to ethanol by alcohol dehydrogenase, and this reaction is essential for sugar metabolism by the yeast. Aldehyde dehydrogenase oxidizes acetaldehyde to promote the synthesis of acetic acid. The synthesis of lactones and their alternative forms is still unknown, but their biochemical synthesis involves chemical cyclization and glutamic acid as a precursor, thus promoting their formation during the alcohol fermentation process [31].

Several factors during alcohol fermentation can modulate acetaldehyde formation, such as yeast strain, aeration, grape must composition, fermentation temperature, must clarification, and sulfite addition ($SO_2$). In this specific case, the addition of sulfite in wines, with an aim to protect the alcohol fermentation process and the entire winemaking process from other undesirable microorganisms, can directly influence acetaldehyde formation since this compound can bind to $SO_2$ to promote a stable adduct acetaldehyde-hydroxysulfonate form as shown in Figure 6.7.

***Oak Flavor Compounds*** Oak compounds have an excellent contribution to wine aroma and flavor, and complex transformations can be assigned in this step of the winemaking procedure. In general, some wineries perform alcohol fermentation in the oak barrels to provide a unique and pleasant flavor to wines. Other wineries use the oak barrels at the wine stabilization step and the final of the winemaking process and maintain the wines in contact with the oak wood for years. The direct contact between the wine and oak barrel provides the extraction of potent compounds from oak to wine, such as vanillin derivatives, volatile phenols, and lactones. All these volatile compounds promote vanilla, spice, woody, and coconut scents.

The wine fermented in oak barrels contains essential compounds that were extracted from the wood and were transformed by the yeast as a result of the alcohol fermentation reaction. Commonly, these reactions are reductive, and a typical example is the reduction of the vanillin odorant to its vanillic alcohol. This alcohol is related to the decrease in the woody wine character, which promotes a pleasant and soft wine flavor. Another common reaction is the formation of furfuryl alcohol and furfuryl mercaptan (2-furanmethanethiol) from its precursor furfural–the latter

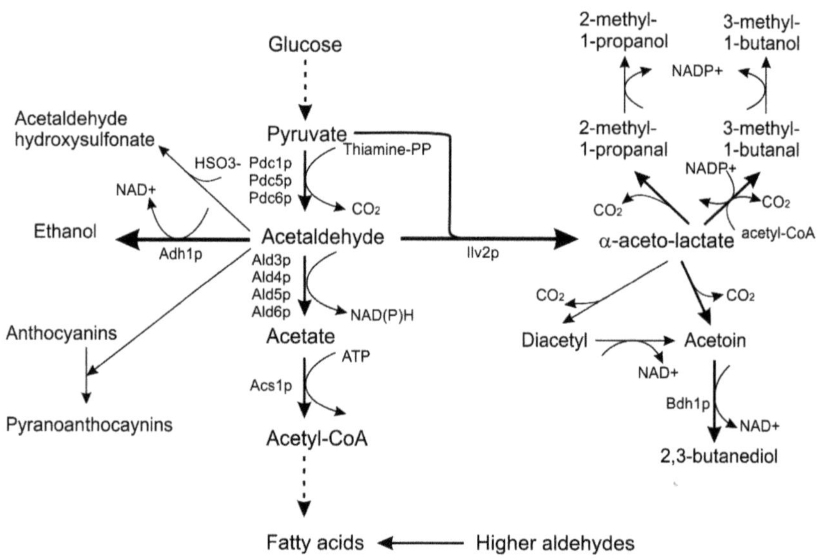


**Figure 6.7** The biochemical pathway for aldehyde metabolism, resulting in higher aldehydes, diacetyl, and acetoin. Source: Adapted from Ugliano and Henschke [31]/ Springer Nature.

is catalyzed by cysteine desulfydrase, an enzyme produced by the fermentation yeasts–thereby releasing $H_2S$ from cysteine under low assimilable nitrogen conditions and promoting a coffee aroma and flavor [46].

### 6.4.2 Mezcal

According to the Official Mexican Standards [47], mezcal is defined as a Mexican alcohol beverage, 100% agave, distilled from the fermented juice of cooked mature maguey plant stems (*Agave* genus). This spirit may contain an alcohol concentration between 35% and 55% v/v at 20 °C. This alcoholic beverage can be produced only in the territory protected by the Appellation of Origin Mezcal (AOM), which includes the states of Guanajuato, Durango, Guerrero, Michoacán, Oaxaca, San Luis Potosí, Tamaulipas, Zacatecas, and Puebla (NOM-070-SCFI-2016). The NOM-070 allows the use of different species of maguey whose biological development and harvest have taken place in the states included in the AOM (NOM-070-SCFI-2016). Mexico has nine genera and 261 species of *Agave*, out of the nine genera and 340 species that make up the *Agavaceae* family [48].

Maguey is a perennial, rosette, monocarpic plant that lives for many years and dies after fruiting [49]. These plants reach maturity, and they are ready to be cut for mezcal production when they are 8–25 years old. In this stage, maguey presents the formation of the "quiote" (stem where flowering occurs), which is removed (a process known as capping) to prevent flowering, and the decrease in the reserve of carbohydrates, which are stored as fructans. These molecules are a complex mixture of branched structures formed by fructosyl bonds $\beta$ 2→1 and fructosyl $\beta$ 2→6, and an internal or external glucosyl residue [50]. Piña (true stem and base where the leaves are attached) is the part of the maguey plant where these carbohydrates are stored, and it is obtained by eliminating their leaves and roots.

Mezcal is produced using mainly *Agave angustifolia* (maguey espadín), *A. potatorum* (maguey tobalá), *A. cupreata* (maguey chino), *A. durangensis* (maguey cenizo), *A. rhodacantha* (maguey mexicano), *A. marmorata* (maguey tepeztate), and *A. karwinskii* (maguey barril) (Mezcal Regulatory Council, 2019). According to the NOM-070 (NOM-070-SCFI-2016), there are three categories of Mezcal: Ancestral Mezcal, Artisanal Mezcal, and Mezcal. These categories are assigned according to their production process, which involves the level of techniques of its procedures and equipment used in different stages of production (cooking maguey, milling cooked maguey, fermentation, and distillation). Its artisanal process characterizes the production of this spirit; thus, in 2018, this category represented around 92% of the total mezcal production.

The carbohydrate content depends on the part of the plant and the stage of maturity [51]. For example, *Agave tequilana* Weber blue variety at maturity stage (seven years) contains approximately 710 mg of fructans, 14 mg of sucrose, 11.7 mg of fructose, 4.3 mg of glucose, and 0.58 mg of starch per gram of "piña" by dry weight. In contrast, a 2-year-old plant has 328 mg of fructans, 39 mg of sucrose, 19 mg of fructose, 14 mg of glucose, and 4.9 mg of starch per gram of "piña" in dry weight [52]. To obtain fermentable sugars (fructose units), "piñas" are cooked. The maturity of the maguey plants is not only necessary for obtaining more fermentable sugars but also to prevent the production of high methanol content. Methanol concentration depends on maguey age; this alcohol derived from methoxylated pectins is present in a higher amount in immature maguey plants, while pectins from mature maguey have fewer methoxyl groups [53], which will result in drinks with less methanol content.

Maguey cooking is the first step that generates classic caramel and smoked sensory notes. Compounds such as furans (furfural, 2-acetyl furan, 5-methyl furfural, 5-hydroxymethylfurfural, and 2-methyl-tetrahydrofuran) [49], alcohols (furfuryl alcohol), aldehydes (vanillin), ketones (2,3-pentanedione, 2-ethyl-cyclobutanone, 2-methyl tetrahydrofuran-3-one), esters (furfuryl acetate, methyl 2-furoate), phenols (carvacrol, thymol), and terpenes (cis-linalool oxide, trans-linalool oxide) [54] are responsible for these aroma scents. Complete thermal hydrolysis of fructans allows the production of furans, which are compounds that at concentrations of $1.5–4.5\,g\,l^{-1}$ in maguey juices can decrease the yield of ethanol production due to their inhibitory effect on the growth of yeasts. It is recommended to conduct thermal hydrolysis of fructans at temperatures between 106 and 116 °C and a cooking time of around 6 and 14 hours to avoid these problems. Under these conditions, the production of furans is reduced [55]. However, these conditions are only met using autoclaves and masonry ovens, in which the temperature is uniform inside and is controlled by injecting steam. Nevertheless, for artisanal mezcal production, the piñas are cooked in pit ovens, where the temperature is not uniform inside and is not controlled. During the maguey cooking process, 5-hydroxymethylfurfural is formed [49], and it is allowed up to 5 mg/100 ml of anhydrous alcohol since it is considered a toxic compound, regulated by the Official Mexican Standards (NOM-070-SCFI-2016). To reduce the concentration of furfural in mezcal, Mezcal Regulatory Council [56] recommends avoiding overcooking of maguey and removing the burnt parts of maguey before the final milling.

Natural fermentation of maguey juices is a complex process in which yeasts carry out the transformation of sugars into ethanol, carbon dioxide, higher alcohols, esters, and aldehydes, among other metabolites [57]. In ancestral and artisanal mezcal production, pulp and bagasse are placed in wooden vats and left to stand over two or three days, followed by the addition of warm water (40 °C) [58]. This mixture is known as a must, which undergoes natural fermentation. Pulp and fiber are poured into a wooden vat and left to stand for 24 hours, and water at approximately 25 °C is then added in the ratio of 2:1 (pulp and fiber: water). After 24 hours, the vat content is stirred. In other cases, maguey pulp and fiber are placed into the vat and left to stand for several hours. Subsequently, warm water (38 °C) is added together with 10 l of a wild inoculum. After 24 hours of fermentation, the vat content is stirred and kept covered during the fermentation process [57]. Water, pulp, and fiber form a pasty consistency which is used by producers to form a kind of coating on the surface of the vat, which hinders oxygen transfer and maintains anaerobic conditions at the bottom of the vat and aerobic conditions at the top. During the fermentation process, must reaches temperatures around 40 °C, and it can take from 8 to 20 days to end the fermentation process. When the taste of the must is no longer sweet, and the production of $CO_2$ is over, the fermentation is finished.

The physicochemical properties of the maguey must depend on various factors such as the species, age, and chemical composition of the maguey and the traditional practices of each producer, among other aspects. Thus, it has been found that, between lots and between distilleries, the concentration of direct reducing sugars at which juices are prepared varied in an approximate range of 120–270 $g\,l^{-1}$ [57]. At the end of fermentation, sugar content ranged from 7–60 $g\,l^{-1}$. Fermentation began with pH values close to 4 and 4.5. During this stage, various metabolites were produced in different concentrations. For example, acetic acid in the fermented juice of *A. angustifolia* and *A. potatorum* reached concentrations of 587 and 213 $mg\,l^{-1}$, respectively [59]. At the end of the natural fermentation, the juice contained 37–50 $g\,l^{-1}$ of ethanol [57]. The final concentration of ethanol after the fermentation depended on the initial fermentable sugar content. Also, fusel alcohols, esters, organic acids, aldehydes, and terpenes are found in the final step of fermentation. Isobutanol (1–20 $mg\,l^{-1}$), propanol (5–15 $mg\,l^{-1}$), and amyl alcohols (1–80 $mg\,l^{-1}$) are produced. These alcohols are formed by the metabolism of the amino acids.

The production of isobutyl and isoamyl alcohol depends on the content of valine and leucine in the must, respectively [32, 60]). Nolasco-Cancino et al. [58] suggest that higher alcohol concentration in maguey juice fermentation depends on the must composition, yeast species, and their interaction. Ethyl esters of fatty acids impart sensory characteristics such as fruity odor, depending on the length of their hydrocarbon chain. In mezcal, the predominant ethyl esters are ethyl octanoate, ethyl decanoate, ethyl dodecanoate, ethyl hexadecanoate, ethyl *cis*-9-octadecenoate (ethyl oleate), and ethyl *cis*-9,12-octadecadienoate (ethyl linoleate) [61]. The most abundant short-chain esters are ethyl acetate and ethyl lactate. Other compounds, coming from the agave plant, such as terpenes like α-terpineno, *p*-cimene, limonene, linalool, 4-terpineol, nerol, geraniol, cedrol, α-bisobolol, α-curcumene, among others, which depend on the maguey species [62, 63]).

Spontaneous fermentation is a natural process that occurs when substrate availability and environmental conditions are appropriate for microbial growth [64]. In this process, microorganisms come from several sources, such as the skin of the harvested maguey from different fields; sweet cooked maguey, which promotes the indirect transition of yeast, bacterial, and fungus through insect vectors; cooked maguey's milling stage; human manipulation; utensils and equipment used during the whole process, and microorganisms adhered to the walls of the wooden vats as well as the microorganisms in each distillery's surroundings [58, 65].

The main yeasts species involved in natural fermentation comprise *S. cerevisiae*, *Saccharomyces exiguous*, *Kluyveromyces marxianus*, *Torulaspora delbrueckii*, *Pichia membranefaciens*, *Pichia kluyvery*, *Pichia fermentans*, *Zygosaccharomyces bailii*, *Zygosaccharomyces rouxii*, *Zygosaccharomyces bisporus*, *Clavispora lusitaniae*, *Candida ethanolica*, and *Candida diversa* [57, 58, 66, 67]. Most studies on mezcal microbiota have reported that *S. cerevisiae* and *K. marxianus* were the predominant yeasts [57, 58, 66, 67].

Verdugo-Valdez et al. [66] reported that *S. cerevisiae* is the predominant yeast species in the fermentation of *Agave salmiana* juices. The population of non-*Saccharomyces* yeasts (*K. marxianus*, *Pichia kluyveri*, *Zygosaccharomyces bailli*, *C. lusitaniae*, *T. delbrueckii*, *C. ethanolica*, *Saccharomyces exiguus*, *Pichia membranifaciens*, and

*Rhodotorula mucilaginosa*) experience a succession of species at the beginning of fermentation. Similarly, Páez-Lerma et al. [67] found that at the beginning of the fermentation of *A. duranguensis* juices, the non-*Saccharomyces* yeast population is high and at the end of the process only *S. cerevisiae* predominates; however, in a second region, it was found that the predominant yeast at the end of fermentation is *T. delbrueckii*. After *S. cerevisiae*, the most frequent species were *K. marxianus*, *Z. rouxii*, *Z. bisporus*, *T. delbrueckii*, and *Pichia membraniefaciens*. In addition to *S. cerevisiae* and *K. marxianus*, Nolasco-Cancino et al. [58] found that *Pichia kudriavzevii* and *P. manshurica* are also predominant in the natural fermentation of the artisanal mezcal in the Oaxaca state. Both yeast species appear to originate exclusively from the Oaxaca state, while *S. cerevisiae* and *K. marxianus* are two common species in mezcal fermentation [58].

The predominance of non-*Saccharomyces* in mezcal fermentation could be associated with the temperature (38–40 °C) of the vat's ecosystem. During maguey juice fermentation, wooden vats reach temperatures around 40 °C as a result of the microbial activity, and warm water is added to the bagasse for the sugar extraction. This environmental condition supports the growth of *P. kudriavzevii* and *K. marxianus*. Both the yeast species can ferment at 40–42 °C [68–71]. The predominance of *K. marxianus* in mezcal and tequila fermentation has also been associated with its fructanase activity [72], which allows them to use maguey fructans as a substrate.

*Kluyveromyces marxianus* also has saponinase activity. Saponins are secondary metabolites present in maguey, and they have antifungal activity, i.e. can inhibit the growth of yeasts. It has been shown that *K. marxianus* can hydrolyze these compounds and grow in that environment, while more considerable damage occurs in the cell membrane of *S. cerevisiae*, and its growth becomes inhibited [73]. Different indigenous strains of *S. cerevisiae* from mezcal can tolerate different stress conditions such as low pH (3), high ethanol content (15% of ethanol), high temperature (42 °C), and high glucose concentration (30% w/v) [74]. Therefore, these strains play an essential role in maguey fermentation, as they are better adapted to the environmental conditions prevalent in distilleries.

The predominance of yeasts in mezcal fermentation depends on their ability to adapt to the stress factors of fermentation vats; these factors are diverse, but we can classify them into three types: cultural, environmental, and composition of maguey. The first depends on the empirical productive practices that each producer has developed and inherited for generations; thus, they are deeply rooted (times, tools, procedures, and quantities); the second depends on the geographical location, which is determined by environmental conditions such as altitude, soil type, temperature, humidity, wind, light, and water composition.

The identification and characterization of the native yeast strains strengthen the concept of Denomination of Origin, which has allowed the development of mezcal, thus giving it a more significant technological and cultural identity. Capozzi et al. [75], indicate that native strains are responsible for the relationship between the beverages and the history of the production area. Therefore, the predominant yeast population in each distillery could be considered as a distinctive signature that gives a particular characteristic to the Mezcal produced there [58].

## 6.5 Conclusions

Flavor is a sensory attribute that is directly related to the chemistry of the beverage, and the beverage chemistry composition depends on the biochemical phenomena that occur during the alcohol fermentation process. It was possible to observe all the biochemical phenomena that play an essential role in the chemical composition and, consequently, in the sensory profile of fermented alcoholic beverages. The alcohol fermentation process is responsible for the synthesis of typical volatile compounds that impart flavor to fermented alcoholic beverages, and this reaction is considered a crucial reaction that leads to the sensory uniqueness and quality of the beverage. Several biochemical pathways still need to be elucidated, and the study of these pathways will be useful for all the producers and industries of fermented beverages. The study and analysis of the biochemical pathways related to the synthesis of flavor compounds are essential to assure the quality of the final product and for producing a beverage with safety for consumers.

## References

1 Sánchez-Rodríguez, L., Syd Ali, N., Cano-Lamadrid, M. et al. (2019). Flavor and Aromas. In: *Postharvest Physiology and Biochemistry of Fruits and Vegetables* (ed. E.M. Yahia and A. Carrillo-López), 385–404. United Kingdom: Elsevier.
2 Jackson, R.S. (2009). *Wine Tasting: A Professional Handbook*. California: Academic Press.
3 De Castilhos, M.B.M., Maia, J.D.G., Gómez-Alonso, S. et al. (2016). Sensory acceptance drivers of pre-fermentation dehydration and submerged cap red wines produced from *Vitis labrusca* hybrid grapes. *LWT Food Sci. Technol.* 69: 82–90.
4 Clemente-Jimenez, J.M., Mingorance-Cazorla, L., Martínez-Rodríguez, S. et al. (2004). Molecular characterization and oenological properties of wine yeasts isolated during spontaneous fermentation of six varieties of grape must. *Food Microbiol.* 21: 149–155.
5 Pretorius, I.S., Bartowsky, E.J., de Barros Lopes, M. et al. (2006). The tailoring of designer grapevines and microbial starter strains for a market-directed and quality-focused wine industry. In: *Handbook of Food Science, Technology and Engineering*, Food Technology and Food Processing, vol. 4 (ed. H.Y. Hui, E. Castell-Perez, L.M. Cunha, et al.). New York: CRC Press.
6 Swiegers, J.H. and Pretorius, I.S. (2007). Modulation of volatile sulfur compounds by wine yeast. *Appl. Microbiol. Biot.* 74: 954–960.
7 Jackson, R.S. (2014). *Wine Science: Principles and Application*, 4e. San Diego: Academic Press.
8 Zamora, F. (2010). Biochemistry of alcoholic fermentation. In: *Wine Chemistry and Biochemistry* (ed. M.V. Moreno-Arribas and M.C. Polo), 3–26. New York: Springer.
9 Loureiro, V. and Malfeito-Ferreira, M. (2006). Spoilage activities of *Dekkera/Brettanomyces* spp. In: *Food Spoilage Microorganisms* (ed. C. Blackburn), 354–398. Cambridge: Woodhead Publishers.

10 Crauwels, S., Steensels, J., Aerts, G. et al. (2015). *Brettanomyces bruxellensis*, essential contributor in spontaneous beer fermentations providing novel opportunities for the brewing industry. *Brewing Sci.* 68: 110–121.

11 Steensels, J., Daenen, L., Malcorps, P. et al. (2015). *Brettanomyces* yeast – from spoilage organisms to valuable contributors to industrial fermentations. *Int. J. Food Microbiol.* 206: 24–38.

12 Ribéreau-Gayon, P., Glories, Y., Maujean, A., and Dubourdieu, D. (2006). *Handbook of Enology, The Chemistry of Wine Stabilization and Treatments*, 2e. Chichester: Wiley.

13 Del Nobile, M.A., D, Amato, D., Altieri, C. et al. (2003). Modeling the yeast growth-cycle in a model wine system. *J. Food Sci.* 68: 2080–2085.

14 Riou, C., Nicaud, J.M., Barre, P., and Gaillardin, C. (1997). Stationary-phase gene expression in *Saccharomyces cerevisiae* during wine fermentation. *Yeast* 13: 903–915.

15 Hallsworth, J.E. (1998). Ethanol-induced water stress in yeasts. *J. Ferment. Bioeng.* 85: 125–137.

16 Viegas, C.A., Almeida, P.F., Cavaco, M., and Sá-Correia, I. (1998). The H+-ATPase in the plasma membrane of *Saccharomyces cerevisiae* is activated during growth latency in octanoic acid-supplemented medium accompanying the decrease in intracellular pH and cell viability. *Appl. Environ. Microbiol.* 64: 779–783.

17 Astray, G., García-Río, L., Mejuto, J.C., and Pastrana, L. (2007). Chemistry in food: flavours. *Electron. J. Env. Agric. Food Chem.* 6: 1742–1763.

18 Hanig, D.P. (1901). Zur Psychophysik des Geschmacksinnes. *Philosophische Studien.* 17: 576–623.

19 Brodal, A. (1981). *Neurological Anatomy in Relation to Clinical Medicine*, 3e. New York: Oxford University Press.

20 Rolls, E.T., Critchley, H.D., Browning, A., and Hernadi, I. (1998). The neurophysiology of taste and olfaction in primates, and umami flavor. *Ann. NY. Acad. Sci.* 30: 426–437.

21 Gilbertson, T.A. and Boughter, J.D. (2003). Taste transduction: appetizing times in gustation. *NeuroReport* 14: 905–911.

22 Margolskee, R.F., Dyer, J., Kokrashvili, Z. et al. (2007). T1R3 and gustducin in gut sense sugars to regulate the expression of Na+-glucose cotransporter 1. *Proc. Natl. Acad. Sci. U.S.A.* 104: 15075–15080.

23 Tahara, Y. and Toko, K. (2013). Electronic tongues. A review. *IEEE Sens. J.* 13: 3001–3011.

24 Grosch, W. (2001). Evaluation of key odorants of foods by dilution experiments, aroma models and omission. *Chem. Senses* 26: 533–545.

25 Jelén, H.H., Majcher, M., and Dziadas, M. (2012). Microextraction techniques in the analysis of food flavor compounds: a review. *Anal. Chim. Acta* 738: 13–26.

26 De Castilhos, M.B.M., Del Bianchi, V.L., Gómez-Alonso, S. et al. (2019). Sensory descriptive and comprehensive GC-MS as suitable tools to characterize the effects of alternative winemaking procedures on wine aroma: Part I: BRS Carmem and BRS Violeta. *Food Chem.* 272: 462–470.

**27** De Castilhos, M.B.M., Del Bianchi, V.L., Gómez-Alonso, S. et al. (2020). Sensory descriptive and comprehensive GC-MS as suitable tools to characterize the effects of alternative winemaking procedures on wine aroma. Part II: BRS Rúbea and BRS Cora. *Food Chem.* 311: 126025.

**28** Sánchez-Palomo, E., Delgado, J.A., Ferrer, M.A., and González Viñas, M.A. (2019). The aroma of La Mancha Chelva wines: chemical and sensory characterization. *Food Res. Int.* 119: 135–142.

**29** Ubeda, C., Kania-Zelada, I., del Barrio-Galán, R. et al. (2019). Study of the changes in volatile compounds, aroma and sensory atributes during the production process of sparkling wine by traditional method. *Food Res. Int.* 119: 554–563.

**30** Romano, P., Fiore, C., Paraggio, M. et al. (2003). Function of yeast species and strains in wine flavour. *Int. J. Food Microbiol.* 86: 169–180.

**31** Ugliano, M. and Henschke, P.A. (2010). Yeasts and wine flavour. In: *Wine Chemistry and Biochemistry* (ed. M.V. Moreno-Arribas and M.C. Polo), 313–392. New York: Springer.

**32** Swiegers, J.H., Bartowsky, E.J., Henschke, P.A., and Pretorius, I.S. (2005). Yeast and bacterial modulation of wine aroma and flavour. *Aust. J. Grape Wine R.* 11: 139–173.

**33** De Castilhos, M.B.M., Cattelan, M.G., Conti-Silva, A.C., and Del Bianchi, V.L. (2013). Influence of two different vinification procedures on the physicochemical and sensory properties of Brazilian non-*Vitis vinifera* red wines. *LWT Food Sci. Technol.* 54: 360–366.

**34** De Castilhos, M.B.M., Corrêa, O.L.S., Zanus, M.C. et al. (2015). Pre-drying and submerged cap winemaking: Effects on polyphenolic compounds and sensory descriptors. Part II: BRS Carmem and Bordô (*Vitis labrusca L.*). *Food Res. Int.* 76: 697–708.

**35** De Castilhos, M.B.M., Conti-Silva, A.C., and Del Bianchi, V.L. (2012). Effect of grape pre-drying and static pomace contact on physicochemical properties and sensory acceptance of Brazilian (Bordô and Isabel) red wines. *Eur. Food Res. Technol.* 235: 345–354.

**36** Escudero, A., Campo, E., Fariña, L. et al. (2007). Analytical characterization of the aroma of five premium red wines. Insights into the role of odor families and the concept of fruitiness of wines. *J. Agric. Food Chem.* 55: 4501–4510.

**37** Lilly, M., Bauer, F.F., Lambrechts, M.G. et al. (2006). The effect of increased yeast alcohol acetyltransferase and esterase activity on the flavour profiles of wine and distillates. *Yeast* 23: 641–659.

**38** Mason, A.B. and Dofour, J.P. (2000). Alcohol acetyltransferases and the significance of ester synthesis in yeast. *Yeast* 16: 1287–1298.

**39** Plata, C., Mauricio, J.C., Millan, C., and Ortega, J.M. (2005). Influence of glucose and oxygen on the production of ethyl acetate and isoamyl acetate by a Saccharomyces cerevisiae strain during alcoholic fermentation. *World J. Microbiol. Biot.* 21: 115–121.

**40** Saerens, S.M.G., Delvaux, F., Verstrepen, K.J. et al. (2008). Parameters affecting ethyl ester production by Saccharomyces cerevisiae during fermentation. *Appl. Env. Microbiol.* 74: 454–461.

41 Molina, A.M., Swiegers, J.H., Varela, C. et al. (2007). Influence of wine fermentation temperature on the synthesis of yeast-derived volatile aroma compounds. *Appl. Microbiol. Biotechnol.* 77: 675–687.

42 Regenberg, B., During-Olsen, L., Kielland-Brandt, M.C., and Holmberg, S. (1999). Substrate specificity and gene expression of the amino-acid permeases in *Saccharomyces cerevisiae*. *Curr. Genet.* 36: 317–328.

43 Hazelwood, L.A., Daran, J.-M., van Maris, A.J.A. et al. (2008). The Ehrlich pathway for fusel alcohol production: a century of research on yeast metabolism. *Appl. Environ. Microbiol.* 74: 2259–2266.

44 Martins, N., Garcia, R., Mendes, D. et al. (2018). An ancient winemaking technology: exploring the volatile composition of amphora wines. *LWT – Food Sci. Technol.* 96: 288–295.

45 Geffroy, O., Lopez, R., Serrano, E. et al. (2015). Changes in analytical and volatile compositions of red wines induced by pre-fermentation heat treatment of grapes. *Food Chem.* 187: 243–253.

46 Tominaga, T., Blanchard, L., Darriet, P., and Dubourdieu, D. (2000). A powerful aromatic volatile thiol, 2-furanmethanethiol, exhibiting roast coffee aroma in wines made from several *Vitis vinifera* grape varieties. *J. Agric. Food Chem.* 48: 1799–1802.

47 NOM-070-SCFI-2015 (2016). *Bebidas alcohólicas – Mezcal – especificaciones.* Secretaria de economía: Publicada en el Diario Oficial de la Federación.

48 García-Mendoza A. (2011) Agavaceae. Flora del Valle de Tehuacán-Cuicatlán. Instituto de Biología, Universidad Nacional Autónoma de México, México, D. F. 88: 1–95.

49 Villanueva-Rodríguez, S.J., Rodríguez-Garay, B., Prado-Ramírez, R., and Gschaedler, A. (2016). Tequila: raw material, classification, process, and quality parameters. In: *Encyclopedia of Food and Health*, 283–289. San Diego: Elsevier.

50 López, M.G., Mancilla-Margalli, N.A., and Mendoza-Diaz, G. (2003). Molecular Structures of Fructans from *Agave tequilana* Weber var. azul. *J. Agric. Food Chem.* 51: 7835–7840.

51 Michel-Cuello, C., Juárez-Flores, B.I., Aguirre-Rivera, J.R., and Pinos-Rodríguez, J.M. (2008). Quantitative characterization of nonstructural carbohydrates of mezcal agave (*Agave salmiana* Otto ex Salm-Dick). *J. Agric. Food Chem.* 56: 5753–5757.

52 Mellado-Mojica, E. and López, M.G. (2012). Fructan metabolism in *A. tequilana* Weber blue variety along its developmental cycle in the field. *J. Agric. Food Chem.* 60: 11704–11713.

53 Pinal, L., Cornejo, E., Arellano, M. et al. (2009). Effect of *Agave tequilana* age, cultivation field location and yeast strain on tequila fermentation process. *J. Ind. Microbiol. Biot.* 36: 655–661.

54 Prado-Jaramillo, N., Estarrón-Espinosa, M., Escalona-Buendía, H. et al. (2015). Volatile compounds generation during different stages of the Tequila production process. A preliminary study. *LWT Food Sci. Technol.* 61: 471–483.

55 García-Soto, M.J., Jiménez-Islas, H., Navarrete-Bolaños, J.L. et al. (2011). Kinetic study of the thermal hydrolysis of *Agave salmiana* for Mezcal production. *J. Agr. Food Chem.* 59: 7333–7340.

**56** Consejo Regulador Mexicano de la Calidad del Mezcal (2019). http://www.crm.org.mx/ (accessed 14 January 2020).

**57** Kirchmayr, M.R., Segura-García, L.E., Lappe-Oliveras, P. et al. (2017). Impact of environmental conditions and process modifications on microbial diversity, fermentation efficiency and chemical profile during the fermentation of mezcal in Oaxaca. *LWT Food Sci. Technol.* 79: 160–169.

**58** Nolasco-Cancino, H., Santiago-Urbina, J.A., Wacher, C., and Ruiz-Terán, F. (2018). Predominant yeasts during artisanal mezcal fermentation and their capacity to fermented maguey juice. *Front. Microbiol.* 9: 2900.

**59** Vera-Guzmán, A.M., Santiago-García, P.A., and López, M.G. (2009). Compuestos volátiles aromáticos generados durante la elaboración de Mezcal de *Agave angustifolia* y *Agave potatorum*. *Rev. Fitotec. Mex.* 32: 273–279.

**60** Dickinson, J.R., Harrison, S.J., and Hewlins, M.J.E. (1998). An investigation of the metabolism of valine to isobutyl alcohol in *Saccharomyces cerevisiae*. *J. Biol. Chem.* 273: 25751–25756.

**61** Nolasco Cancino, H. (2007). *Caracterización y cuantificación de lípidos simples en mezcal y sus productos intermedios por cromatografía de gases capilar (tesis de maestría)*. Ciudad de México, México: Universidad Nacional Autónoma de México.

**62** Martínez-Aguilar, J.F. and Peña-Álvarez, A. (2009). Characterization of five typical agave plants used to produce mezcal through their simple lipid composition analysis by gas chromatography. *J. Agric. Food Chem.* 57: 1933–1939.

**63** Peña-Alvarez, A., Díaz, L., Medina, A. et al. (2004). Characterization of three Agave species by gas chromatography and solid-phase microextraction gas chromatography-mass spectrometry. *J. Cromatogr.* 1027: 131–136.

**64** Navarrete-Bolaños, J.L. (2012). Improving traditional fermented beverages: how to evolve from spontaneous to directed fermentation. *Eng. Life Sci.* 12: 410–418.

**65** Lachance, M.A. (1995). Yeast communities in a natural Tequila fermentation. *Anton. van Lee.* 68: 151–160.

**66** Verdugo-Valdez, A., Segura-Garcia, L., Kirchmayr, M. et al. (2011). Yeast communities associated with artisanal Mezcal fermentations from *Agave salmiana*. *Anton. Van Lee.* 100: 497–506.

**67** Páez-Lerma, J.B., Arias-Garcia, A., Rutiaga-Quiñones, O.M. et al. (2013). Yeasts isolation from the alcoholic fermentation of *Agave duranguensis* during Mezcal production. *Food Biotechnol.* 27: 342–356.

**68** Dhaliwal, S.S., Oberoi, H.S., Sandhu, S.K. et al. (2011). Enhanced ethanol production from sugarcane juice by galactose adaptation of a newly isolated thermotolerant strain of *Pichia kudriavzevii*. *Bioresource Technol.* 102: 5968–5975.

**69** Gallardo, J.C.M., Souza, C.S., Cicarelli, R.M.B. et al. (2011). Enrichment of a continuous culture of *Saccharomyces cerevisiae* with the yeast *Issatchenkia orientalis* in the production of ethanol at increasing temperatures. *J. Ind. Microbiol. Biotechnol.* 38: 405–414.

**70** Hu, N., Yuan, B., Sun, J. et al. (2012). Thermotolerant *Kluyveromyces marxianus* and *Saccharomyces cerevisiae* strains representing potentials for bioethanol production from Jerusalem artichoke by consolidated bioprocessing. *Appl. Microbiol. Biotechnol.* 95: 1359–1368.

**71** Limtong, S., Sringiew, C., and Yongmanitchai, W. (2007). Production of fuel ethanol at high temperature from sugar cane juice by a newly isolated *Kluyveromyces marxianus*. *Bioresour. Technol.* 98: 3367–3374.

**72** Arrizón, J., Morel, S., Gschaedler, A., and Monsan, P. (2012). Fructanase and fructosyltransferase activity of non-*Saccharomyces* yeasts isolated from fermenting must of mezcal. *Bioresource Technol.* 110: 560–565.

**73** Alcázar, M., Kind, T., Gschaedler, A. et al. (2017). Effect of steroidal saponins from Agave on the polysaccharide cell wall composition of *Saccharomyces cerevisiae* and *Kluyveromyces marxianus*. *LWT-Food Sci. Technol.* 77: 430–439.

**74** Ruiz-Terán, F., Martínez-Zepeda, P.N., Geyer-de la Merced, S.Y. et al. (2018). Mezcal: indigenous *Saccharomyces cerevisiae* strains and their potential as starter cultures. *Food Sci. Biotechnol.* 28: 459–467.

**75** Capozzi, V., Garofalo, C., Chiriatti, M.A. et al. (2015). Microbial terroir and food innovation: the case of yeast biodiversity in wine. *Microbiol. Res.* 181: 75–83.

# Part IV

# Food Industry Ingredients

# 7

# The Resinoids: Their Chemistry and Uses

*Daniel J. Strub[1], Maria Strub[2], and Nicolas Baldovini[3]*

[1]Wrocław University of Science and Technology, Faculty of Chemistry, Department of Chemical Biology and Bioimaging, Wyb. Wyspiańskiego 27, 50370, Wrocław, Poland
[2]Liquid Technologies sp. z o.o., ul. Gdańska 13, 50344, Wrocław, Poland
[3]Universite Cote d'Azur, Institut de Chimie de Nice, CNRS UMR 7272, Parc Valrose, F-06108, Nice, France

## 7.1 Introduction

Aromatic plants and aroma materials have been used by humans since ancient times due to their sensory, antibacterial, antifungal, and antiviral properties. In addition, the aromatic resins have been used for religious purposes for thousands of years, and their utilization is mentioned in holy books, e.g. *the Bible*, where myrrh and frankincense were gifts from the magi to Jesus Christ [1, 2], galbanum (gum resin from the plants in the genus *Ferula*), frankincense (gum resin from plants in the genus *Boswellia*), and myrrh (resinous exudate from plants in the genus *Commiphora*) are frequently mentioned in the Bible as ingredients of sacred incense, mastic (dried gumdrops from *Pistacia lentiscus* L.) is widely used in Orthodox churches in eastern Europe [2] and labdanum (oleoresin from *Cistus ladaniferus*) has been mentioned as a perfumery ingredient by Herodotus in the fifth century BC [3]. Plant resins are lipid-soluble mixtures of volatile and nonvolatile compounds belonging generally to the terpene and phenolic families [4], and the majority of commercially important resins are terpenoids. The most important products derived from plant resins are essential oils and resinoids. The former is produced exclusively by physical methods, such as distillation or mechanical pressing, and the latter by extraction. Both types of products can be characterized as a complex mixture of organic compounds, which are often structurally diverse. Resinoids as commercial products are materials derived mainly from plant resins. The definition of a resinoid is ambiguous depending on the source. According to ISO 9235:2013 (aromatic natural raw materials – vocabulary), it is a material "obtained from a dry plant natural raw material by extraction with one or several solvents" with total or partial removal of the solvent(s). Various sources describe

*Natural Flavours, Fragrances, and Perfumes: Chemistry, Production, and Sensory Approach,*
First Edition. Edited by Sreeraj Gopi, Nimisha Pulikkal Sukumaran, Joby Jacob, and Sabu Thomas.

resinoids as extracts from plant exudates: some with a strong emphasis on volatile hydrocarbon solvent as an extractant [5], and some highlighting the wide array of extracting solvents [6, 7]. Commercially available aromatic products called "resinoids" are produced with the aid of solvents of various polarities. In most cases, these aromatic materials are highly viscous and dark-colored. Resinoids shall be distinguished from other extracts, e.g. from herbs and spices, which, according to ISO 9235:2013, shall be labeled as "extracted oleoresins".

In contrast to the very popular volatile natural materials – essential oils, the amount of relevant data regarding the bioactivity of resins and resinoids is very limited. This chapter presents a concise overview of the chemistry, activity, and uses of some of the commercially available resinoids.

### 7.1.1 Asafoetida (*Ferula assa-foetida*)

The term "asafoetida" mostly refers to the exudate of *Ferula assa-foetida*. The plant grows up to 2 m high, and the resin is isolated from 4 to 5 year old specimens by cutting their roots and collecting their thick, milky juice, which hardens over time. The main producers of asafoetida are Iran and Afghanistan. The resin is commonly composed of 40–64% resinous matter, 25% gum, 10–17% volatile oil, and 1.5–10% ash [8]. The main classes of compounds present in chloroform and dichloromethane extracts belong to the coumarin, phenol, and sulfide families (Figure 7.1) [9–12]. Especially the first one is structurally diverse with extensive variations of the alkylated side chain of umbelliferone. However, the most important from a sensorial point of view is the sulfide family. The major sulfurous components that are mostly responsible for the putrid onion notes include 1-(methylthio)propyl 1-propenyl disulfide, 1-methylpropyl (3-methylthio-2-propenyl) disulfide, and dimethyl trisulfide.

The main application areas of asafoetida due to its sensory profiles (Table 7.1) are seasonings and savory flavors [14]. Common asafoetida's products include pure and compounded (diluted/mixed with starch or Arabic gum) resin, tincture, fluid extract, and essential oil, while the resinoid is less present on the market. Though the asafoetida products find use in many flavor creations, they can also be found in fragrance compositions. Classical *Ma Griffe* (Carven, 1946) compounded by the famous Jean Carles, *Cabochard* (Gres, 1959, perfumer: Bernard Chant), *Tendre Poison* (Christian Dior, 1994, perfumer: Edouard Flechier), and the more recently launched *Resine Precieux* (Sultan Pasha Attars, 2015, perfumer: Sultan Pasha) are good examples of the use of this sulfurous-smelling material. The biological activities of asafoetida extracts have not been extensively studied (Table 7.2, entries 1–4). However, the most promising double-blind studies of ethanolic extract for the treatment of the irritable colon have shown that it was effective at the 1% confidence level. It also exhibits moderate cytotoxic activity on Dalton's lymphoma ascites tumor cells (entry 2). Methanolic extract of asafoetida resin has shown very good antifertility activity (entry 3) in rats, reaching 80% effectiveness at a dose of 400 mg $kg^{-1}$. Despite several studies showing no mutagenic properties of asafoetida, it has been pointed out that its ethanolic extract exhibits weak mutagenic properties on two strains of *Salmonella typhimurium* TA98.

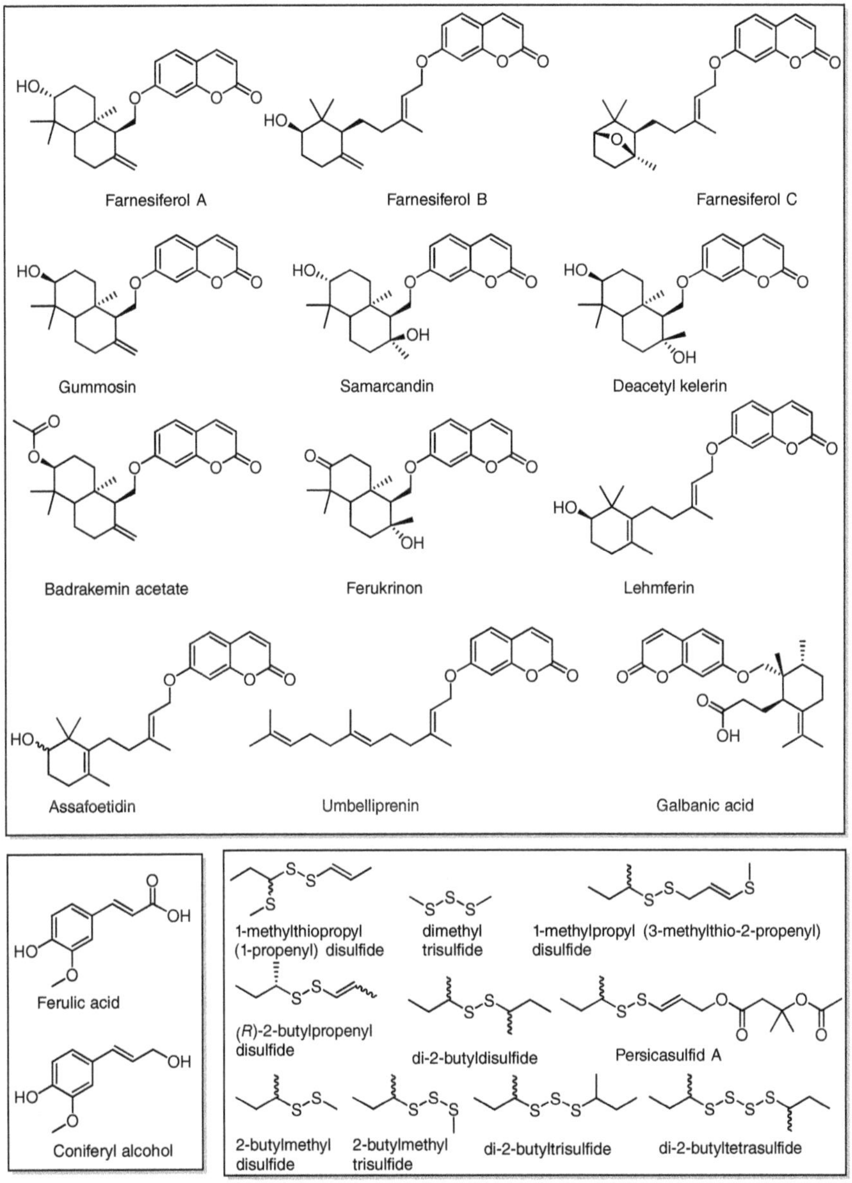


**Figure 7.1** The main classes of compounds present in asafoetida resinoid.

### 7.1.2 Galbanum (*Ferula gummosa*)

Galbanum is an exudate collected from several *Ferula* species, but the official plant is *Ferula gummosa* (syn. *Ferula galbaniflua*), which like *F. assa-foetida* belongs to the family of giant fennels. These resinous plants are widespread in the Middle East. Resinous material is collected similarly to asafoetida, after hardening into separate tears or yellow–green masses [4]. The major producer of this aromatic material is

**Table 7.1** Flavor and odor profile of some commercially important resinoids.

| Resinoid | Flavor profile | Odor profile |
|---|---|---|
| Asafoetida | Sulfurous, onion, garlic, meaty | Sulfurous, onion, garlic |
| Galbanum | Vegetative, green, peppery, herbal, seedy, and earthy [13] | Intensely rich-green, woody-balsamic with a dry undertone, typical "green peppers" foliage-like note [7] |
| Benzoin Siam | Floral, medicinal, spicy | Long-lasting, chocolate-like, sweet, balsamic notes [6] |
| Benzoin Sumatra | Sweet, vanillic, balsamic, spicy | Warm, powdery, sweet, balsamic notes [6] |

Iran (15 tonnes annually) [22]. There are two types of commercial galbanum resins – "hard" (Persian) or "soft" (Levant). Resinoid is produced mainly from the Levant-type galbanum. The exudate is composed of 6–26% essential oil [7, 8], up to 67% resin, 19% gum, and 8% foreign matter (sand, wood splinters, dirt). The resinoid is obtained by extraction of crude galbanum with a nonpolar solvent with a yield of up to 50% [23]. Similar to asafoetida resinoid, galbanum resinoid comprises a diverse family of coumarin derivatives (Figure 7.2a). The most abundant compounds in the volatile fraction (Figure 7.2b) are pinenes (α- and β-) and Δ-(+)-3-carene [24]. This material's key aroma contributors (Figure 7.2c) are thioester *sec*-butyl 3-methyl-2-butenethioate, disubstituted pyrazine (2-(*sec*-butyl) 3-methoxypyrazine), and two trienones ((6*Z*,8*E*)-undeca-6,8,10-trien-3-one, (6*Z*,8*E*)-undeca-6,8,10-trien-4-one), which are responsible for the typical fruity-green odor of galbanum [25].

The main application area of galbanum resinoid is perfumery. It provides characteristic green top notes (Table 7.1) and has good fixative properties. It can be found in perfumes like *Chanel N°19* and *Egoiste Platinum* (Chanel, 2011 and 1993, respectively, perfumer: Jacques Polge), *(untitled)* (Maison Martin Margiela, 2010, perfumer: Daniela Andrier), and *Nude* (Bill Blass, 1991, perfumer: Sophia Grojsman). The biological effects of galbanum extracts (Table 7.2) are not outstanding. They possessed rather weak antimicrobial properties against *Corynebacterium striatum* and *Enterococcus faecalis* (Table 7.2, entry 6), and weak AChE inhibitory properties (Table 7.2, entry 5). However, the petroleum extract of galbanum resin exhibits good spasmolytic properties (Table 7.2, entry 7) on isolated rat ileum contractions induced by KCl and acetylcholine (ACh).

### 7.1.3 Elemi (*Canarium luzonicum*)

The term "Elemis" does not refer to a single specific material but is used to name resins obtained from various species of the Burseraceae family (*Canarium*, *Dacryodes*, *Protium*). They are generally strongly odorant semisolid materials, containing large amounts of triterpenoids (like α- and β-amyrin) [4]. The botanical

**Table 7.2** Some important biological activities of commercially available resinoids.

| Entry | Biological effect | Solvent used | Results |
|---|---|---|---|
| | | *Asafoetida* | |
| 1 | Treatment of the irritable colon [15] | Ethanol | Effective at the 1% confidence level |
| 2 | *In vitro* cytotoxicity on Dalton's lymphoma ascites tumor cells [16] | Ethanol | $EC_{50} = 0.60\ mg\,ml^{-1}$ |
| 3 | Antifertility activity in Sprague-Dawley rats [17] | Methanol | Prevention of pregnancy in 80% specimens (dose $400\ mg\,kg^{-1}$) |
| 4 | *In vitro* mutagenicity toward *Salmonella typhimurium* TA98 #510 and #4 [18] | Ethanol | Weak mutagenicity with a dose-related response |
| | | *Galbanum* | |
| 5 | AChE inhibition [19] | Dichloromethane | $IC_{50} = 88.44\ \mu g\,ml^{-1}$ |
| 6 | Antimicrobial properties [20] | Methanol | *Corynebacterium striatum*<br>$MIC = 156\ \mu g\,ml^{-1}$<br>$MBC = 625\ \mu g\,ml^{-1}$<br>*Enterococcus faecalis*<br>$MIC = 312\ \mu g\,ml^{-1}$<br>$MBC = 1250\ \mu g\,ml^{-1}$ |
| 7 | Relaxant effect on rat's ileum [21] | Diethyl ether | $IC_{50}\ (KCl) = 550 \pm 40\ \mu g\,ml^{-1}$<br>$IC_{50}\ (ACh) = 10 \pm 2.2\ \mu g\,ml^{-1}$ |
| | | *Labdanum* | |
| 8 | Antimicrobial properties [20] | Ethanol | *Escherichia coli*<br>$1000\ ppm \leq MIC \leq 10\,000\ ppm$<br>*Staphylococcus aureus*<br>$1000\ ppm \leq MIC \leq 10\,000\ ppm$<br>*Bacillus megaterium*<br>$1000\ ppm \leq MIC \leq 2000\ ppm$<br>*Aspergillus niger*<br>$MIC \geq 10\,000\ ppm$<br>*Botrytis cinerea*<br>$MIC \geq 10\,000\ ppm$<br>*Verticillium albo-atrum*<br>$MIC \geq 10\,000\ ppm$<br>*Mucor racemosus*<br>$1000\ ppm \leq MIC \leq 10\,000\ ppm$ |

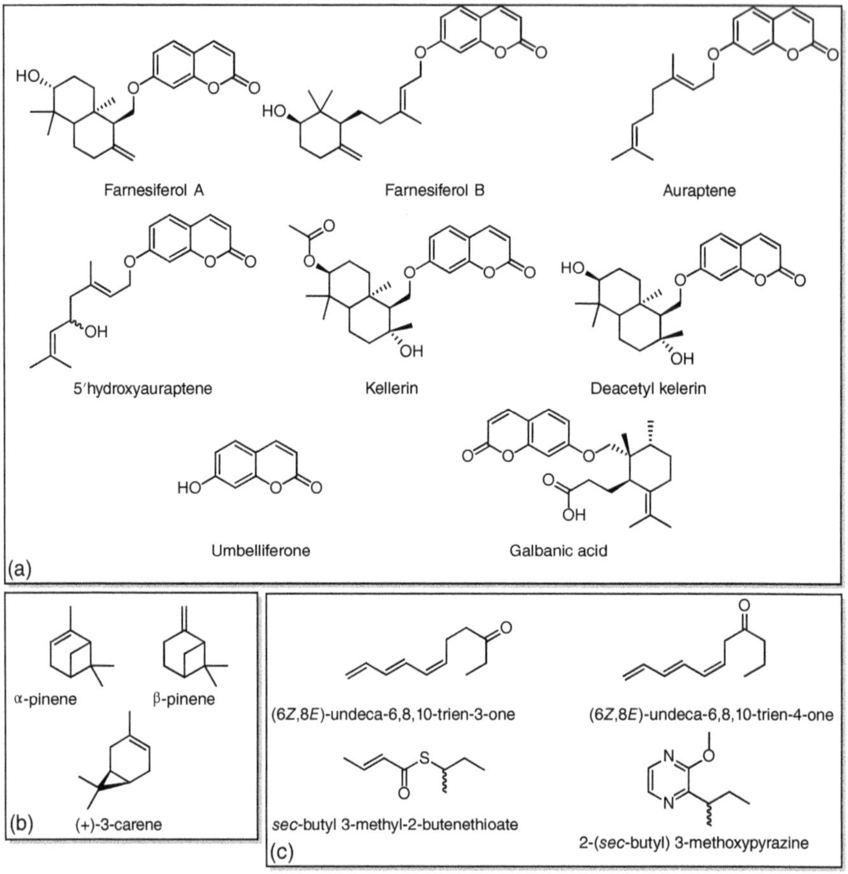


**Figure 7.2** Characteristic compounds of galbanum resinoid. (a) Coumarin derivatives; (b) Most abundant volatile constituents; (c) Key aroma compounds.

species from which the most common elemi is obtained is *Canarium luzonicum*, a tree native to the Philippine archipelago. This species produces the "Manila elemi," sometimes used in local medicine for its antirheumatic and antitussive properties [4]. In the past, it was used for the preparation of varnishes, and nowadays it still has some applications in the perfume industry. The chemical composition of this material has received little attention. It was shown to contain triterpenic acids [26], and about 15–25% of essential oil containing mostly α-phellandrene and limonene [27, 28]. It also contains elemol and elemicin, two compounds named after their isolation from elemi (Figure 7.3).

### 7.1.4 Styrax (*Liquidambar orientalis* Mill. and *Liquidambar styraciflua*)

Styrax (often named storax) is a resinous exudate of deciduous trees of the *Liquidambar* genus (*Hamamelidaceae*). In perfumery, two varieties have commercial importance: the "oriental" type coming from *Liquidambar orientalis* Mill. (mostly originating from Turkey), and the American one produced by trees native

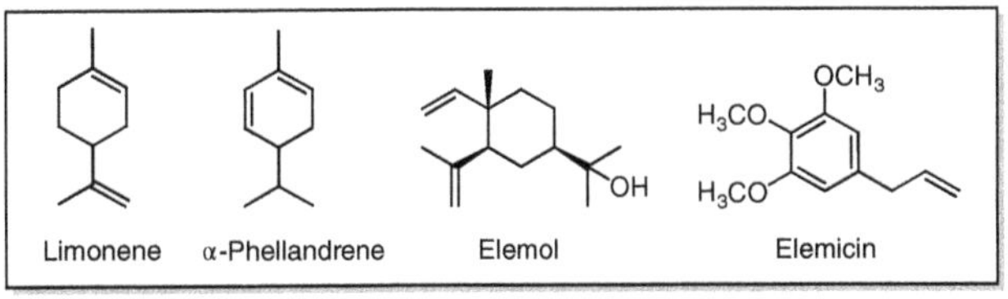


**Figure 7.3** Main volatile constituents of elemi.

to Central America (Honduras, Mexico, Guatemala) and the United States of America: *Liquidambar styraciflua* L. Care should be taken to avoid confusion with *Styrax officinalis* [2], an eastern Mediterranean plant, which is botanically totally different from *Liquidambar*, and not a resin producing tree.

Traditionally, Styrax was a remedy in Turkish folk medicine against ulcers. In perfumery, it is used either as an essential oil or as a resinoid for perfume formulations. Styrax's chemical composition has not been thoroughly investigated [29, 30]. It seems that the main constituent of both American and oriental essential oil types is styrene, which name illustrates the fact that it was first isolated from styrax. Styrene content is sometimes as high as 70% in the essential oil of the oriental type [30], although because of its volatility and its ability to polymerize, its percentage strongly depends on the freshness of the material. Its characteristic sharp and penetrating odor contributes to the specific odor of styrax. Other volatiles isolated from this material are α- and β-pinene, and β-caryophyllene (in the American variety). Some aromatic constituents are also present, such as cinnamic acid and its esters (3-phenylpropyl and cinnamyl cinnamates, Figure 7.4), but because of their low volatility, their content in the essential oil is generally low, and much higher in the resinoid. The secondary metabolites of other parts of the tree (leaves, bark) have been investigated in some studies, which have revealed additional heavy constituents that may be actual constituents of the resinoid, even if up to now, no study is concentrated on the nonvolatile

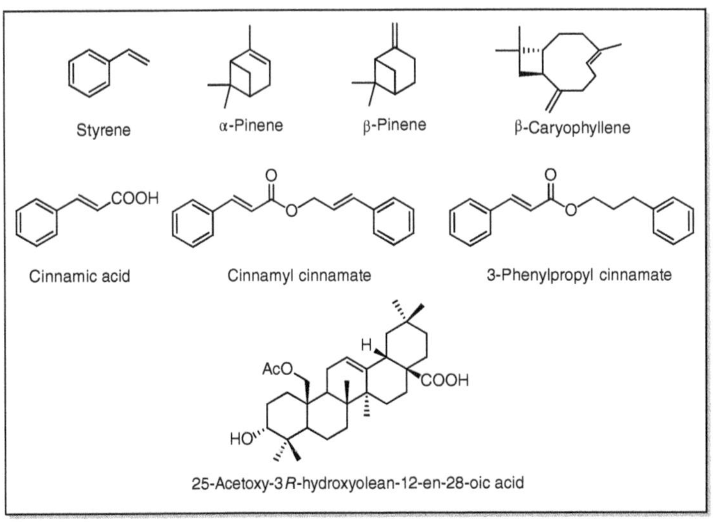


**Figure 7.4** Some constituents of styrax.

part of styrax resinoids. Hence, 25-acetoxy-3*R*-hydroxyolean-12-en-28-oic acid and several structurally related other triterpenoids were isolated from the cuticles of *Liquidamber styraciflua*, and the former compound showed strong cytotoxicity against various human cancer cell lines [31, 32]. Flavonoids and hydrolyzable tannins with anti-inflammatory activity were also isolated from the stems and leaves of this species [33], but these compounds are probably not apolar enough to be found in the resinoid.

## 7.2 Benzoin Siam (*Styrax tonkinensis craib ex hartwiss*) and Benzoin Sumatra (*Styrax benzoin*)

Benzoin resinoids are most commonly obtained from tree exudates widespread in Thailand, Laos, Cambodia, Vietnam (benzoin Siam), and the island of Sumatra (benzoin Sumatra) [6]. Benzoin Sumatra resin has a bigger market share (~4000 tonnes) than benzoin Siam resin (~70 tonnes) [34], even given the fact that the latter is more valued internationally for perfume and pharmaceutical purposes. Benzoin resinoids are obtained by extraction with various solvents. Yields vary substantially with the grade of resins – 85–95% (benzoin Siam), and 65–95% (benzoin Sumatra) [7]. Benzyl benzoate is the main constituent of both resinoids. These two materials can be differentiated by the presence of benzoic acid (benzoin Siam, Figure 7.5a) derivatives or cinnamic acid (benzoin Sumatra, Figure 7.5b) derivatives [7, 35].

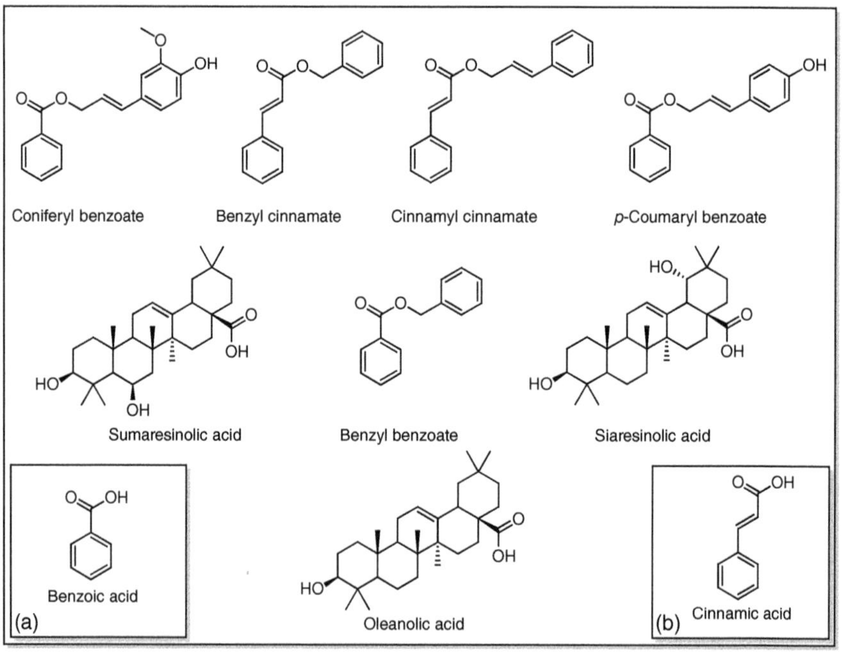


**Figure 7.5** Main compounds of benzoin resinoids. (a) Characteristic compound for the benzoin Siam resinoid; (b) Characteristic compound for the benzoin Sumatra resinoid.

Benzoin resinoids are excellent fixatives and find application in perfumery. Due to their low price and agreeable sensory properties (Table 7.1), they are often used in soap perfumes. Benzoin Siam resinoid is valued and used in fine perfumery as a fixative in light, floral fragrances [5]. It is also preferred by the food industry as a flavoring. Benzoin Siam resinoid can be found, for example, in "*Replica*" *Coffee Break* (Maison Margiela, 2019, perfumer: Jacques Cavallier) and "*Replica*" *Whispers in the Library* (Maison Margiela, 2019, perfumer: Marie Salamagne). Applications of benzoin resins are wider than resinoids. The same applies to the number of biological studies. Nevertheless, it has been proven that benzoin Siam resinoid can be potentially considered for aromatherapy purposes, as it exhibits very good inhibitory properties against human leukocyte elastase ($0.1 \leq IC_{50} \leq 1$ ppm), which contributes to the pathogenesis of inflammatory disorders [36].

## 7.3 Labdanum (*Cistus ladaniferus*)

*Cistus ladaniferus* is a pyrophoric plant and a source of highly prized labdanum gum. When the plant's flowers begin to fade, the shrub develops leafy twigs. Its branches are covered with secretory hairs, which release abundant quantities of gum with an amber-like fragrance. The gum was historically harvested by combing goats' beards that moved amid the shrubs [4]. Nowadays, cistus branches are gathered and processed on a large scale. Hydrodistillation or steam diffusion of cistus branches results in the traditional cistus essential oil, which has a lemony, fresh, and amber-like olfactory profile. The addition of an extract from distillation water to the traditional oil gives a complete cistus essential oil with predominant amber-like, warm, and gourmand notes. Extraction of cistus twigs and branches with a hydrocarbon-based solvent leads to a cistus concrete, which can be processed via ethanolic extraction that gives a traditional cistus absolute with dry wood and sweet notes or via molecular distillation, resulting in an absolute with balsamic, smoky, and hot notes. On the other hand, young cistus branches can be dipped in the hot solution of sodium carbonate. This solution is acidified with sulfuric acid and raw labdanum gum is obtained. Its hydrodistillation results in labdanum oil with intensive amber-like woody notes. The extraction of gum with a hydrocarbon solvent leads to labdanum concrete, which after ethanolic extraction gives labdanum absolute that possesses alcoholic, balsamic, and mild notes. Ethanolic extraction of a raw gum leads to a labdanum resinoid with a balsamic, sweet, and amber-like olfactory profile. The most characteristic constituents of labdanum resinoid are labdane-type acids (Figure 7.6).

Labdanum resinoid is a high-viscosity dark-brown material that has properties desired by the perfume industry. It has wide applications in soap perfumes, where it acts as a fixative, blender, and sweetener [7]. It is also used in the food industry in ice creams, beverages, sweets, and baked goods. Data on the biological activities of labdanum resinoid is very scarce (Table 7.2), and the available results are not promising as this material showed very weak properties against selected important microorganisms.

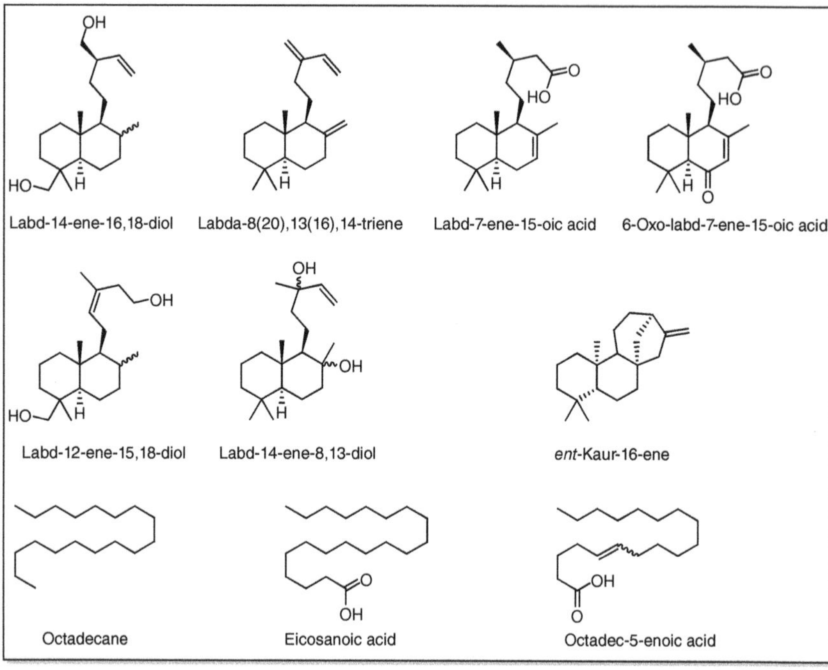


**Figure 7.6** Characteristic compounds of labdanum resinoid.

## 7.4 Myrrh (*Commiphora myrrha*)

The genus *Commiphora* is composed of almost 200 species native to the Arabian peninsula and Eastern Africa (Somalia, Ethiopia, Kenya), but also present in South America. These plants belong to the Burseraceae family, which includes resin-producing trees. The most common examples of commercial resinous material produced from Burseraceae are frankincense and myrrh (respectively from *Boswellia* and *Commiphora* genus). Unfortunately, the traceability of such resins sold by local producers is often extremely difficult to control, and consequently, the actual botanical identity of the trees from which the material was collected is often unclear. The voluntary mixing of different types of resin is also common and brings a lot of confusion in phytochemical studies. Therefore, a lot of inconsistencies can be found in the literature concerning the chemical composition of these resins. "True" myrrh (Heerabol myrrh, medicinal myrrh, bitter myrrh) is considered to be produced by *Commiphora myrrha* Engl. (Syn. *Commiphora molmol* Engl. ex Tschirch), which are shrub-like trees, growing in countries of the Horn of Africa, in the South of the Arabian Peninsula, and India and Pakistan. China is the largest market for myrrh resins, mainly because of its use in traditional Chinese medicine.

*Commiphora myrrha* Engl. has a very characteristic sweet, warm-balsamic, licorice, and mushroom-like odor. It is widely used in perfumery and aromatherapy, as well as a cure for many diseases in traditional medicines. A lot of biological activities of its constituents have been demonstrated [37], like antimicrobial [38–41], anti-quorum sensing [39], anti-inflammatory [42], lipogenesis inhibition effect [42], analgesic, antifungal [41], cytotoxic [43, 44], renoprotective [45], neuroprotective [46, 47], and dermoprotective properties [48]. Not surprisingly, as a result of its considerable importance in traditional medicine (especially Chinese), the chemical composition of myrrh resin has been thoroughly investigated to better characterize the constituents responsible for the activities cited above [49].

A significant part (30–60%) of myrrh resin consists of a water-soluble gum containing mostly sugars and polysaccharides. The remaining alcohol-soluble portion (hence constituting the resinoid) is composed of terpenic constituents, of which 10–50% are volatile enough to be extracted by hydrodistillation when myrrh essential oil is prepared. This latter material is indeed the main product of the transformation of the resin since it contains all of the odoriferous constituents, as well as a large portion of the bioactive products, and it has been extensively investigated [41, 50–53]. The particularity of myrrh essential oil is its low content of monoterpene compounds and its very high amount of characteristic furanosesquiterpenoids. Generally, furanodiene, curzerene, lindestrene, and furanoeudesma-1,3-diene [54] are the main components (Figure 7.7) and the sum of these three constituents represents more than half of the total amount. Many other furanosesquiterpenoids are also present in the resin, and a lot of them show interesting bioactivities.

The studies reporting on the odor-donating compounds of myrrh are scarce because most phytochemical studies are far from perfumistic interests, and thus generally neglect to mention the olfactory properties of isolated samples. Nevertheless, in one of the earliest investigations on myrrh essential oil, Brieskorn and Noble reported that furanoeudesma-1,3-diene possessed "an intensive fragrance" [55] Later, Wilson and Mookherjee [56] (from IFF company) mentioned that lindestrene had a "deep rich leathery, incensey, warm, balsamic, sweet, very typical myrrh character". Among many other odorant compounds isolated in this study, dihydropyrocurzerenone was present at a low amount (c. 0.1%), and displayed a very characteristic "very heavy resinous myrrh odor with rich, sweet incense note." In another study describing the olfactory properties of myrrh components, Zhu reported the first isolation and structure determination of myrrhone ("weak floral with relatively strong animal-like note odor"), *epi*-curzerenone "floral and animal-like odor", furanogermacra-1*E*,10(15)-dien-6-one "floral and somewhat animal-like note", τ-cadinol and eudesm-4(15)-ene-1β,6a-diol, both showing "very interesting animal and castoreum-like odor quality" [57]. In addition to the volatile part, heavier components of myrrh resin are also expected to be present in the resinoid, and many publications reported the presence of di- and tri-terpenoids [38, 58, 59]. This nonvolatile part is indeed rather complex, as shown by the high number of components identified in a series of publications conducted during the last 10 years: polyfunctional and nitrogen-containing sesquiterpenoids (lactams, diols, hydroxyesters, hydroxyketones. . .) [42, 46, 47, 58, 60], as well as unusual dimers of sesquiterpenoids (Figure 7.7) [38, 42, 43, 61].

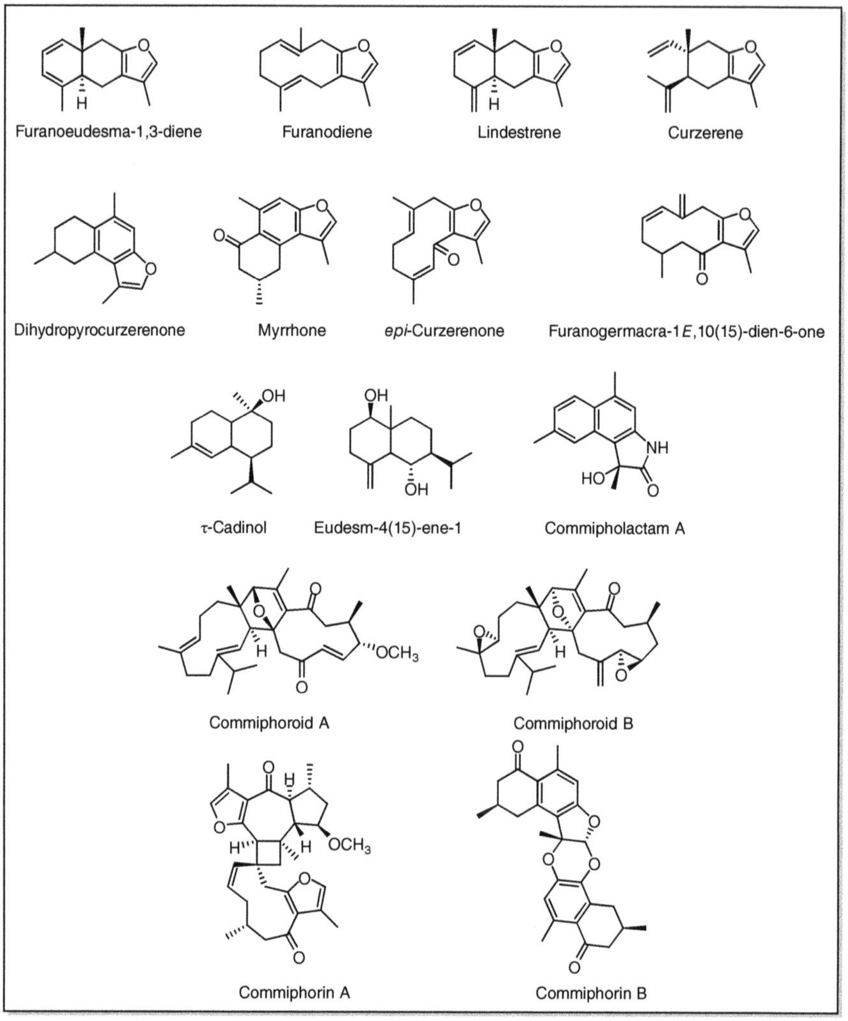


**Figure 7.7** Selected constituents of myrrh resinoid: fragrant furanosesquiterpenoids, other odorants, a nitrogen compound, and sesquiterpenoid dimers.

## 7.5 Conclusions

The chemistry of resinoids is a complex subject, especially for the nonvolatile part. Their volatile constituents have been quite thoroughly studied because most of the commercially available resins are also steam- or hydro-distilled to the respective essential oils. Despite the long history of the use of resinoids, the data on their biological activity is scarce. Solubility issues of the resinoid matrix in various media might be partially responsible for the limited popularity of *in vitro* studies. The second reason is that these materials are niche, compared to widely studied essential oils. This brief overview has shown that resinoids might have some application potential besides the common use as flavor and fragrance additives, but more studies need to be undertaken, particularly *in vivo* experiments.

## References


1 Sell, C. (2019). *Perfume in the Bible*. Royal Society of Chemistry.

2 Niebler, J. (2017). Incense materials. In: *Springer Handbook of Odor* (ed. A. Buettner), 13–14. Cham: Springer International Publishing.

3 Ohloff, G. and Pickenhagen, W. (2012). *Scent and Chemistry*. Wiley.

4 Langenheim, J.H. (2003). *Plant Resins: Chemistry, Evolution, Ecology, and Ethnobotany*. Timber Press.

5 Curtis, A. and Williams, D.G. (2001). *Introduction to Perfumery*. Micelle Press.

6 Surburg, H. and Pantent, J. (2016). Natural raw materials in the flavor and fragrance industry. *Common Fragrance Flavor Mater.* 193–264.

7 Arctander, S. (2017). *Perfume and Flavor Materials of Natural Origin*. Lulu.com.

8 George, C.K. (2012). Chapter 8 – Asafoetida. In: *Handbook of Herbs and Spices*, 2e (ed. K.V. Peter), 151–165. Woodhead Publishing.

9 Farhadi, F., Asili, J., Iranshahy, M., and Iranshahi, M. (2019). NMR-based metabolomic study of asafoetida. *Fitoterapia* 139: 104361.

10 Abd El-Razek, M.H., Ohta, S., Ahmed, A.A., and Hirata, T. (2001). Sesquiterpene coumarins from the roots of *Ferula assa-foetida*. *Phytochemistry* 58 (8): 1289–1295.

11 Degenhardt, A., Liebig, M., Kohlenberg, B. et al. (2012). Novel insights into flavor chemistry of asafetida. In: *Recent Advances in the Analysis of Food and Flavors. ACS Symposium Series. 1098*, 167–175. American Chemical Society.

12 Rajanikanth, B., Ravindranath, B., and Shankaranarayana, M.L. (1984). Volatile polysulphides of asafoetida. *Phytochemistry* 23 (4): 899–900.

13 Burdock, G.A. (2009). *Fenaroli's Handbook of Flavor Ingredients*, 6e. Taylor & Francis.

14 Wright, J. (2018). *Flavor Creation*, 3e. Blurb, Incorporated.

15 Rahlfs, V.W. and Mössinger, P. (1979). Asa foetida bei Colon irritabile. *Deut. Med. Wochenschr.* 104 (04): 140–143.

16 Unnikrishnan, M.C. and Kuttan, R. (1988). Cytotoxicity of extracts of spices to cultured cells. *Nutr. Cancer* 11 (4): 251–257.

17 Keshri, G., Lakshmi, V., Singh, M.M., and Kamboj, V.P. (1999). Post-coital antifertility activity of ferula assafoetida extract in female rats. *Pharm. Biol.* 37 (4): 273–276.

18 Shashkanth, K.N. and Hosono, A. (1986). In vitro mutagenicity of tropical spices to streptomycindependent strains of *Salmonella typhimurium* TA98. *Agri. Biol. Chem.* 50 (11): 2947–2948.

19 Adhami, H.-R., Scherer, U., Kaehlig, H. et al. (2013). Combination of bioautography with HPTLC–MS/NMR: a fast identification of acetylcholinesterase inhibitors from Galbanum†. *Phytochem. Anal.* 24 (4): 395–400.

20 Abedini, A., Roumy, V., Mahieux, S. et al. (2014). Antimicrobial activity of selected Iranian medicinal plants against a broad spectrum of pathogenic and drug multiresistant micro-organisms. *Lett. Appl. Microbiol.* 59 (4): 412–421.

21 Sadraei, H., Asghari, G.R., Hajhashemi, V. et al. (2001). Spasmolytic activity of essential oil and various extracts of *Ferula gummosa* Boiss. on ileum contractions. *Phytomedicine* 8 (5): 370–376.

22 Axtell, B.L. and Fairman, R.M. (1992). *Minor Oil Crops*. Food and Agriculture Organization of the United Nations.

23 Food and Agriculture Organization of the United Nations (1992). Galbanum natural oleoresin. http://www.fao.org/3/X5043E/x5043E0h.htm (accessed on 04 May 2020).

24 Moyler, D.A. and Clery, R.A. (2005). The aromatic resins: their chemistry and uses. In: *Flavours and Fragrances* (ed. K.A.D. Swift), 96–115. Woodhead Publishing.

25 Miyazawa, N., Nakanishi, A., Tomita, N. et al. (2009). Novel key aroma components of galbanum oil. *J. Agric. Food. Chem.* 57 (4): 1433–1439.

26 Cotterrell, G.P., Halsall, T.G., and Wriglesworth, M.J. (1970). The chemistry of triterpenes and related compounds. Part XLVII. Clarification of the nature of the tetracyclic triterpene acids of elemi resin. *J. Chem. Soc. C* 5: 739–743.

27 Brieskorn, C.H. and Krauß, G. (1986). Monoterpene, Sesquiterpene und dimere Phellandrene aus Manila-Elemi*. *Planta Med.* 52 (04): 305–307.

28 Villanueva, M.A., Torres, R.C., Başer, K.H.C. et al. (1993). The composition of Manila elemi oil. *Flavour Fragrance J.* 8 (1): 35–37.

29 Chalchat, J.-C., Garry, R.-P., and Mathieu, J.-P. (1994). Composition of the volatile fraction from honduras styrax, *Liquidambar styraciflua* L. *J. Essent. Oil Res.* 6 (1): 73–75.

30 Fernandez, X., Lizzani-Cuvelier, L., Loiseau, A.-M. et al. (2005). Chemical composition of the essential oils from Turkish and Honduras Styrax. *Flavour Fragrance J.* 20 (1): 70–73.

31 Sakai, K., Fukuda, Y., Matsunaga, S. et al. (2004). New cytotoxic oleanane-type triterpenoids from the cones of *Liquidamber styraciflua*. *J. Nat. Prod.* 67 (7): 1088–1093.

32 Fukuda, Y., Yamada, T., Wada, S.-i. et al. (2006). Lupane and oleanane triterpenoids from the cones of *Liquidamber styraciflua*. *J. Nat. Prod.* 69 (1): 142–144.

33 Mancarz, G.F.F., Laba, L.C., da Silva, E.C.P. et al. (2019). *Liquidambar styraciflua* L.: a new potential source for therapeutic uses. *J. Pharm. Biomed. Anal.* 174: 422–431.

34 Fernández, C.G. (2004). Benzoin, a resin produced by Styrax trees in North Sumatra Province, Indonesia. Center for International Forestry Research.

35 Fernandez, X., Lizzani-Cuvelier, L., Loiseau, A.-M. et al. (2003). Volatile constituents of benzoin gums: Siam and Sumatra. Part 1. *Flavour Fragrance J.* 18 (4): 328–333.

36 Baylac, S. and Racine, P. (2004). Inhibition of human leukocyte elastase by natural fragrant extracts of aromatic plants. *Int. J. Aromather.* 14 (4): 179–182.

37 Nomicos, E.Y.H. (2007). Myrrh: medical marvel or myth of the Magi? *Holis. Nurs. Prac.* 21 (6): 308–323.

38 Zhu, C.-Z., Hu, B.-Y., Liu, J.-W. et al. (2019). Anti-*Mycobacterium tuberculosis* terpenoids from *Resina Commiphora*. *Molecules* 24 (8): 1475.

39 Rubegeta, E., Ahmad, A., Kamatou, G.P.P. et al. (2019). Headspace analysis, antimicrobial and anti-quorum sensing activities of seven selected African Commiphora species. *S. Afr. J. Bot.* 122: 522–528.

40 Alhussaini, M.S., Saadabi, A.M., Alghonaim, M.I., and Ibrahim, K.E. (2015). An evaluation of the antimicrobial activity of *Commiphora myrrha* Nees (Engl.) Oleo-gum Resins from Saudi Arabia. *J. Med. Sci.* 15 (4): 198–203.

41 Ljaljević Grbić, M., Unković, N., Dimkić, I. et al. (2018). Frankincense and myrrh essential oils and burn incense fume against micro-inhabitants of sacral ambients. Wisdom of the ancients? *J. Ethnopharmacol.* 219: 1–14.

42 Dong, L., Qin, D.-P., Di, Q.-Q. et al. (2019). Commiphorines A and B, unprecedented sesquiterpenoid dimers from *Resina Commiphora* with striking activities on anti-inflammation and lipogenesis inhibition. *Org. Chem. Front.* 6 (23): 3825–3833.

43 Hu, B.-Y., Qin, D.-P., Wang, S.-X. et al. (2018). Novel terpenoids with potent cytotoxic activities from *Resina Commiphora*. *Molecules* 23 (12): 3239.

44 Gao, W., Su, X., Dong, X. et al. (2015). Cycloartan-24-ene-1α,2α,3β-triol, a cycloartane-type triterpenoid from the resinous exudates of *Commiphora myrrha*, induces apoptosis in human prostatic cancer PC-3 cells. *Oncol. Rep.* 33 (3): 1107–1114.

45 Dong, L., Luo, Q., Cheng, L.-Z. et al. (2017). New terpenoids from *Resina Commiphora*. *Fitoterapia* 117: 147–153.

46 Xu, J., Guo, Y., Li, Y. et al. (2011). Sesquiterpenoids from the resinous exudates of *Commiphora myrrha* and their neuroprotective effects. *Planta Med.* 77 (18): 2023–2028.

47 Xu, J., Guo, Y., Zhao, P. et al. (2012). Four new sesquiterpenes from *Commiphora myrrha* and their neuroprotective effects. *Fitoterapia* 83 (4): 801–805.

48 Auffray, B. (2007). Protection against singlet oxygen, the main actor of sebum squalene peroxidation during sun exposure, using *Commiphora myrrha* essential oil. *Int. J. Cosmet. Sci.* 29 (1): 23–29.

49 Shen, T., Li, G.-H., Wang, X.-N., and Lou, H.-X. (2012). The genus *Commiphora*: a review of its traditional uses, phytochemistry and pharmacology. *J. Ethnopharmacol.* 142 (2): 319–330.

50 Baser, K.H.C., Demirci, B., Dekebo, A., and Dagne, E. (2003). Essential oils of some *Boswellia* spp., Myrrh and Opopanax. *Flavour Fragrance J.* 18 (2): 153–156.

51 Marongiu, B., Piras, A., Porcedda, S., and Scorciapino, A. (2005). Chemical composition of the essential oil and supercritical $CO_2$ extract of *Commiphora myrrha* (Nees) Engl. and of *Acorus calamus* L. *J. Agric. Food. Chem.* 53 (20): 7939–7943.

52 Morteza-Semnani, K. and Saeedi, M. (2003). Constituents of the essential oil of *Commiphora myrrha* (Nees) Engl. var. molmol. *J. Essent. Oil Res.* 15 (1): 50–51.

53 Dekebo, A., Dagne, E., and Sterner, O. (2002). Furanosesquiterpenes from *Commiphora sphaerocarpa* and related adulterants of true myrrh. *Fitoterapia* 73 (1): 48–55.

54 Maradufu, A. and Warthen, J.D. (1988). Furanosesquiterpenoids from *Commiphora myrrh* oil. *Plant Sci.* 57 (2): 181–184.

55 Brieskorn, C.H. and Noble, P. (1983). Two furanoeudesmanes from the essential oil of myrrh. *Phytochemistry* 22 (1): 187–189.

56 Wilson, R.A. (1983). Characterization of aroma donating compounds of myrrh. *IXth International Essential Oil Congress*. Singapore.

57 Zhu, N., Sheng, S., Sang, S. et al. (2003). Isolation and characterization of several aromatic sesquiterpenes from *Commiphora myrrha*. *Flavour Fragrance J.* 18 (4): 282–285.

58 Zhu, S.-S., Qin, D.-P., Wang, S.-X. et al. (2019). Commipholactam A, a cytotoxic sesquiterpenoidal lactam from *Resina Commiphora*. *Fitoterapia* 134: 382–388.

**59** Shen, T., Wan, W.-Z., Wang, X.-N. et al. (2009). A triterpenoid and sesquiterpenoids from the resinous exudates of *Commiphora myrrha*. *Helv. Chim. Acta* 92 (4): 645–652.

**60** Dong, L., Cheng, L.-Z., Yan, Y.-M. et al. (2017). Commiphoranes A–D, carbon skeletal terpenoids from *Resina Commiphora*. *Org. Lett.* 19 (1): 286–289.

**61** Liu, J.-W., Liu, Y., Yan, Y.-M. et al. (2018). Commiphoratones A and B, two sesquiterpene dimers from *Resina Commiphora*. *Org. Lett.* 20 (8): 2220–2223.

# 8

# Seasoning, Herbs, and Spices

*Anjali Anil*[1]*, Józef T. Haponiuk*[2]*, and Sumith Kunnirikka*[1]

[1] *Tropical flavors (P) Ltd., Kolenchery, Cochin, Kerala, 682311, India*
[2] *Gdansk University of Technology, Department of Polymer Technology, Gabriela Narutowicza 11/12, Gdańsk, 80-233, Poland*

## 8.1 Introduction

We cannot even imagine having food without any flavor or having several foods with the same flavor. Fortunately, each food has its own natural flavoring. However, sometimes it needs to be intensified or improved. Hence, the use of seasoning and flavoring in foods is relevant here.

Seasoning is an ingredient that intensifies or improves the taste and flavor of the food without altering its natural flavor. The individual flavor of seasoning in the food product cannot be tasted, if seasoning is used accurately. This skill can be developed by tasting the food throughout the cooking process. Flavoring is an ingredient that actually alters the natural flavor of the food to which it is added. Seasoning brings out the flavors in the food and adds complementary tastes to enhance the eating experience. Generally, we can season food at any time of the cooking process, but in some cases, it is necessary to add seasoning at a particular time of cooking process. For example, we can add seasoning during the entire cooking process of food like soup, but it is necessary to add seasoning at the beginning of the cooking process such as roast, because it allows sufficient time for seasoning to be absorbed effectively throughout the food. Dried seasoning should be added prior to the fresh seasoning in the cooking process.

A wide range of ingredients can be used as seasonings. The most common seasonings are salt, pepper, and lime juice. Seasoning powders also contain sugar, acids, flavor enhancers, anticaking agents, fillers, preservatives, and antioxidants. However, the main seasoning ingredients are classified as herbs and spices.

Spices and herbs are the building blocks of flavor and taste in food applications because they improve and strengthen taste and flavor profile of the food. The herbs

*Natural Flavours, Fragrances, and Perfumes: Chemistry, Production, and Sensory Approach,*
First Edition. Edited by Sreeraj Gopi, Nimisha Pulikkal Sukumaran, Joby Jacob, and Sabu Thomas.

are usually fresh or dried leaves of plants that contains aromatic compounds responsible for their characteristic flavor and taste. Spices come from the bark, bud, fruit, root, flower, or seed of various plants or trees; they contain volatile oils and active components that give them their characteristic aroma, taste, and pungency. Herbs tend to have milder flavor, while spices tend to be more intensified. Both herbs and spices play a critical role in appearance and texture of the food. Spices and herbs are available in many forms, including fresh, dried, frozen, whole, ground, crushed, pureed, and extracts, and their forms are chosen by the food developer according to the specific application, processing methods/parameters, and shelf life.

## 8.2 Spices as Seasoning Ingredient

Spices are one of the key ingredients in seasoning mix that provide characteristic aroma, taste, and color to the food. The word spice originated from the Latin word "species," which means a specific kind that reflects the parts of the plants, including seed, berry, bark, kernel, aril, rhizome, root, flower, bulb, fruit, and flower bud cultivated for their aromatic, fragrant, pungent, and coloring properties. The sensations of sweet, salty, piney, sour, bitter, spicy, sulfury, earthy, and pungent are derived from an overall combination of aroma and taste in spices due to the presence of volatile components and nonvolatile components, respectively. Crunchiness, smoothness, or chewiness of spices in seasonings influence the texture of food, which adds overall flavor perception. Some common spices that are used to season foods are listed below.

### 8.2.1 Ajwain (*Trachyspermum ammi*)

Ajwain is native to Eastern Mediterranean, and it is now cultivated in South India, Europe, Pakistan, Afghanistan, Iran, and Egypt. Ajwain is a small, ridged, oval, light brown to purple red in color, caraway-like seed with slightly spicy and bitter taste with thyme-like aroma [1]. Ajwain seeds contains 3–4% essential oils, which contain mainly thymol, paracymene, γ-terpinene, β-pinene, dipentene, camphene, and myrcene. Ajwain seeds contain calcium, phosphorus, and iron [2]. Ajwain is widely used in North Indian, North African, Pakistani, and Iranian cuisine, including berbere, pakora, panchphoron, and chat masala [1].

### 8.2.2 Asafoetida (*Ferula asa-foetida*)

Asafoetida is indigenous to Iran, India, Afghanistan, and Pakistan, and it is also cultivated in China and Russia. It is a coagulated, dark brown, resin-like gum obtained from the juice of rhizome of the ferula. Asafoetida has strong garlic-like unpleasant aroma with bitter taste [1]. The essential oil content of asafoetida ranges from 10% to 17% and contains sulfur compounds like 2-butyl-1-propenyl disulfide, 1-(methyl thio) propyl 1-propenyl disulfide, and 2-butyl-3-(methyl thio)-2-propenyl disulfide; terpenoids; and farnesiferoles [3]. Asafoetida contains vitamin B complex, calcium, phosphorus, and iron [4]. Asafoetida is an important ingredient in Indian and Iranian dishes, include chewda, sambar, and chat masala [1].

### 8.2.3 Black Pepper (*Piper nigrum*)

Black pepper is known as the king of spices, and it is commonly called black gold. It is indigenous to southwest Malabar coast of India, and it is now cultivated in Vietnam, Sri Lanka, Thailand, Malaysia, Indonesia, Singapore, and Brazil. Black pepper is a spherical, wrinkled, dried, unripe berry with an outer shell. It has woody and pungent aroma with a biting taste [1]. Piperine gives pungency, and its essential oil gives characteristic aroma to black pepper. White pepper is obtained by removing the outer pericarp of black pepper before drying by the water steeping and retting technique [5]. The essential oil content of black pepper varies from 2% to 5%, and it mainly contains constituents like α and β pinene, α- phellandrene, limonene, β-caryophellene, piperonal, dihydroocerveol, ρ-cymene, sabinene, camphene, myrcene, and γ-terpinenes. Black pepper contains vitamin B complex, vitamin A, calcium, potassium, sodium, phosphorus, and iron [2]. Black pepper is an important ingredient in North American, European, Southeast Asian, and Arabian dishes, including cajun, creole, pepperoni, jerk, curry powders, berbere, ras-el-hanout, baharat, galat dagga, and zhoug [1].

### 8.2.4 Celery (*Apium graveolens*)

Celery seed is indigenous to Eastern and Southern Europe and the Mediterranean region, and it is also cultivated in India, France, Britain, China, Japan, United States, and Hungary. It is a dark brown, dried, tiny spherical stiff seed with harsh spicy aroma with warm bitter taste [1]. Celery seeds are a rich source of vitamin C, potassium, calcium, and magnesium [6]. Celery seeds contain 2–3% essential oils that include limonene, β-selinene, *n*-butylidene phthalide, and myrcene. These seeds are commonly used in North American, European, and Southeast Asian dishes, including bouquet garni, gumbo blends, stuffing blends, pickles, curries, and tomato juice blends [1].

### 8.2.5 Chili (*Capsicum annum*)

Chili pepper are native to Caribbean, Mexico, and Central and South America, and it is now cultivated in India, China, US, Africa, and Southeast Asia. The shape, size, color, and pungency differ with respect to their variety. The chili fruit is used ripe or unripe, dried or fresh, ground, crushed, or whole. The aroma of chili fruit can be hot, sweet, fruity, earthy, smoky, and floral. Chili pepper contains 0.2% to 2% capsaicinoids that are responsible for their pungency and include capsaicin, dihydrocapsaicin, and nordihydrocapsaicin. The red color of paprika chili is due to the presence of constituents such as carotenoids, capsanthin, capsorbin, zeaxanthin, cryptoxanthin, lutein, and α- and β-carotene. The characteristic aroma of chili is due to the compound 3-isobutyl-2-methoxy pyrazine [1]. Chili contains vitamin A, vitamin C, vitamin B6, vitamin K, potassium, and copper [7]. Heat chili and color chili are an important ingredient in all cuisines. It is used in a wide variety of dishes, including jerk, peri-peri, romesco, berbere, jhoug, kimchi, kochujang, sambal, pebre, nuoc cham, mole negro, nam prik, ma pla, goulash, ras-el-hanout, curry powders, baharat, and chermola [1].

### 8.2.6 Cinnamon (*Cinnamomum cassia*)

Cinnamon is commonly known as sweet/Chinese wood. It is indigenous to China, Vietnam, and Burma, and it is also cultivated in Malaysia and Indonesia. Chinese cinnamon has a thick bark with coarse surface, while Ceylon cinnamon has a thin bark with smooth surface. Cinnamon has woody and spicy aroma with sweet and bitter taste. It contains calcium, potassium, manganese, iron, magnesium, and vitamin C [1]. The essential oil content of cinnamon bark is usually 0.9–7%, and it contains four major components, including cinnamaldehyde, cinnamic acid, cinnamyl alcohol, and coumarin [8]. Cinnamon is commonly used in Asian, Latin American, European, and Mediterranean region. It is a key ingredient of dishes like garam masala, mole negro, panang curry, baharat, berbere, Indian curry powders, pho bo, ras-el-hanout, and Chinese five spices [1].

### 8.2.7 Clove (*Syzyium aromaticum*)

Cloves are the dried unblossomed flowering bud of clove trees; they are picked as rose-colored buds and become dark reddish brown when dry. They are indigenous to the Moluccas spice island of Indonesia, and they are also cultivated in India, Brazil, Sri Lanka, Malaysia, and Jamaica. Cloves consist of spicy, woody, and musty aroma with sharp, pungent, sweet, and bitter taste [1]. The dried clove buds contain about 15–20% of essential oils, including mainly eugenol; acetyl eugenol; β-caryophyllene; cratagolic acid; tannins; galatonnic acid; methyl salicylate; flavonoids such as eugenin, kaempferol, rhamnetin, and eugenitin; and terpenoids such as oleanolic acid [9]. Clove contains calcium, vitamin C, vitamin A, sodium, manganese, potassium, and magnesium. Clove is an important ingredient in American, North African, Indian, Chinese, and European dishes such as Chinese five spices, berbere, baharat, ras-el-hanout, rendang, gulais, garam masala, curry powders, Worcestershire sauce, and ketchup [1].

### 8.2.8 Coriander (*Coriandrum sativum*)

Coriander is indigenous to Asia, and it is also cultivated in North and South America. It is a dried, round-shaped ribbed seed having brownish yellow color with sweet, spicy, and nutty flavor and bitter and citrus taste. Coriander contains calcium, phosphorus, sodium, and magnesium. Depending upon their variety, coriander seeds contain 0.2–2% essential oils, including mainly linalool, α-pinene, γ-terpinene, geranyl acetate, and camphor [1, 10]. It is commonly used in Southeast Asia, North Africa, Caribbean, and Turkey dishes, which include curry powder, pickle blends, berbere, zhoug, baharat, and garam masala [1].

### 8.2.9 Cumin (*Cuminium cyminum*)

Cumin is native to the Mediterranean region, Indonesia, India, and China, and it is also cultivated in Middle East, Mexico, and Argentina. Cumin seeds are dried, stiff, dark brown or brownish yellow color, elongated, oval-shaped seeds with nutty, earthy, spicy, and bitter taste and warm aroma. Cumin contains 2.5–4.5% essential oils, including mainly cuminic aldehyde, β-pinene, terpinine, ρ-cymene, cuminyl alcohol, and β-farnesene. Cumin contains vitamin A, calcium, potassium, sodium, iron, magnesium, and phosphorus. Cumin is commonly used in Mexican, Southeast Asian, Caribbean, Middle East, and North African dishes, including garam masala, ras-el-hanout, sambar, curry powder, achiote, jhoug, panchphoron, chili con carne, and baharat [1].

### 8.2.10 Fennel (*Foneiculum vulgare*)

Fennel is indigenous to the Mediterranean region, and it is also cultivated in Europe, India, Iran, and US. Fennel seed is elongated, oval-shaped, ridged with pale or bright green to yellowish brown in color. Fennel seed is also called sweet cumin, and it has sweet, fresh, camphoraceous, anise-like aroma with sweet and bitter taste [1]. Fennel seeds contain vitamin B complex, vitamin C, calcium, iron, magnesium, manganese, phosphorus, potassium, sodium, and zinc [11]. Fennel seeds contain 1–6% essential oil, including fenchone, trans-anethole, limonene, camphene, and α-pinene [12]. Fennel seeds are widely used in Mediterranean, European, and Asian dishes, including curry powders, Chinese five spices, panchphoron, mirepoix, and herbes de provence [1].

### 8.2.11 Fenugreek (*Trigonella foenum graecum*)

Fenugreek is indigenous to the Mediterranean region, and it is now cultivated in Asia, Europe, and North and South America. Fenugreek seeds also called Greek hay and cow's horn because the seeds are enclosed in a sickle-shaped envelope. The seeds are cuboid, hard with smooth surface, yellow to brown in color with bitter taste, and sweet hay-like aroma [1]. Fenugreek seeds contain 0.02–0.05% essential oils, including mainly components like *n*-hexanol, heptanoic acid, dihydroactiniolide, dihydrobenzofuran, tetradecane, α-murolene, pentadecane, and β-elemene. Fenugreek seeds contain calcium, phosphorus, iron, sodium, potassium, and vitamins including vitamin A, C, B1, B2, and niacin [2]. Fenugreek is an important ingredient in Indian, Armenian, Mediterranean, Iranian, and Arabian dishes, including fish curry, sambar, panchphoron, berbere, zhug, hilbeh, aboukht, and chemen blends [1].

### 8.2.12 Garlic (*Allium sativum*)

Garlic is native to central Asia, and it is now cultivated in Europe, North and South America. Garlic is a white bulb with a plump and egg shaped bulblets covered by a transparent white or purple outer skin. Garlic can be used as fresh or dried. Fresh garlic has strong pungency and flavor, which become milder during cooking. Dried garlic has very strong persistent aroma and taste [1]. Garlic contains 0.1–0.25% essential oil. Garlic has an enzyme called alliinase that acts on alliin to produce allicin, which breaks down to diallyl disulfide and contribute to the strong penetrating sulfur-type aroma. Garlic powder contains vitamin A, vitamin B complex, calcium, phosphorus, iron, sodium, and potassium [5]. Garlic is an important ingredient in European, Mediterranean, Middle Eastern, Asian, Latin American, and American dishes, including tabil, rouille, adobos, sofritos hummus, moles, mojo, tzatziki, chimmichuri, aioli, and refogado blend [1].

### 8.2.13 Ginger (*Zingiber officinale*)

Ginger is native to Southern India and Southeast Asia, and it is cultivated in Nigeria, Japan, Jamaica, Mexico, Sri Lanka, United States, and Australia. Indian, Chinese, Jamaican, and African are the major varieties of ginger. Ginger is a rhizome that can be used in fresh and dried forms. Fresh ginger is firm, plump, not shriveled, tan or pale brown, juicy, spicy, and sweet with strong aroma. Dried ginger is fibrous, shriveled, hard, and less pungent. The essential oil content of ginger ranges from 1% to 4% and contains zingiberene, curcumene, α-pinene, sabinene, limonene, borneol, zingerone, shogaol, and paradol. Ginger contains magnesium, potassium, phosphorus, calcium, sodium, niacin, vitamin A, and manganese. Ginger is commonly used in Mediterranean, Asian, European, Caribbean, Australian, and North American dishes, including curry powders, berbere, Chinese five spices, stir fry, ras-el-hanout, congee, laksa, ginger ale, and quatre epices [1].

### 8.2.14 Green Cardamom (*Elletaria cardamomum*)

Green cardamom is known as the queen of spices, and it is commonly called true cardamom or small cardamom, which is indigenous to South India [1]. It is also cultivated in Sri Lanka, Nepal, Indonesia, and Tanzania [13]. Cardamom fruit is usually 1–2 cm in length, ellipsoidal, indehiscent, fleshy, and green in color, and it turns yellow and leathery when dry [14]. It contains calcium, potassium, magnesium, and manganese [1]. The essential oil content of green cardamom ranges from 6% to 14% with respect to their genotype and processing method [15]. Essential oils consist of mostly monoterpene constituents that include 1,8-cineole, α-pinene, α-terpineol, linalool, linalyl acetate, nerolidol, and the ester component α-terpinyl acetate [16–18]. The characteristic sweet and spicy flavor and the taste of green cardamom enhance sweet and savory dishes. It is commonly used in Asian, Arabian and North African dishes, including garam masala, berbere, curry powder, baharat, ras-el-hanout, korma, kabsah, biriyani, zhoug, and coffee [1].

### 8.2.15 Nutmeg and Mace (*Myristica fragrans*)

Nutmeg/Mace is native to Indonesia, and it is cultivated in South India, Sri Lanka, Caribbean islands, and Malaysia. Nutmeg is a dried, light brown/gray in color, wrinkled seed enclosed by a hard, smooth, blackish brown nut. It has spicy, sweet, and bitter taste with a terpene and camphor-like aroma. Mace is the interlaced, smooth and shiny, deep red/reddish brown, leathery covering of a nutmeg seed. It has strong terpene-like aroma and more bitter taste than nutmeg [1]. The essential oil content of nutmeg ranges from 3.9% to 16.5%, while that of mace varies from 7% to 14%. The volatile oil of both nutmeg and mace contains aroma compounds, including mainly sabinene, α- and β-pinene, myrcene, 1,8-cineole, myristicin, limonene, safrole, and 4-terpineol. Mace contains vitamin A, phosphorus, potassium, magnesium, sodium, and calcium, and nutmeg contains vitamin B complex, and iron along with prior nutrients [2]. Both nutmeg and mace are important ingredients of Asian, Caribbean, European, and North African dishes, including curry powders, garam masala, ras-el-hanout, baharat, galat dagga, rendang, bechamel sauce, quatre epices, and jerk [1].

### 8.2.16 Onion (*Allium cepa*)

Onion is indigenous to central Asia, and it is also cultivated in Europe, Southeast Asia, North and South America, and Egypt. Onion is a brownish red or yellowish white bulb. Onion can be used as a fresh or dried powder form. Fresh-cut onion has a strong, pungent, and penetrating odor that becomes sweet and pleasant when cooked. Onion contains 0.01–0.015% essential oils, which consist of methyl propyl disulfide, methyl propyl trisulfide, and dipropyl trisulfide released through the action of allinase enzyme [1]. Onion contains vitamin B complex, vitamin C, vitamin D, calcium, zinc, phosphorus, potassium, sodium, magnesium, and iron [5]. Onions are good for sauce, stew, and soup. Onion is an important ingredient in American, Asian, and European dishes, including cili boh, rojak, curry powders, pierogi filling, and stir fry blends [1].

### 8.2.17 Star Anise (*Illicium verum*)

Star anise is indigenous to China and Vietnam, and it is also cultivated in India, Philippines, Japan, and Korea. Star anise is a reddish brown, hard, irregular star-shaped fruit with eight carpels fused around a central core. Star anise has a sweet and pungent flavor with bitter taste [1]. Star anise contains vitamin B complex, calcium, iron, phosphorus, potassium, manganese, zinc, and magnesium. The essential oil content of star anise varies from 2.5% to 5%, and it mainly consists of compounds like α- and β-pinene, camphene, cis- and trans-anethole, linalool, safrole, anisaldehyde, and acetoanisole [2]. Star anise is an important ingredient in Southeast Asian, Indian, Chinese, and European dishes, including chettinad chicken, Chinese five spices, hoisin, Singapore pork curry, and barbecue blend [1].

### 8.2.18 Turmeric (*Curcuma domestica*)

Turmeric is native to South and Southeast Asia, and it is cultivated in India, Sri Lanka, China, Peru, Jamaica, and Malaysia. It is also called Indian saffron and yellow ginger. Turmeric is a rhizome with bright orange flesh and brownish yellow skin. It has musky and earthy aroma with bitter and gingery/peppery taste. Madras turmeric is bright yellow in color, while Alleppy turmeric is brownish yellow in color [1]. Turmeric contains 1.5–6% essential oils consisting mainly of sesquiterpenes, zingiberene, sabinene, cineole, borneol, and α- phellandrene. The coloring component of turmeric is curcuminoids that includes curcumin, demethoxy curcumin, and bis-demethoxy curcumin [19]. Turmeric contains vitamin C, iron, sodium, potassium, phosphorus, magnesium, and calcium. It is widely used in Indian and Southeast Asian dishes, including curry powders, laksa, pulao, nasi kuning, and bumbu blend [1].

## 8.3 Herbs as Seasoning Ingredient

Herbs are plants with savory and aromatic properties that are used for flavoring and garnishing of food. Culinary herbs are generally defined as the parts of plant, including leaf, stem, and stalk that provide specific flavors, aromas, colors, and visual appeal to the food. A few common culinary herbs used in seasonings are listed below.

### 8.3.1 Basil (*Osimum basilicum*)

Basil is indigenous to Southeast Asia, Europe and India, and they are also cultivated in Iran, Africa, Japan, and US. Sweet basil has bright green leaves with floral, sweet anise-like aroma, and cooling undertone, and it also has delicate, fresh, and slight minty taste [1]. The essential oil of basil contains components such as α- and β-pinene, methyl chavicol, 1,8-cineole, linalool, ocimene, borneol, geraneol, β-caryphyllone, *n*-cinamate, and eugenol. Basil contains magnesium, calcium, potassium, sodium, iron, copper, manganese, and zinc [20]. Basil leaves are widely used in Italian, Mediterranean, and Southeast Asian foods, including pizza sauce, pestos, pistous, Thai green, Vietnamese phos, umeboshi, bouquet garni, and Malaysian kurma [1].

### 8.3.2 Oregano (*Origanum vulgare*)

Oregano is native to the Mediterranean region, India, Mexico, and South America. Oregano has dark green leaves with herbaceous, phenolic, slightly floral, and bitter taste and slight lemony and pungent note. The aroma strength varies with different subspecies. Dried oregano has more potent flavor than fresh oregano [1]. The essential oil content of oregano ranges from 1% to 4%, and it contains compounds, including mainly thymol, carvacrol, ρ-cymene, sabinene, geraniol, borneol, camphene, linalool, β-myrcene, β-caryophyllene, akhdarenol and betulin. Oregano contains vitamin B complex, vitamin E, potassium, calcium, magnesium, phosphorus, zinc, manganese, iron, copper, sulfur, chlorine, iodine, and selenium [21]. Oregano is widely used in Italian, Spanish, Mediterranean, and Latin American foods, including pizza sauce, pasta sauce, zatar, chili con carne, frijole, and sante fe seasoning [1].

### 8.3.3 Parsley (*Petroselinum sativum*)

Parsley is indigenous to the Mediterranean region, Greece, Turkey, and Sardina, and it is now cultivated in Europe, Middle East, and US. Parsley leaves have a mild, herbaceous, green taste with a slight lemony aroma. Dried leaves have no flavor and taste, and therefore, they are used as a garnish. Essential oils of parsley leaves mainly include myristicin, apiole, α-pinene, β-phellandrene, ρ-mentha-1,8-triene, α-ρ-dimenthylstyrene, myrcene, and limonene [1]. Parsley leaves are rich in vitamin A, vitamin C, iron, iodine, and magnesium [22]. Parsley leaves are widely used in European, Middle Eastern, Turkish, Iranian, and Lebanese dishes, including bouquet garni, persillade, gremolada, pesto, and chimmichurri blend [1].

### 8.3.4 Rosemary (*Rosmarinus offinialis*)

Rosemary is indigenous to the Mediterranean region, and it is also cultivated in Europe and United States. Rosemary leaves are leathery, shiny, pine-like, small, narrow, and dark green in color. Dried leaves are curved with brownish green color. Fresh-cut rosemary leaves have strong, piney, tea-like fragrance with sweet, minty, peppery, camphor-like taste and bitter and woody aftertaste. Dried leaves have strong persistent aroma [1]. The essential oil content of rosemary ranges from 0.5% to 2.5%, and it contains mainly 1,8-cineole, α-pinene, camphene, and camphor, while the nonvolatile fraction consists of phenolic compounds such as rosmarinic acid, carnosic acid, carnosol, and rosmanol [23]. Rosemary contains vitamin C, vitamin A, potassium, sodium, calcium, magnesium, iron, and phosphorus. Rosemary is widely used in Mediterranean and European dishes, including bouquet garni, herbs de provence, pizza sauce, lamb tagine, and chicken roast marinade [1].

### 8.3.5 Thyme (*Thymus vulgaris*)

Thyme is native to the Mediterranean range, and it is also cultivated in Europe, Caribbean islands, and United States. Thyme leaves are green, slightly rolled, pined, broad/narrow, oval, and covered in fine hairs. They are used fresh and dried. They have piney, smoke, bitter taste with herbaceous and slightly floral aroma [1]. The essential oil content of thyme ranges from 1.5% to 5% that contains mainly thymol, terpinen-4-ol, carvacrol, ρ-cymene, pinene, camphene, myrcene, 1,8-cineole, terpinene, D-linalool and flavanoids such as apigenin, naringenin, luteolin, and thymonin [24]. Thyme contains vitamin A, vitamin B complex, potassium, sodium, phosphorus, calcium, and magnesium. Thyme is an important ingredient in

European, Caribbean, and Middle East foods, including bouquet garni, dukkah, zahtar, cajun spice, herb de provence, and jerk seasoning [1].

## 8.4 Seasoning Blends

Seasoning blends are created from a mixture of different spices and herbs that are skillfully mixed together to bring unique flavors from different cultures, styles of cooking and even simple combinations that improves the fresh taste of the food. Some of the most popular seasoning blends and their ingredients from all over the world are given below (Table 8.1).

## 8.5 Future Aspects

It seems like the produce of the future will be optimized for flavor. It is estimated that 20% tasting experience comes from the taste that is from the tongue. Whereas 80% comes from the smell and aroma.

One of the biggest developing fields in food industry is "food pairing." The basic concept behind this is the pairing of certain elements mainly based on their texture and common flavor they are sharing. This procedure makes it easy to discover new ingredient combinations.

A common pairing we found is wine and food. The backbone of this pairing includes flavor (whether it is fruity, herbal, etc.), taste (sweet, acidic, sour, bitter, etc.), and texture (which includes body, weight, and structure). While selecting wine and food for pairing, we have to understand the concept of flavor present in these items, such as the weak identified flavor, intensity of flavor, and whether these include any spicy characteristics. There are five primary sensations such as sweetness, sourness, saltiness, bitterness, and umami. Usually if the food is sweet, it will decrease the sweetness of wine served by sensory adaptation. Likewise, foods that are rich in amounts of acidity will decrease sourness of wine and makes the wine taste richer. In contrast, food with bitter taste will decrease the bitterness of a wine served with it [25]. Hence, there are many techniques in food industry to extract bitterness from many foods during cooking. Likewise, umami taste in wine can make the food bitter and metallic in taste. Temperature also plays a role in pairing of wine and food because the colder the wine, the less is the sweetness and the more is the acidity.

Flavors that create healthy dishes without compromising the taste and color of dishes are more important and will grow in popularity in the future.

**Table 8.1** Ingredient list of different types of seasoning.

| Seasoning | Origin | Type | Ingredients |
|---|---|---|---|
| Adobo seasoning | Spain, Portugal | Sauce | Salt (34.28%), paprika powder (17.14%), black pepper powder (11.43%), onion powder (8.57%), oregano dried (8.57%), cumin powder (8.57%), garlic powder (5.72%), chili powder (5.72). |
| Baharat seasoning | Middle East | Condiment | Paprika powder (24.59%), blackpepper powder (19.67%), cumin powder (14.75%), coriander powder (9.84%), cinnamon powder (9.84%), clove powder (9.84%), cardamom powder (4.92), star anise powder (4.92), nutmeg powder (1.63%). |
| Barbecue seasoning | United States | Condiment | Smoked paprika powder (30%), sugar (20%), celery salt (10%), garlic powder (10%), mustard powder (10%), black pepper powder (5%), salt (15%). |
| Berbere seasoning | Middle East | Spice mix | Mexico chili powder (50.79%), paprika powder (25.39%), cayenne pepper (6.35%), onion powder (2.12%), ginger powder (2.12%), cumin powder (2.12%), coriander powder (2.12%), cardamom powder (2.12), fenugreek powder (2.11%), garlic powder (1.06%), cinnamon powder (1.06%), nutmeg powder (0.53%), pimento powder (1.06%), clove powder (1.06%). |
| Cajun seasoning | Louisiana | Condiment | Salt (16.67%), garlic powder (16.67%), paprika powder (20.83%), black pepper powder (8.33%), onion powder (8.33%), cayenne pepper (8.33%), oregano dried (10.42%), thyme dried (10.42%). |
| Chimichurri seasoning | Argentina | Sauce | Oregano dried (18.36%), basil dried (18.36%), parsley dried (12.24%), thyme dried (12.24%), salt (12.25%), black pepper powder (6.12%), savory dried (6.12%), smoked paprika powder (6.13%), garlic powder (4.08%), chili crushed (4.08%). |
| Chinese five spice seasoning | China | Spice mix | Cinnamon powder (20%), clove powder (20%), fennel powder (20%), star anise powder (20%), Schezwan pepper powder (20%). |
| Creole seasoning | Louisiana | Condiment | Onion powder (10%), garlic powder (10%), oregano dried (10%), basil dried (10%), thyme dried (5%), black pepper powder (5%), white pepper powder (5%), cayenne pepper crushed (5%), paprika powder (25%), salt (15%). |

(Continued)

**Table 8.1** (Continued)

| Seasoning | Origin | Type | Ingredients |
|---|---|---|---|
| Fajita seasoning | Mexico | Condiment | Corn starch (30.77%), chili powder (20.51%), salt (10.26%), paprika powder (10.25%), sugar (10.26%), onion powder (5.13%), garlic powder (5.13%), cayenne pepper powder (2.56%), cumin powder (5.13%). |
| Garam masala | India | Spice mix | Cumin powder (31.58%), coriander powder (15.79%), cardamom powder (15.79%), black pepper powder (15.79%), cinnamon powder (10.53%), clove powder (5.26%), nutmeg (5.26%). |
| Harissa seasoning | Tunisia | Paste | Minced chili (48.98%), smoked paprika powder (12.24%), cumin powder (12.25%), coriander powder (12.25%), salt (6.12%), garlic powder (6.12%), caraway powder (2.04%). |
| Herbs de Province seasoning | France | Spice mix | Rosemary dried (11.54%), fennel (5.77%), savory dried (11.54%), thyme dried (11.54%), basil dried (11.54%), marjoram dried (11.54), lavender dried (11.54%), parsley dried (11.54%), oregano dried (5.77%), tarragon dried (5.77%), bay leaf powder (1.91%). |
| Italian herb seasoning | US | Spice mix | Basil dried (20%), oregano dried (20%), rosemary dried (20%), thyme dried (20%), marjoram dried (20%). |
| Jerk seasoning | Jamaica | Marinade | Onion powder (14.12%), garlic powder (14.12%), cayenne pepper powder (9.41%), salt (9.41%), black pepper powder (9.41%), thyme dried (9.41%), sugar (9.41%), pimento powder (4.71%), parsley dried (4.71%), paprika powder (4.71%), chili flakes (2.35%), cinnamon powder (2.35%), nutmeg powder (2.35%), clove powder (2.35%), cumin powder (1.18%). |
| Panchphoron | Bangladesh | Spice mix | Cumin powder (20%), mustard powder (20%), fenugreek powder (20%), fennel (20%), nigella (20%). |
| Peri-peri seasoning | South Africa | Sauce | Peri-peri pepper powder (6.25%), paprika powder (25%), oregano dried (12.5%), cardamom powder (12.5%), ginger powder (12.5%), onion powder (12.5%), garlic powder (12.5%), salt (6.25%). |
| Pumpkin pie spice seasoning | US | Spice mix | Cinnamon (61.53%), nutmeg powder (15.38%), ginger powder (15.38%), clove powder (7.71%). |
| Ras-el-hanout | North Africa | Spice mix | Cumin powder (16.66%), ginger powder (16.66%), salt (16.66%), black pepper powder (12.5%), cinnamon powder (8.34%), coriander powder (8.34%), cayenne pepper powder (8.34%), pimento powder (8.34%), clove powder (4.16). |

| Seasoning | Origin | Type | Ingredients |
|---|---|---|---|
| Shichimi togarashi seasoning | Japan | Spice mix | Orange peel dried (26.08%), cayenne pepper powder (34.79%), sesame seeds (17.39%), ginger powder (8.69%), poppy seeds (8.69%), white pepper (2.18%), nori (2.18%). |
| Taco seasoning | Mexico | Condiment | Chili powder (37.5%), garlic powder (3.125%), onion powder (3.125%), crushed chili (3.125%), oregano dried (3.125%), paprika powder (6.25%), cumin powder (18.75%), salt (12.5%), black pepper powder (12.5%). |
| Zaatar seasoning | Middle East | Spice mix | Cumin powder (20%), thyme dried (20%), sumac (20%), toasted sesame (20%), marjoram (6.67%), salt (6.67%), black pepper powder (6.66%). |

## References


1 Raghavan, S. (2006). *Handbook of Spices, Seasonings, and Flavourings*. CRC press.

2 Parthasarathy, V.A., Chempakam, B., and Zachariah, T.J. (ed.) (2008). *Chemistry of Spices*. CABI.

3 Amalraj, A. and Gopi, S. (2017). Biological activities and medicinal properties of asafoetida: a review. *J. Trad. Complementary Med.* 7 (3): 347–359.

4 Mahendra, P. and Bisht, S. (2012). *Ferula asafoetida*: traditional uses and pharmacological activity. *Pharmacogn. Rev.* 6 (12): 141.

5 Peter, K.V. (2001). *Handbook of Herbs and Spices*. Woodhead Publishing Limited.

6 Sowbhagya, H.B. (2014). Chemistry, technology, and nutraceutical functions of celery (*Apium graveolens* L.): an overview. *Crit. Rev. Food Sci. Nutr.* 54 (3): 389–398.

7 Whelton, P.K. and He, J. (2014). Health effects of sodium and potassium in humans. *Curr. Opin. Lipidol.* 25 (1): 75–79.

8 He, Z.D., Qiao, C.F., Han, Q.B. et al. (2005). Authentication and quantitative analysis on the chemical profile of cassia bark (cortex cinnamomi) by high-pressure liquid chromatography. *J. Agric. Food. Chem.* 53 (7): 2424–2428.

9 Bhowmik, D., Kumar, K.S., Yadav, A. et al. (2012). Recent trends in Indian traditional herbs *Syzygium aromaticum* and its health benefits. *J. Pharmacogn. Phytochem.* 1 (1): 13–22.

10 Diederichsen, A. (1996). *Coriander: Coriandrum sativum L*, vol. 3. Bioversity International.

11 Malhotra, S.K. (2012). Fennel and fennel seed. In: *Handbook of Herbs and Spices* (ed. K.V. Peter), 275–302. Woodhead Publishing.

12 Javed, R., Hanif, M.A., Ayub, M.A., and Rehman, R. (2020). Fennel. In: *Medicinal Plants of South Asia* (ed. M.A. Hanif, H. Nawaz, M.M. Khan and H.J. Byrne), 241–256. Elsevier.

**13** Garg, G., Sharma, S., Dua, A., and Mahajan, R. (2016). Antibacterial potential of polyphenol rich methanol extract of *Cardamom (Amomum subulatum)*. *J. Innov. Biol.* 3 (1): 271–275.

**14** Murugan, M., Dhanya, M.K., Deepthy, K.B. et al. (2016). *Compendium on Cardamom*. Kerala Agricultural University Cardamom Research Station.

**15** Menon, A.N. (2000). Studies on the volatiles of cardamom (*Elleteria cardamomum*). *J. Food Sci. Technol.* 37 (4): 406–408.

**16** Ashokkumar, K., Murugan, M., Dhanya, M.K. et al. (2020). Phytochemical variations among four distinct varieties of Indian cardamom *Elettaria cardamomum* (L.) Maton. *Nat. Prod. Res.* 34 (13): 1919–1922.

**17** Kaskoos, R.A., Ali, M., Kapoor, R. et al. (2006). Essential oil composition of the fruits of *Eletteria cardamomum*. *J. Essent. Oil Bear. Plants* 9 (1): 81–84.

**18** Yashin, A., Yashin, Y., Xia, X., and Nemzer, B. (2017). Antioxidant activity of spices and their impact on human health: a review. *Antioxidants* 6 (3): 70.

**19** Nasri, H., Sahinfard, N., Rafieian, M. et al. (2014). Turmeric: a spice with multifunctional medicinal properties. *J. Herb Med. Pharmacol.* 3: 5–8.

**20** Shahrajabian, M.H., Sun, W., and Cheng, Q. (2020). Chemical components and pharmacological benefits of Basil (*Ocimum basilicum*): a review. *Int. J. Food Prop.* 23 (1): 1961–1970.

**21** Kintzios, S.E. (2012). Oregano. In: *Handbook of Herbs and Spices*, Woodhead Publishing Series in Food Science, Technology and Nutrition (ed. K.V. Peter), 417–436.

**22** Farzaei, M.H., Abbasabadi, Z., Ardekani, M.R.S. et al. (2013). Parsley: a review of ethnopharmacology, phytochemistry and biological activities. *J. Tradit. Chin. Med.* 33 (6): 815–826.

**23** Andrade, M.A., Ribeiro-Santos, R., Bonito, M.C.C. et al. (2018). Characterization of rosemary and thyme extracts for incorporation into a whey protein based film. *Lwt* 92: 497–508.

**24** Hudaib, M., Speroni, E., Di Pietra, A.M., and Cavrini, V. (2002). GC/MS evaluation of thyme (*Thymus vulgaris* L.) oil composition and variations during the vegetative cycle. *J. Pharm. Biomed. Anal.* 29 (4): 691–700.

**25** Koone, R. L. (2012). Predictability of food and wine pairing using a sensory approach. Theses and Dissertations Retrieved from https://scholarworks.uark.edu/etd/603 (accessed 19 June 2021).

# Part V

# Regulations, Consumer Trends, and In Silico Biology

# 9

# Regulatory Aspects for Flavor and Fragrance Materials

*Neha N. Areekal[1] Vaishak Ramachandran[1], Anirudh Jayakumar[1], Józef T. Haponiuk[2], and Roshin Thankachan[2,3]*

[1]*Karunya Institute of Technology and Sciences, Department of Food Processing Engineering, Coimbatore, 641114, Tamil Nadu, India*
[2]*Gdansk University of Technology, Department of Polymer Technology, G. Narutowicza 11/12, Gdańsk, 80233, Poland*
[3]*Tropical flavors (P) Ltd., R&D Centre, Kolenchery, Cochin, Kerala, 682311, India*

## 9.1 Introduction

The amount of additives used in food is an important feature of overall food safety and the quality of the end product. These chemical compounds (additives) can constitute health threats either partially or completely if the total consumption is too high than the standard value. Therefore, exposure assessment is a central aspect of risk management. Dietary exposure to food additives should be measured by integrating concentration data for all food products with details about their intake. The stepwise technique for dietary exposure assessment of food additives is such that, as the precision of dietary exposure tests rises, the expense of gathering sufficient data and equipment required to carry out assessments increases [1]. The methodologies of dietary intake assessment applied to a non-average population, and in particular, to those who eat a comparatively large amount of food commodities containing a higher concentration of chemical compounds (additives) show that when the consumption of such compounds increases beyond the Generally Recognized as Safe (GRAS) level, they potentially contribute to a health risk [2]. In the food industry, flavors put up a wonderful forum. Flavor-incorporated food is important for consumers, and without tastes, the keen interest of the customer for food ceases to exist. In most cases, foodstuffs containing both artificial and synthetic flavors are avoided because the consumer population gives much importance to naturally derived substances, which do not cause any adverse effect on the health condition after their consumption. Components that produce flavor are harmful to their health [3]. Chemically synthesized flavors result in environmentally unfriendly manufacturing practices. It also decreases the selectivity of the substrate, which can also contribute

*Natural Flavours, Fragrances, and Perfumes: Chemistry, Production, and Sensory Approach*, First Edition. Edited by Sreeraj Gopi, Nimisha Pulikkal Sukumaran, Joby Jacob, and Sabu Thomas.

to the creation of undesirable materials, thus lowering process yield and increasing the downstream prices [4].

In recent years, researchers and food processing industries are working on biocatalytic flavor extraction and synthesis due to consumer acceptance of natural flavors rather than chemically derived compounds [5]. These methods of flavor extraction techniques are more precise with fewer side reactions and yield better quality flavors [6]. Clear legislation (European Commission, 1996) sets certain procedures for the establishment of a list of authorized flavoring substances. A registry of around 2700 flavorings used by the food industry in or on food products in member states was introduced for the implementation of this regulation [7–9]. A stepwise method developed by the Joint FAO/WHO Expert Committee on Food Additives (JECFA) is the protocol for the safety assessment of flavorings adopted by the European Commission to create a positive list of these substances [10]. Intake data from existing uses, structure–activity interaction, metabolism, and toxicity are incorporated into the JECFA process. Furthermore, the purity and chemical detail specification is evaluated. The belief that there are levels of consumption that do not cause harmful effects for chemicals with established toxicological profiles is implicit in setting appropriate daily intakes (ADIs). The theory of the threshold of toxicological concern (TTC) expands this definition by suggesting that, below a certain threshold value, the risk to human health is probably very low. Researchers have found that the TTC principles can be applied to low concentrations of chemicals that lack toxicity data in food [11]. The maximized survey-derived daily intake (MSDI) approach is used in the protocol developed by the JECFA for the safety assessment of flavoring substances to measure the intake of a given flavor and equate it with the above thresholds [12]. The intake is estimated for consumers only and is believed to be 10% of the population for all flavoring substances used in the food industry.

The US food supply is one of the best and safest as compared to those of other nations worldwide. A well-established food ingredient regulatory scheme put forward by US Congress in the Federal Food, Drug, and Cosmetic Act, which is implemented and run by the Food and Drug Administration (FDA), assures maximum food safety for the consumers. In recent years, a large amount of research has been placed on food products that are licensed as food additives and GRAS substances in the United States. Many reports highlighted questions related to GRAS conclusions, such as a lack of clarity of protocols, the potential for conflict of interest problems due to industry support by GRAS expert committees, the absence of a required re-evaluation after an initial GRAS conclusion that would account for a rise in potential disclosure or the existence of new, appropriate safety data, and a perceived interpretation [7].

For instance, concerns have also been raised about the status of caffeine as an FDA-listed (21 CFR 182.1180) GRAS substance, with the agency taking action to protect consumers from significant overexposure of caffeine in dietary supplement products [13]. There are relatively few examples of permitted food additives or GRAS substances that induce acute or chronic illness within their expected conditions of use, except illness associated with severe overconsumption of a few food ingredients [14]. The irrational fears about food additives and GRAS substances drive focus from food safety concerns that are of great public health importance, most importantly microbial contamination of a range of foods that are widely eaten. The current chapter aims to provide an updated analysis of scientific opinions, regulatory aspects for flavor and fragrance materials, biosynthesis, etc. of the same and to verify its exposure for estimating the safest intake level for the consumer population.

## 9.2 Biosynthesis of Food Flavors

In the food and fragrance industries, flavors set up a fantastic forum. It is important for customers, and without flavors, the consumer's perception of food will be gone. Artificial foodstuffs containing mostly synthetic flavors are avoided. In the food industry, major flavors and fragrances are currently developed by chemical synthesis acting as artificial flavors or naturally. Identical flavors are not eco-friendly [15]. Biosynthesis provides a large number of compounds with flavors of cherry, almond, strawberry, etc. Flavors that are chemically synthesized will also result in environmentally unfriendly processes of production. It also decreases the selectivity of the substrate, which can also contribute to the creation of undesirable materials, thus reducing the process yield. Extraction of flavors from plants also has various problems such as the presence of a low concentration of the desired compound which makes it more expensive [6, 16].

### 9.2.1 Enzymes Used for Food Flavor Synthesis

The use of enzymes or microorganisms in food preparations is an age-old process. But recently, with the advancement of technology, novel enzymes with a wide range of applications and specificity have been developed, and new application areas are still being explored. Microorganisms such as bacteria, yeast, and fungi and their enzymes are widely used in several food preparations for improving the taste and texture, and they offer huge economic benefits to industries [17]. Precisely, enzyme encapsulation and coenzyme regeneration are highly efficient processes for flavor synthesis. Conversion of 1-phenyl-2-propanone with NADH+ H+-dependent yeast alcohol dehydrogenase into (*S*)-1-phenyl-2-propanol; NAD+ regeneration is an example [5, 18]. The cherry and almond-tasting benzaldehyde is produced from the cyanogenic glycoside amygdalin obtained from the cherry kernels and almond meal by using β-glucosidase. Another example is L-menthol, which is the major constituent of peppermint oil [19].

***Lipase***: Lipase is the most widely used flavor enzyme. It is used in the esterification process in organic solvents to produce flavor esters. Some of these esters are isoamyl acetate, isoamyl butyrate, geranyl acetate, and citronellyl acetate [19, 20]. Lipase catalyzed production of flavor esters is impaired by several esterification reactions. Variables such as alcohol molarity, water addition, agitation speed, temperature, and quantity of enzyme immobilized affect this process[21]. Butyl acetate is added to beverages to provide pineapple flavor. This is popular in food, beverage, cosmetic, and pharmaceutical industries. It is synthesized via the green route of enzymatic reaction synthesis. The synthesis of butyl acetate is demonstrated in several research papers. Similarly, lipase from *Rhizomucor miehei,* immobilized on magnetic nanoparticles uses a solvent-free method as a biocatalyst [22]. Thus, flavor esters are selectively hydrolysed and continuously formed by lipases from fatty acids. Yet, despite this high catalytic efficiency of lipases, the main concerns with most reactions are high production costs and the factors linked to its stability.

***Hydrolase*:** The industrial importance of hydrolases exceeds that of other classes of enzymes. Hydrolases are a group of enzymes that catalyze bond cleavages by reaction with water [23]. These enzymes can catalyze hydrolysis of the ester linkages as in the case of triglycerides, for example, vegetable oil which is a combination of fatty acid and glycerol. Thus, esters will be formed by hydrolases when mixed with acid or alcohols in a nonaqueous system. Various flavoring compounds such as ethyl butyrate, isobutyl butyrate, and isoamyl butyrate are synthesized using this method [21, 24].

***Protease***: Proteases are used widely in the food industry for the production of bread, baked foods, crackers, and waffles, where these enzymes are used to reduce the mixing time, decrease dough consistency and uniformity, regulate the gluten strength in bread, and improve the texture and flavor [17]. Flavor production in soy sauce, fish sauce, nato, tempeh, and miso is achieved by these enzymes. Esterification and transesterification are responsible for flavor production. Protein-mediated hydrolases enhance and add flavors to food by protease-mediated hydrolysis. A key disadvantage of hydroxylated flavors of protein is that they will cause bitterness. This bitterness can be managed by vacuum hydrolyzation that will induce bitterness [25]. The processing treatment, for instance, enzyme treatment of protease for crayfish by-products, permitted the concentration of flavors to be increased, such as benzaldehyde and pyrazine [26]. Savory flavors are created through the heating of protein compounds in acidic pH. Protease is also responsible for cheese production [27].

### 9.2.2 Biosynthesis of Flavors by Fermentation

The most frequent development of flavors by fermentation consists of cheese, wine, and other alcoholic goods. Sources range from various sources to yeast, bacteria, and molds (fungi). This fermentation can sometimes also lead to the production of some beneficial bacteria for humans, like lactic acid bacteria. Though off-flavors are often produced by fermentation, yeast plays a major role in wine production and offers a distinctive flavor when fermented [28]. *Saccharomyces cerevisiae* is one of the microflorae most frequently used in fermentation. Similarly, Basidiomycetes, which comprise almost all edible mushrooms, possess a unique extracellular enzyme system (the so-called secretome) and have already been shown to produce flavors de novo by biotransformation [29]. In the fermentation medium with dissolved oxygen and acetyl CoA, it is also possible for fungi to convert compounds into flavors via the fermentation process. Intracellular enzymes and metabolites are also produced during fermentation. In such cases, yeast as a flavoring agent itself produces a fruity flavor in beer [30]. Volatile acids and esterification processes are used to produce this fruity taste, followed by the formation of volatile esters. Similarly, two strategies are followed in vanillin production. One of the most prevalent is with *Aspergillus niger*, which transforms ferulic acid to vanillic acid, and *Pycnoporus cinnabarinus* further reduces the vanillic acid to vanillin. Similarly, vanillin synthesis is done by bioconversion of eugenol to vanillin via ferulic acid through bacterial strains (Arthrobacter, Corynebacterium, or Pseudomonas). It also appears that bacteria transform eugenol to vanillic acid [31]. The *Pseudomonas putida* strain can catalyze much better than the yeast. Ferulic acid is effectively converted into vanillic acid, and ferulic acid is converted into vanillin by a strain of *Streptomyces setonii* [32].

### 9.2.3 Production of Flavors from Agro Waste

Vanillin can be synthesized from agricultural waste through the use of vanilla bean and vanillin plant caps. It is generated with the support of microorganisms that can transform the precursor such as ferulic acid. Material waste used in the wood pulp industry is a possible source of lignin to generate biovanillin. Study shows that ferulic acid can be obtained from agro waste, which includes lignocellulose waste [33]. The functional feature has been genetically modified by inserting *Pseudomonas fluorescens* BF13 genes for the development of vanillin. There are other agricultural by-products, such as rice bran, which are produced more than 10 000 000 tonnes a year [34]. In China, the refining industry also produces abundant ferulic acid. In addition, arabinofuranosyl residue is esterified to heteroxylans in the cell walls of cereal grains [35]. Ferulic acid has also been found in abundance in rice bran oil (crude) waste residue. The potential of *Ceratocystis fimbriata* to grow on different industrial wastes is used for the manufacture of aroma production and fragrance. The development of fruity aromas is carried out by *C. fimbriata* in solid-state fermentation [36]. Similarly, cassava bagasse and sugar cane bagasse produce a strong fruity aroma with the help of *C. fimbriata*. Glucose introduction into the solid medium produces an intense fragrance, which is influenced by this xylanase enzyme necessary for the release of ferulic acid [37].

### 9.2.4 Production of Flavors through Plant Cells

The main sources of flavor compounds are plant and plant products that contain many compounds such as basic fragrances and oils. Fruit structures such as furanones and pyrones are obtained from many species of tree bark and leaves [38]. These are flavor compounds derived from carbohydrates and consist of a mild odor. Furanones are mainly present alone in the fruit portion, whereas pyrones are obtained from barks and leaves of plant species. Furanones are another flavor compound developed from furanol [39]. Furanol plays a major role in the production of attractive main fruit flavors. Terpenoids are the most common flavoring compounds that also function as a bioactive substance and as a drug for the treatment of different diseases in humans [40]. Sweet basil's glandular trichomes are rich in both monoterpenes, sesquiterpenes, and phenylpropenes. Terpenes are also widely available and commercially used as a compound for flavoring, insecticides, and perfumes, and they are mainly used as flavor raw materials [41]. Production of cinnamaldehyde compounds starts with the assistance of compounds from cinnamon and benzaldehyde from bitter almond oil; this single chemical from the plant is responsible for the whole characteristic flavors and aroma [42]. The entirety or synergism of multiple chemical constituents from the plant is responsible for its typical flavors and fragrances. However, a single compound also contributes to characteristic flavors, for instance, nootkatone found in grapefruit essential oil is responsible for its characteristic flavor [43]. Woody aroma is obtained from a plant called vetiver (*Chrysopogon zizanioides*), where β-vetivone is the bioactive compound responsible for the characteristic woody fragrance of the plant and is one of the key components present in vetiver essential oil [44].

## 9.3 Safety Evaluation of Added Flavors by FDA

Flavors and fragrance materials are applied to food products as mixtures of individual compounds or a group of compounds generically known as "flavoring ingredients." Flavor ingredients can be either substances that impart or modify flavor profile of the processed food or substances that promote the functioning of the flavoring substances in a mixture [12]. Compound flavoring substances may include substances with chemically elucidated fragrance such as vanillin and natural aromatic chemical complexes or as botanical extracts (e.g. vanilla extract) or essential oils (e.g. orange and other citrus oils) [13]. According to the definitions made by FDA, the "flavoring agents and adjuvants" are said to be "compounds that are added to food products which impart or help to impart a taste or aroma to food." In contrast to the terms used in the FEMA GRAS program, Flavor and Extract Manufacturers Association (FEMA)'s flavoring compounds – substances that convey or change the taste – are considered as flavoring agents by FEMA [7]. The flavorings and fragrance materials when incorporated with certain food ingredients can serve multiple functions. If, therefore, a GRAS inference is the reason for the use of that material by the regulatory authority, then only the designated technical consequence, the basic purpose and use, that is GRAS [14, 45].

According to FDA and GRAS conclusions on food flavors and fragrance materials, human consumption estimates are a critical component of the safety assessment [14]. One critique of the GRAS findings is that the possible human consumption of GRAS food products is impossible to accurately predict, a presumed issue compounded by the supposed lack of knowledge of the identity of GRAS substances found in the US food supply [46]. FEMA has performed periodic studies for several years on the volumes of FEMA GRAS flavor ingredients that regularly "disappear" into the US food chain. Thus, a reliable and conservative method of human dietary exposure can be estimated using such survey results. The importance of the FEMA surveys was recognized by the FDA in its final rule on the petition by NGO. FEMA requires applications for GRAS status to include an estimated amount of usage forecast for fresh GRAS flavor ingredients. There is no poundage survey information available for spice ingredients not yet in production [47]. The findings and significance of genotoxicity assays on flavor ingredients have been regularly assessed by the FEMA Expert Team. Moreover, the genotoxic ability of the synthetic flavoring food additives was subjected to the NGO petition and was tested by the FDA. The FDA concluded that in the related tests, five of the compounds (benzophenone, ethyl acrylate, myrcene, pulegone, and pyridine) did not exhibit genotoxic capacity [48].

GRAS tests of flavor additives are performed by the FEMA Expert Panel only under their expected conditions of use in human foods. Food is explained as (i) Items used for food or drinks for humans or for human or other animal eats; (ii) Chewing gum, and; (iii) Articles used as components of any such article. The main feature of a GRAS decision is the terms of the planned application of a food product. The circumstances in which a substance is determined to be GRAS are essential for complying with the regulatory criteria defined in FFDCA Section 201 since it is the application of the substance which is GRAS and not the substance itself (s). There are three elements to the criteria of planned use for the GRAS status of a food ingredient; (i) the technological result designated for the food; (ii) the usage of defined categories of food; and (iii) the levels of usage that correlate to the specified food categories. According to FEMA GRAS status, to apply these categories of conditions for the food applications, flavor and food ingredients are generally classified into two categories: (1) substances that impart or alter flavor are flavoring substances and (ii) adjuvants used for taste, including emulsifiers, preservatives, or solvents that enable the flavoring substances to work in the finished food. Other applications of "FEMA GRAS flavor additives" must have "regulatory authority to use". The food additive status as determined by the FDA can be included by another regulatory authority. Submission to the FDA voluntary notice service for GRAS and acceptance of a letter of "no objection" from the FDA require an individual decision of GRAS that fulfills the FFDCA Section 201(s), the legislative requirements of the GRAS provision, and that follows the rules and policies of the FDA on GRAS conclusions.

The various food authorities have given a green signal to flavors that are widely used in food depending on permissible levels. Consumption of flavors beyond permissible levels result in problems like allergic reaction, food hypersensitivity, worsening of asthmatic systems, abdominal pain, diarrhea, and vomiting [49]. Some of the flavors that are commonly used in F&F industries and permission levels are briefed in Table 9.1.

Diacetyl (2,3-butanedione) is a well-known volatile aromatic compound that is mainly present in butter, cheese, etc. which is very significant in the food industry. It has many advantages like a more consistent quality product, buttermilk flavor, etc. This is also an important compound for microwave popcorn, but the American conference of governmental industrial hygienists has found out that inhalation of this can cause lung diseases [50]. In beer, this can impart a butterscotch flavor. The

**Table 9.1** Table for permissible limits of various flavors.

| Flavors | Permissible limits | Application | Remarks |
|---|---|---|---|
| Diacetyl [50] | 0.02–16.7 ppm | Important aroma compounds in butter, cheese, margarine, etc. | |
| Isoamyl acetate [51] | 100 ppm | Used to confer banana flavor in food | Masks the flavor of banana |
| Benzaldehyde [52] | | Provide almond flavor to fruit juices, baked goods | Masks the odor of almond |
| Cinnamic Aldehyde [53] | 1100–6400 ppm | Replacement of chemical synthetic preservative | |
| Ethyl propionate [49] | | Used in ready to drink beverages | Masks the flavor of pineapple |
| Methyl anthranilate [54] | 17.5–33 ppm | Used in soft drinks, candies, chewing gum, etc. | Masks grape flavor |
| Limonene [55] | 31–68 ppm | Used in beverages, chewing gum, etc. | Masks lime flavor |
| Ethyl decadienoate [56] | 5 ppm | | Masks the flavor of green apple |
| Allyl hexanoate [57] | 500 ppm | Used in pineapple juice, beverages, and yogurts containing pineapple | Masks the flavor of pineapple |
| Ethyl maltol [58] | 20–110 ppm | Used in candies, ice creams, dairy products, etc. | Masks the odor of caramelized sugar |
| Ethylvanillin [59] | 100–800 ppm | Used in chocolates, sweets, ice creams, etc. | Provides vanilla flavor to the products |
| Methyl salicylate [60] | 10–5000 ppm | Used in chewing gums, baked goods, candies, etc. | Masks the flavor of wintergreen |

fermentation temperature, contamination of bacteria, and aeration level are some of the factors that influence the level of beer. In home-brewed beer, the level can be 0.05 to greater than 1 ppm [61]. Isoamyl acetate, also known as isopentyl acetate, is an organic compound that produces odor in fruits like bananas and pears. When this is added to the bacterial growth media, it can efficiently be imported into the cells of bananas, and it gets converted to isoamyl acetate which in turn imparts flavor [51]. A high concentration of this can cause headache, drowsiness, dizziness, and fatigue. Studies also suggested that extremely high levels can result in death. Benzaldehyde is yet another aromatic aldehyde that can be used as a flavoring and a fragrance agent, which gives an almond-like odor [52].

Natural benzaldehyde can be obtained from the reaction of cinnamaldehyde, i.e. the hydrolysis of cinnamaldehyde with water undergoes retro-aldol condensation, which is a reverse reaction between aldol condensation and aldol reaction with acid or base catalyst in an aqueous medium [53]. It can also be obtained from kernels of apricot, peaches, etc. by various kinds of reactions. Natural benzaldehyde obtained from this can be used as a flavor enhancer and also as an ingredient for other kinds of fruity flavors [52]. These compounds are also known to produce various kinds of health hazards to humans. Oral intake of high doses can create problems like kidney damage. Studies have reported that the flavor of cinnamon can be due to the presence of oils. One of the main compounds in oil is cinnamaldehyde. The IUPAC name of this compound is 3-phenyl prop-2-enol with a boiling point of 246 °C [53]. As the name implies this gives cinnamon its flavor and odor. The occurrence of these flavors can be seen in trees like cinnamon, cassia, and camphor. This can also be used as a flavoring agent in food items like ice cream, candy, and beverages. [61]. In baby foods, it is allowed in the range of 2000 ppm, whereas in baked goods and breakfast cereals, it is allowed in the range of 3500 and 2200 ppm, respectively. Studies have proven that 180 000 kg is consumed in food each year, and its usage has increased by 3% per year [53]. The toxicity of this is comparatively low. A joint FAO/WHO expert committee on food additives set an ADI value for this as 0.7 mg per kg body weight. In some people, this can create allergy with symptoms like swells, and itching [62].

Ethyl propionate is also a flavoring agent with a transparent and colorless liquid with a melting point of −73 °C. The pineapple-like odor of this compound makes this perfect for use as a flavoring agent in fruit syrups. It is mainly used for creating both rum and fruity notes in products. In the gas chromatography method, the maximum peak area of this compound in the production of commercial rum is 7.476% [56]. Methyl anthranilate, also known as methyl 2-aminobenzoate, is an ester of anthranilic acid that can be used as the main contributor in fruit juices because of its unique and pleasant fruity grape smell. It is mainly used in products like candy, chewing gum, soft drinks, and alcoholic drinks. [63]. It can occur in essential oils of various plant species like neroli, jasmine, and bergamot, and it is commonly produced by chemical methods like petroleum-based processes. Apart from flavoring agents, it can also be used as an oxidative inhibitor in pharmaceuticals. Even though it is known to be present in nature extensively, it cannot be extracted directly or conveniently because of its low level. A joint FAO/WHO expert committee on food additives has set an ADI value of 0–1.5 mg per kg body weight [62]. Limonene is an oil extracted from citrus fruits like oranges. This comes under the compound known as terpenes which protect the plant from its predators. The two isomers of limonene are D- and L-limonene. The main compound D-limonene is a colorless liquid with a lemon-like odor that is extensively used as a flavoring agent [55].

Several studies have proven that temperature is the major factor in the antimicrobial activity of limonene. The presence of other compounds can also increase this activity in limonene. The FDA considers this a safe food additive, but consumption of 20 g of D-limonene can cause diarrhea and can increase the protein level of urine in humans. Ethyl decadienoate is an organic chemical compound that can be used as a flavoring and fragrance agent because of its pear-like taste and odor. Allyl hexanoate is an organic compound found to occur naturally in pineapples. Trace levels of allyl hexanoate are found in fruits like mango, acerola, and some varieties of mushrooms. Allyl hexanoate can also be used as a flavoring agent in chewing gums, candies, and baked goods. JECFA has given an ADI value of 0–130 mg per kg body weight [57]. Ethyl maltol is an organic compound that is used as a flavoring in confectionaries like candies, ice creams, and dairy products. This can also be used to cover the odor of highly concentrated products. When compared with maltol, this compound has a sweeter caramel-like odor and fruity taste. The caramel-like presence of this compound in sweet and cotton candy is found to be 4.2 mg $ml^{-1}$. JECFA in 1974 has set an ADI value of 0.2 mg per kg body weight, and they also concluded that there is no safety concern when taken at an estimated level as a flavoring substance [58].

## 9.4 Conclusion

Usage of flavors and fragrances is increasing day by day in the production of various food products. The main idea underlying the incorporation of flavors is to deliver appealing food products by masking the unpleasant odor. Various regulations made by FDA and FEMA are said to be some of the important parameters that should be known before the addition of certain flavors and fragrances. Knowledge of various flavors and their permissible limits helps in preventing health problems that arise through the addition of flavors in excess dosage. More research has to be done to understand the interactions of flavors in different foods, as long as flavors play an important role in foods.

## References

1 Skovgaard, N. (2003). Safety evaluation of certain mycotoxins in food: Prepared by the fifty-sixth meeting of the joint FAO/WHO Expert Committee on Food Additives (JECFA). WHO Food Additives Series: 47; FAO Food and Nutrition Paper: 74; IPCS-International Programme on Chemical Safety, WHO Geneva, 2001, 701 pages; WHO ISBN 92 4 166047 3, *Int. J. Food Microbiol.*, 81(2). 176, ISSN 0168-1605, https://doi.org/10.1016/S0168-1605(02)00160-5.

2 EFSA Panel on Food Additives and Nutrient Sources Added to Food (2003, 2003). Opinion of the scientific panel on food additives, flavourings, processing aids and materials in contact with food (AFC) related to L-carnitine-L-tartrate for use in foods for particular nutritional purposes. *EFSA J.* 1 (12): 19–13. https://doi.org/10.2903/j.efsa.2003.19.

3 Teixeira, M.I., Andrade, L.R., Farina, M., and Rocha-Leão, M.H.M. (2004). Characterization of short chain fatty acid microcapsules produced by spray drying. *Mater. Sci. Eng., C* 24 (5): 653–658, ISSN 0928-4931, https://doi.org/10.1016/j.msec.2004.08.008.

4 Krings, U. and Berger, R. (1998). Biotechnological production of flavours and fragrances. *Appl. Microbiol. Biotechnol.* 49: 1–8. https://doi.org/10.1007/s002530051129.

5 Vandamme, E.J. and Soetaert, W. (2002). Bioflavours and fragrances via fermentation and biocatalysis. *J. Chem. Technol. Biotechnol.* 77: 1323–1332. https://doi.org/10.1002/jctb.722.

6 Janssens, L., De Pooter, H.L., Schamp, N.M., and Vandamme, E.J. (1992). Production of flavours by microorganisms. *Process Biochem.* 27 (4): 195–215, ISSN 1359-5113, https://doi.org/10.1016/0032-9592(92)80020-4.
7 Neltner, T.G., Alger, H.M., O'Reilly, J.T. et al. (2013). Conflicts of interest in approvals of additives to food determined to be generally recognized as safe: out of balance. *JAMA Intern. Med.* 173 (22): 2032–2036. https://doi.org/10.1001/jamainternmed.2013.10559, PMID: 23925593.
8 European Commission, (1999). Commission decision 1999/217/EC of 23 February 1999 adopting a register of flavouring substances used in or on foodstuffs. Official Journal of the European Communities, L 84, pp. 1–137. http://data.europa.eu/eli/dec/1999/217/oj (accessed 11 January 2022)
9 European Commission, (2002). Commission Decision 2002/113/EC of 23 January 2002 amending Commission Decision 1999/217/EC as regards the register of flavouring substances used in or on foodstuffs. Official Journal of the European Communities, L 49. pp. 1–160. https://www.fao.org/faolex/results/details/en/c/LEX-FAOC042626 (accessed 11 January 2022).
10 Joint FAO/WHO Expert Committee on Food Additives (1996: Geneva, Switzerland), World Health Organization & Food and Agriculture Organization of the United Nations (1997). Evaluation of certain food additives and contaminants: forty-sixth report of the Joint FAO/WHO Expert Committee on Food Additives. World Health Organization. https://apps.who.int/iris/handle/10665/41962 (accessed 11 January 2022).
11 Kroes, R., Renwick, A.G., Cheeseman, M. et al. (2004 Jan). European branch of the international Life Sciences Institute. Structure-based thresholds of toxicological concern (TTC): guidance for application to substances present at low levels in the diet. *Food Chem. Toxicol.* 42 (1): 65–83. https://doi.org/10.1016/j.fct.2003.08.006. PMID: 14630131.
12 Munro, I.C., Kennepohl, E., and Kroes, R. (1999). A procedure for the safety evaluation of flavouring substances. Joint FAO/WHO Expert Committee on Food Additives. *Food Chem. Toxicol.* 37 (2–3): 207–232. https://doi.org/10.1016/s0278-6915(98)00112-4. PMID: 10227744.
13 FDA News Release (April 12, 2018). FDA takes step to protect consumers against dietary supplements containing dangerously high levels of extremely concentrated or pure caffeine. https://www.fda.gov/news-events/press-announcements/fda-takes-step-protect-consumers-against-dietary-supplements-containing-dangerously-high-levels (accessed 11 January 2022).
14 Bendig, P., Maier, L., and Vetter, W. (2012). Brominated vegetable oil in soft drinks – an underrated source of human organobromine intake. *Food Chem.* 133 (3): 678–682, ISSN 0308-8146, https://doi.org/10.1016/j.foodchem.2012.01.058.
15 Forti, L., Di Mauro, S., Cramarossa, M.R. et al. (2015). Non-conventional yeasts whole cells as efficient biocatalysts for the production of flavors and fragrances. *Molecules* 20 (6): 10377–10398. https://doi.org/10.3390/molecules200610377.
16 Ly, M.H., Aguedo, M., Goudot, S. et al. (2008). Interactions between bacterial surfaces and milk proteins, impact on food emulsions stability. *Food Hydrocolloids* 22: 742–751.

**17** Raveendran, S., Parameswaran, B., Ummalyma, S.B. et al. (2018). Applications of microbial enzymes in food industry. *Food Technol. Biotechnol.* 56 (1): 16–30. https://doi.org/10.17113/ftb.56.01.18.5491.

**18** Kragl, U., Kruse, W., Hummel, W., and Wandrey, C. (1996). Enzyme engineering aspects of biocatalysis: cofactor regeneration as example. *Biotechnol. Bioeng.* 52 (2): 309–319. https://doi.org/10.1002/(SICI)1097-0290(19961020)52:2<309::AID-BIT11>3.0.CO;2-E. PMID: 18629898.

**19** Schreier, P. (1997). Enzymes and flavour biotechnology. In: Berger, R. G. (Volume Editor). In: *Biotechnology of Aroma Compounds. Advances in Biochemical Engineering/Biotechnology*, vol. 55 (ed. W. Babel, H.W. Blanch, C.L. Cooney, et al.). Berlin, Heidelberg: Springer https://doi.org/10.1007/BFb0102062.

**20** Gatfield, I.L. (1999). Biotechnological production of natural flavor materials. In: *Flavor Chemistry* (ed. R. Teranishi, E.L. Wick and I. Hornstein). Boston, MA: Springer https://doi.org/10.1007/978-1-4615-4693-1_19.

**21** Singh, A., Naik, V., Lal, S. et al. (2011). Improving the efficiency of healthcare delivery system in underdeveloped rural areas. In: *2011 Third International Conference on Communication Systems and Networks (COMSNETS 2011)*, IEEE, 1–6. https://doi.org/10.1109/COMSNETS.2011.5716519.

**22** Güvenç, A., Kapucu, N., and Mehmetoğlu, Ü. (2002). The production of isoamyl acetate using immobilized lipases in a solvent-free system. *Process Biochem.* 38 (3): 379–386, ISSN 1359-5113, https://doi.org/10.1016/S0032-9592(02)00099-7.

**23** Alcántara, A.R., Hernaiz, M.-J., and Sinisterra, J.-V. (2011). Chapter 3.28 – Biocatalyzed production of fine chemicals. In: *Comprehensive Biotechnology*, 2e (ed. M. Moo-Young), 309–331. Academic Press, ISBN 9780080885049, https://doi.org/10.1016/B978-0-08-088504-9.00225-7.

**24** Welsh, F.W., Murray, W.D., Williams, R.E., and Katz, I. (1989). Microbiological and enzymatic production of flavor and fragrance chemicals. *Crit. Rev. Biotechnol.* 9 (2): 105–169. https://doi.org/10.3109/07388558909040617.

**25** Wilmon, M. (1983). 4361586 Vacuum enzymatic digestion of protein material. *Biotechnol. Adv.* 1 (1): 153–154, ISSN 0734-9750, https://doi.org/10.1016/0734-9750(83)90520-7.

**26** Baek, H.H. and Cadwallader, K.R. (1996). Volatile compounds in flavor concentrates produced from crayfish-processing byproducts with and without protease treatment. *J. Agric. Food. Chem.* 44 (10): 3262–3267. https://doi.org/10.1021/jf960023q.

**27** Aaslyng, M.D., Stephen Elmore, J., and Mottram, D.S. (1998). Comparison of the aroma characteristics of acid-hydrolyzed and enzyme-hydrolyzed vegetable proteins produced from soy. *J. Agric. Food. Chem.* 46 (12): 5225–5231. https://doi.org/10.1021/jf9806816.

**28** Liguori, L. and Minisci, F. (2000). High selectivity in the oxidation of mandelic acid derivatives and in O-methylation of protocatechualdehyde: new processes for synthesis of vanillin, iso-vanillin, and heliotropin. *Org. Process Res. Dev.* 4: 534–543.

**29** Bosse, A.K., Fraatz, M.A., and Zorn, H. (2013). Formation of complex natural flavours by biotransformation of apple pomace with basidiomycetes. *Food Chem.* 141 (3): 2952–2959. https://doi.org/10.1016/j.foodchem.2013.05.116, PMID: 23871045.

**30** Vasantha Rupasinghe, H.P., Joshi, V.K., Smith, A., and Parmar, I. (2017). Chapter 3 – Chemistry of fruit wines. In: *Science and Technology of Fruit Wine Production* (ed. M.R. Kosseva, V.K. Joshi and P.S. Panesar), 105–176, ISBN 9780128008508. Academic Press https://doi.org/10.1016/B978-0-12-800850-8.00003-X.

**31** Zheng, L., Zheng, P., Sun, Z. et al. (2007). Production of vanillin from waste residue of rice bran oil by *Aspergillus niger* and *Pycnoporus cinnabarinus*. *Bioresour. Technol.* 98 (5): 1115–1119. https://doi.org/10.1016/j.biortech.2006.03.028. PMID: 16782330.

**32** Gallage, N.J. and Møller, B.L. (2015). Vanillin–bioconversion and bioengineering of the most popular plant flavor and its de novo biosynthesis in the vanilla orchid. *Mol. Plant* 8 (1): 40–57, ISSN 1674-2052. https://doi.org/10.1016/j.molp.2014.11.008.

**33** Stentelaire, C., Lesage-Meessen, L., Oddou, J. et al. (2000). Design of a fungal bioprocess for vanillin production from vanillic acid at scalable level by *Pycnoporus cinnabarinus*. *J. Biosci. Bioeng.* 89 (3): 223–230, ISSN 1389-1723, https://doi.org/10.1016/S1389-1723(00)88823-4.

**34** Havkin-Frenkel, D. and Belanger, F.C. (2016). Chapter 5 – Biotechnological production of vanillin. In: *Biotechnology in Flavor Production* (ed. D. Havkin-Frenkel and N. Dudai). Wiley. ISBN: 978-1-118-35405-6.

**35** Nieman, K.M., Rains, T.M., Dicklin, M.R., and Maki, K.C. (2014). Chapter 5 – Wheat fiber in postprandial metabolic profile and health. In: *Wheat and Rice in Disease Prevention and Health* (ed. R.R. Watson, V.R. Preedy and S. Zibadi), 59–66. Academic Press, ISBN 9780124017160, https://doi.org/10.1016/B978-0-12-401716-0.00005-2.

**36** Bramorski, A., Soccol, C.R., Christen, P., and Revah, S. (1998). Fruity aroma production by *Ceratocystis Fimbriata* in solid cultures from agro-industrial wastes, industrial microbiology. *Rev. Microbiol.* 29 (3): https://doi.org/10.1590/S0001-37141998000300012.

**37** McAndrew, B.A. (1992). *Perfumes: Art, Science and Technology* (ed. P.M. Muller and D. Lamparsky). London, New York: Elsevier Applied Science, 1991. No. of pages: VIII + 658, price US$190.00. £95.00. ISBN 1–85166–573–0. Flavour Fragr. J., 7: 239–240. https://doi.org/10.1002/ffj.2730070414.

**38** Turner, G.W. and Croteau, R. (2004). Organization of monoterpene biosynthesis in Mentha. Immunocytochemical localizations of geranyl diphosphate synthase, limonene-6-hydroxylase, isopiperitenol dehydrogenase, and pulegone reductase. *Plant Physiol.* 136 (4): 4215–4227. https://doi.org/10.1104/pp.104.050229.

**39** Guentert, M. (2007). The flavour and fragrance industry—past, present, and future. In: *Flavours and Fragrances* (ed. R.G. Berger). Berlin, Heidelberg: Springer https://doi.org/10.1007/978-3-540-49339-6_1.

**40** Modzelewska, A., Sur, S., Kumar, S.K., and Khan, S.R. (2005). Sesquiterpenes: natural products that decrease cancer growth. *Curr. Med. Chem. Anticancer Agents* 5 (5): 477–499. https://doi.org/10.2174/1568011054866973. PMID: 16178774.

**41** Iijima, Y., Davidovich-Rikanati, R., Fridman, E. et al. (2004). The biochemical and molecular basis for the divergent patterns in the biosynthesis of terpenes and phenylpropenes in the peltate glands of three cultivars of basil. *Plant Physiol.* 136 (3): 3724–3736. https://doi.org/10.1104/pp.104.051318. PMID: 15516500; PMCID: PMC527170.

**42** Remaud, G., Debon, A.A., Martin, Y.-l. et al. (1997). Authentication of bitter almond oil and cinnamon oil: application of the SNIF-NMR method to benzaldehyde. *J. Agric. Food. Chem.* 45 (10): 4042–4048. https://doi.org/10.1021/jf970143d.

**43** Paula Dionísio, A., Molina, G., Souza de Carvalho, D. et al. (2012). Chapter 11 – Natural flavourings from biotechnology for foods and beverages. In: *Woodhead Publishing Series in Food Science, Technology and Nutrition, Natural Food Additives, Ingredients and Flavourings* (ed. D. Baines and R. Seal), 231–259. Woodhead Publishing, ISBN 9781845698119, https://doi.org/10.1533/9780857095725.1.231.

**44** Gupta, P. and Phulara, S. (2021). Chapter 3 – Terpenoids: types and their application. In: *Biotechnology of Terpenoid Production from Microbial Cell Factories* (ed. P. Gupta and S. Phulara), 47–78. Academic Press, ISBN 9780128199176, https://doi.org/10.1016/B978-0-12-819917-6.00006-5.

**45** Hallagan, J.B. and Hall, R.L. (2009). Under the conditions of intended use – new developments in the FEMA GRAS program and the safety assessment of flavor ingredients. *Food Chem. Toxicol.* 47 (2): 267–278. https://doi.org/10.1016/j.fct.2008.11.011. PMID: 19041920.

**46** Neltner, T.G., Kulkarni, N.R., Alger, H.M. et al. Navigating the U.S. Food Additive Regulatory Program. *Compr. Rev. Food Sci. Food Saf.* 10: 342–368. https://doi.org/10.1111/j.1541-4337.2011.00166.x.

**47** Smith, T.R.L., Adams, T.B., Cohen, S.M. et al. (2004). Safety evaluation of natural flavour complexes. *Toxicol. Lett.* 149 (1–3): 197–207, ISSN 0378-4274, https://doi.org/10.1016/j.toxlet.2003.12.031.

**48** Gooderham, N.J., Cohen, S.M., Eisenbrand, G. et al. (2020). The safety evaluation of food flavoring substances: the role of genotoxicity studies. *Crit. Rev. Toxicol.* 50 (1): 1–27. https://doi.org/10.1080/10408444.2020.1712589.

**49** Gries, G., Gries, R., Perez, A.L. et al. (1994). Ethyl propionate: synergistic kairomone for African palm weevil, *Rhynchophorus phoenicis* L. (Coleoptera: Curculionidae). *J. Chem. Ecol.* 20 (4): 889–897. https://doi.org/10.1007/BF02059585. PMID: 24242203.

**50** Clark, S. and Winter, C.K. (2015). Diacetyl in foods: a review of safety and sensory characteristics. *Compr. Rev. Food Sci. Food Saf.* 14: 634–643. https://doi.org/10.1111/1541-4337.12150.

**51** Furukawa, S. (2012). Chapter 8 – Sake: quality characteristics, flavour chemistry and sensory analysis. In: *Woodhead Publishing Series in Food Science, Technology and Nutrition, Alcoholic Beverages* (ed. J. Piggott), 180–195. Woodhead Publishing, ISBN 9780857090515, https://doi.org/10.1533/9780857095176.2.180.

**52** Donald, J.B. (1976). Benzaldehyde, food and cosmetics toxicology. 14 (Supplement): 693–698, ISSN 0015-6264, https://doi.org/10.1016/S0015-6264(76)80079-X.

**53** Sun, Q., Li, J., Sun, Y. et al. (2020). The antifungal effects of cinnamaldehyde against *Aspergillus niger* and its application in bread preservation. *Food Chem.* 317: 126405, ISSN 0308-8146, https://doi.org/10.1016/j.foodchem.2020.126405.

**54** Scott, R.D. (1923). Methyl anthranilate in grape beverages and flavors. *Ind. Eng. Chem.* 15 (7): 732–733. https://doi.org/10.1021/ie50163a031.

**55** Ravichandran, C., Badgujar, P.C., Gundev, P., and Upadhyay, A. (2018). Review of toxicological assessment of d-limonene, a food and cosmetics additive. *Food Chem. Toxicol.* 120: 668–680. https://doi.org/10.1016/j.fct.2018.07.052. PMID: 30075315.

**56** Gatfield, I. and Kindel, G. (1998). Process for the Preparation of trans-2, cis 4-Decadienic acid ethylester. EP 0770685, assigned to Haarman & Reimer.

**57** Raffo, A., D'Aloise, A., Magrì, A.D., and Leclercq, C. (2012). Quantification of allyl hexanoate in pineapple beverages and yogurts as a case study to characterise a source of uncertainty in dietary exposure assessment to flavouring substances. *Food Addit. Contam., Part A* 29 (1): 43–53. https://doi.org/10.1080/19440049.2011.623285.

**58** Kocadağlı, T. and Gökmen, V. (2019). Caramelization in foods: a food quality and safety perspective. In: *Encyclopedia of Food Chemistry* (ed. L. Melton, F. Shahidi and P. Varelis), 18–29. Academic Press, ISBN 9780128140451, https://doi.org/10.1016/B978-0-08-100596-5.21630-2.

**59** Vijayalakshmi, S., Disalva, X., Chittaranjan, S., and Arun, A. (2019). Vanilla-natural vs artificial: a review. *Res. J. Pharm. Technol.* 12 (6): 3068–3072. https://doi.org/10.5958/0974-360X.2019.00520.1.

**60** Greene, T., Rogers, S., Franzen, A., and Gentry, R. (2017). A critical review of the literature to conduct a toxicity assessment for oral exposure to methyl salicylate. *Crit. Rev. Toxicol.* 47 (2): 98–120. https://doi.org/10.1080/10408444.2016.1236071.

**61** Lee, H.H., Lee, K.T., and Shin, J.A. (2017). Analytical method validation and monitoring of diacetyl in liquors from Korean market. *Food Sci. Biotechnol.* 26 (4): 893–899. https://doi.org/10.1007/s10068-017-0124-0.

**62** WHO (2021). Joint FAO/WHO Expert Committee On Food Additives, Ninety-first meeting (Safety evaluation of certain food additives, and contaminants), Virtual meeting.

**63** Api, A.M., Belsito, D., Bhatia, S. et al. (2015). RIFM fragrance ingredient safety assessment, ethyl anthranilate, CAS registry number 87-25-2. *Food Chem. Toxicol.* 82 (Supplement): S97–S104, , ISSN 0278-6915. https://doi.org/10.1016/j.fct.2015.03.017.

# 10

# Sensory Science and its Perceptual Properties

*Constantina Tzia, Virginia Giannou, Tryfon Kekes, Charikleia Chranioti, and Maria Katsouli*

*National Technical University of Athens, Laboratory of Food Chemistry and Technology, School of Chemical Engineering, 5 IroonPolytechniou St., Polytechnioupoli, Zografou, 15772, Athens, Greece*

## 10.1 Introduction

Consumers choose foods based on a number of factors, which in sum can be thought of as "quality." Quality of foods, in specific, includes those attributes of the product, which influence its selection and acceptability by the consumers, such as the sensorial, nutritional, safety and healthy/functional characteristics, shelf life, and quantitative factors (weight and composition). However, the central core of foods' quality is designated by their sensorial characteristics such as appearance, shape, size, color, texture, kinesthetic, taste, odor, and flavor, since they are the first to be perceived. Sensorial characteristics are easily evaluated by the human senses, thereby affecting the products' acceptability in the food market. Among all the sensorial characteristics, flavor, which is defined as the combination of taste and odor (aroma), is of greater interest for consumers' acceptance [1].

The sensorial or organoleptic attributes of foods are evaluated through sensory analysis. Sensory testing has been developed as a formalized, structured, and codified methodology using trained assessors and specific sensory tests. Especially for flavor, analytical techniques are also used for the identification and quantification of aroma compounds.

Quality is of primary importance to both the food industry and consumers. Food industry, being one of the most powerful industries nowadays, designs products' quality and determines their specifications with the sensorial attributes being of primal importance [2]. Sensory analyses are performed for the needs of the quality control and sensory techniques are also used in products' research and development [3].

*Natural Flavours, Fragrances, and Perfumes: Chemistry, Production, and Sensory Approach*, First Edition. Edited by Sreeraj Gopi, Nimisha Pulikkal Sukumaran, Joby Jacob, and Sabu Thomas.

## 10.2 Sensorial Characteristics

The sensorial attributes or characteristics are those perceived by the visual, olfactory, gustatory, tactile/kinesthetic, and auditory senses [4–7]. Perceived quality includes sensorial and non-sensorial factors, which in turn are subject to individual differences and situational factors. However, the taste, smell, texture, flavor, and appearance of food contribute greatly to the consumers' perception of its quality comprising its "sensorial quality." Consumed foods and drinks should not only be nutritious but also be appealing to the consumers. Consumers' experiences from food consumption govern future selection and expectation over food products. Sensorial attributes in agricultural fresh produce vary depending on different factors, while postharvest deterioration affects them negatively. Processing and packaging of foods aim at delaying, inhibiting, or reducing the deterioration processes and maintaining the sensorial properties for an extended time period. The sensorial attributes are of decisive importance for the definition of the shelf life of food products [8]. For all the above reasons, the mechanisms for the perception of the sensorial attributes have been investigated, and a specialized scientific "sensory evaluation" (i.e. trained assessors, sensory tests/methods) methodology has been developed for their measurement. The particular unique attributes of food products should be defined, standardized, and evaluated by the food industry.

The sensorial attributes are often classified according to the senses they involve and the order in which the consumer tends to perceive them. In fact, in addition to the five senses (vision/sight, smell, taste, touch, and hearing), several other senses must be included for the complete classification, such as kinesthetic, temperature, deep pressure, and pain [2, 7]. A sensory evaluation procedure generally follows the steps given below [3]: (i) Consumers/assessors, at first, look at the color and the general appearance of a food, perceiving other additional visual characteristics such as size, shape, integrity, gloss/shine, form, surface texture, and defects for solid products or transparency/opacity, viscosity/consistency, and carbonation for liquid products. (ii) They then touch the food perceiving its handfeel/tactile texture (softness, firmness, elasticity, adhesiveness, etc.) or use a tool (knife, spoon, folk) to obtain a portion of food for tasting. In liquid products, they can detect their rheological properties (viscous, fluidity, thickness, etc.) by pouring in a suitable container. (iii) Thereafter, by holding the product, either food or drink, near their mouth, ready to taste it, they can perceive its odor. Odor is also called aroma and often has a desirable sensation. (iv) The circulation of a drink around the oval cavity or the consumption of a food through biting and mastication and decomposition of its components by saliva allows them to perceive both the kinesthetic characteristics through the tongue and hearing (chewability, smoothness, body, juiciness, softness, grittiness, stiffness, crispness), and the taste (sweetness, sourness, salty, bitterness). The volatile and non-volatile constituents are also released during this step. Thus, apart from the taste, the texture perceiving/evaluation is also completed in the tongue/mouth, and the sensation is known as mouthfeel. (v) During food mastication or drink consumption, one can eventually perceive the taste, odor, and/or texture (in the tongue/mouth) at the same time, namely the flavor of a product. More specifically, flavor is primarily the combined impression perceived by the chemical senses (odor and taste) from a product inside the oral cavity. Aromatics represent the volatile constituents that originate from food in the mouth and are perceived by the olfactory system via the posterior nares. Oral-somatosensation also results through thermal, tactile, and irritation sensations. (vi) After swallowing a food or drink, an aftertaste and/or off-flavor remains. (vii) The entire experience results in the overall acceptance of the food or drink.

Concluding, the sensorial attributes, characteristics, or factors involved in the perception of the quality of foods or drinks are presented in Table 10.1. For certain sensorial attributes, known as composite sensations, more than one sense are required for their expression and evaluation such as texture (touch, kinesthetic on tongue/mouth, and hearing) and flavor (taste and odor).

Sensorial attributes may be further analyzed in qualitative dimensions (or sensory qualities) associated with different sensations. For example, color can be defined as red, green, or blue; the taste as sweet, salty, sour, or bitter; and the odor as fragrant,

**Table 10.1** Classification of sensorial characteristics/attributes/properties.

| | |
|---|---|
| Appearance (vision/sight) | Color (hue, value, and chroma)<br>Size, shape, form, and integrity<br>Defects<br>Geometric properties: gloss/shine, clarity (liquids)<br>Surface texture, consistency |
| Odor (smell) | Qualities: fragrant/flowery/fruity, acid/sharp, burned/tarry/scorched, and caprylic/goat-like<br>Primary odors: camphoraceous, pungent, ethereal, floral, pepperminty, musky, putrid |
| Taste | Qualities: sweet, salty, sour, bitter, umami<br>Chemesthesis: pungency, astringency, metallic taste |
| Texture | Visual texture/appearance |
| i. Visual (sight) | Mechanical properties (nonmastication): (by sampling, slicing, spreading, pouring, etc.) |
| ii. Auditory (hearing) | Mechanical properties (mastication): cohesiveness, viscosity, elasticity, adhesiveness particle size/shape and air cell size/shape properties |
| iii. Tactile/handfeel (touch or by tool) | Disintegration properties (melt down on palate) |
| iv. Oral (tongue)<br>Kinesthetic (tongue)<br>Phase change<br>Mouthfeel (tactile, pain, temperature) | |
| Flavor (odor–taste–temperature–pain–touch) | Taste, odor, somatosensations (irritation, thermal tactile) |

acid, burned, and caprylic [2]. Texture can be analyzed into primary and secondary characteristics depending on the food. For each sensation of the sensorial attributes, quantitative dimensions also exist to describe differences in the magnitude or intensity (i.e. pale to dark for color, low to high-intensity for taste, etc.).

The "perception" process consists of the selection, organization, and interpretation of the information provided by the sensory systems and results from cognitive processes, understanding, and experience [2, 9]. The perceived food quality is the result of the interactions between the physicochemical properties of foods and human sensory receptor organs; thus, the primary sensorial experiences of the consumer are obtained. It is influenced by memories, personality, mood, knowledge, and expectations, further resulting in more complex perceptions of quality and/or acceptability. Understanding the interactions that elicit specific sensations of quality and intensity is attained by psychophysics terms and can contribute to consumers' perception of food quality. Regarding the quantitative aspects of sensorial experience, the threshold representing the minimal stimulus energy required to elicit a sensation and the intensity scale that can be used to describe its magnitude are of interest. The just noticeable difference (JND), which corresponds to the smallest additional amount that can be perceived, is also used. A basic overview of the individual effect of each sensorial attribute on the sensorial quality and perception as well as their interpretation is provided below.

### 10.2.1 Appearance

Appearance plays an important role in the perception of food quality and may be the first sensory evaluation made on its quality [10]. The visual appearance of food is determinant for the choice and consumption of foods, and it may often be the only attribute on which the decision to purchase is based. Visual perception is the result of the stimulation of the receptors in the retina of the eye by electromagnetic radiation. General appearance covers all visible characteristics of food quality such as color, size, shape, particle size, distribution of pieces, glitter, sheen and gloss, and wholeness (Table 10.1) [2, 10]. Choice is, in fact, governed by a hierarchy of appearance properties that creates expectations. However, color is considered as the most important from all the visible characteristics [11]. Visual characteristics also include the surface texture for solids, the gross texture or consistency for fluids and the clarity for solids or liquids [6]. Of the sensorial attributes, those related to appearance may be most susceptible to objective measurement, but they are important to consumers. Therefore, food producers should seek an attractive appearance for food products to increase their quality perception.

### 10.2.2 Color

Color is an appearance property attributed to the spectral distribution of light. It is a characteristic of light measurable in terms of intensity and wavelength [12]. Color perception is the brain's response to a stimulus of the retina that results from the detection of light after it has interacted with the food [13]. In color perception, both physical and psychological components are involved. Wavelengths in the visible portion of the electromagnetic spectrum not absorbed by the viewed food are seen by the eye and interpreted by the brain as color. In particular, blue is perceived in wavelengths between 400 and 500 nm, green and yellow between 500 and 600 nm, and red between 600 and 800 nm [6]. The three physical properties of light: wavelength, intensity, and purity are associated with three psychological dimensions: hue, brightness, and saturation [2, 13]. Color is commonly expressed in terms of three components (tri-stimulus colorimetry): hue (predominant wavelength reflected that determines the perceived color), value (lightness or darkness), and chroma (intensity strength of the color) (Table 10.1) [14, 15].

Color is a critical aspect of the appearance of a food, perhaps the most recognizable amongst the visual attributes of food. It is easily perceptible and correlated to quality of either fresh produce or processed foods; hence, it is used as a quality index for a number of foods [16]. The quality of most food products is associated with a certain color and is used as an index for its evaluation. Other visual properties, associated with light, are also important for the quality of foods. Glossiness, transparency, haziness, and turbidity, referred to as geometric properties, are appearance properties related to the geometric manner in which light is reflected and transmitted (Table 10.1) [7, 12]. The way light is reflected in a food creates different perceptions. For example, haziness or glossiness (shine or polish) result when light is reflected in all directions or in only a single direction, respectively, translucency when the light is transmitted though the food, or turbidity when the transmitted light is scattered by particles within the food.

### 10.2.3 Shape-Size

Size and shape are important visual attributes of foods that may be easily accessed by consumers. In fruits and vegetables, the size and geometric shape vary depending on the product, while they are related to their variety, maturation stage, etc. or quality grading; their weight changes along with their size. So, consumers select them based on their experience and look for fruits and vegetables with the desirable geometric characteristics [14]. Quality grading of the raw materials by size/shape (i.e. fruits and vegetables) is attained in the food industry to facilitate food processing and obtain uniform and desirable final products [12]. In general, consumers expect processed food products to be of uniform size and shape. The wholeness of a food can be visually perceived and contributes to the quality of some products, i.e. canned fruits, potato chips. In some cases, the size and shape are designed (i.e. in new formulated food products), or in others, they are well defined (i.e. in biscuits). Surface texture can also be important, especially in fibrous products (meats and fish) in which fiber geometry is observable, or in any other products where a smooth surface texture is desirable.

### 10.2.4 Defects

Defects of foods are in most cases related to color, consistency, and other appearance attributes, and therefore, they can easily be determined by consumers. Defects or imperfections result due to either the absence of something necessary for perfection or the presence of something that distracts from perfection. Defects of foods are studied and analyzed in respect to their significance; a number of defects, depending on their significance, can render a product defective. Food products are graded for defects, and the tolerance is established in terms of maximum numbers of defective units allowable.

### 10.2.5 Odor

Odor of a food is the second most important sensorial attribute that contributes immeasurably to the pleasure of consuming. Odor, like appearance, may be a valuable index to the quality of a food and even to its wholesomeness and edibility [7, 16].

Smell is the sensation resulting from the stimulation of receptors in the olfactory epithelium of the nose by volatile compounds. A reaction eventually takes place between the odor molecules and the chemoreceptors, thereby producing a neural pulse that reaches the brain where the information is stored [17]. The odor of a food or a beverage is perceived by the olfactory system either when its volatiles enter the nasal passage by active or passive sniffing through the nares (orthonasal olfaction) or through the mouth-inhaled air (rostronasal olfaction). The volatiles of a product are also released during chewing or swishing around in the mouth and are passed into the nasal cavity through the nasopharynx (retronasal olfaction) [6, 10]. The olfactory mechanism is both complex and more sensitive than the gustation process. The amount of volatiles released from a food product depends on the temperature, the nature of the compounds (volatility), and the surface properties of the food (i.e. softness/hardness, porosity, and watery-ness/dryness). Some odors can be released as a result of an enzymatic reaction, as in the case of the freshly cut surface of an onion [6].

Perceptible odors of foods can be described by the threshold (the lower concentration that creates an odor impression) and the odor quality. It is remarkable that humans can recognize thousands of odors. The odor quality describes the particular character of the aroma of a food, attributed to its aromatics.

Numerous theories have been proposed (vibrational, chemical, chemical–vibrational, electrochemical, enzymatic, informational, physical, and physicochemical), and many classifications of odor have been made to explain the odor quality aspects [7]. Odor description in terms of certain sensory qualities and of odor classification using multidimensional scaling approaches has been achieved. Notwithstanding, there is no internationally standardized odor terminology. Many terms may be ascribed to a single compound, and a single term may be associated with many compounds [6, 17].

Thus, to define and classify odors, four basic odor components (qualities) have been suggested: (i) fragrant or flowery or fruity, (ii) acid or sharp, (iii) burned or tarry or scorched, and (iv) caprylic or goatlike (Table 10.1). Additionally, a four-digit numeric system has been developed to classify odors and a nine-point scale ($0 =$ absence of any component and $8 =$ maximum intensity) has been proposed to indicate the magnitude of each of the four components in a particular odorous compound [2, 16]. So, rate 0000 represents an odorless (anosmic) substance and rate 8888 represents a substance being as intensely fragrant, acid, burned, and caprylic as possible, while any odor can be expressed according to this system, for example the odor of coffee is 7683.

Many attempts have been made to explain the relationship between molecular structure and odor sensation and classify odors based on the molecule geometry (size and shape); compounds that differ markedly in structure may have similar odors, or conversely, small differences in structure may give molecules with markedly different odors. According to this theory, seven primary odors are suggested to sufficiently characterize various odors. These primary odors are as follows: camphoraceous, pungent, ethereal, floral, pepperminty, musky, and putrid (Table 10.1) [16, 17]. Another theory (membrane puncturing) suggests an odor spectrum dependent on the molecule cross-sectional area of the odorant and the ease that it can pass from air to moisture on the olfactory receptor surface and be adsorbed across the interface of the thin lipid membrane underneath. According to this theory, groups of related odors occupy distinct areas in plots of molecule cross-sectional area vs. free energy of adsorption, which are related to the shape, size, functional groups and their distribution and position. The vibrational theory relates the odorous character to the vibrational specificity of the molecules (Raman spectra area is related to the vibrations of chains and flexing or twisting of bonds between groups of atoms in the molecule) [16, 17].

Food odors, natural or artificial, are usually the result of the presence of mixtures of many different odorous compounds. Regarding the particular aroma of a certain food, it can be either related mainly to the presence of one compound (i.e. citral for lemon) or to a limited number of compounds (named contributory aroma compounds), which create the impression of certain foods (i.e. esters for fruits, thiosulfonates and thiosulfinates for onion) [6, 17]. Consumers are commonly able to recognize the contributory odors and aroma characteristic of foods, and they generally expect pleasant odor and aroma from all the food products they consume. Therefore, any deviation from the definite odor of a certain food or any unexpected odor does not prepossess the consumer, while any unpleasant odor can be the reason for the rejection of a product. Odor supplements taste and encourages the consumption facilitating the ingestion of nutritious foods and activating their digestion. Moreover, odor may play, in several cases, a protective role against spoiled foods, foods processed under poor sanitary conditions, or even contaminated foods as they usually have a detectable unpleasant odor [7]. Aroma and odor of food products are influenced by their processing and storage; therefore, it is often attempted to enhance them, especially in foods where odor is one of their most attractive features (i.e. in fruit juices). In other cases, special processing steps are associated with the development of a desirable odor (i.e. ripening of cheese or aging of wine) or with the elimination of undesirable odors (i.e. deodorization of vegetable oils). Finally, herbs and spices, natural or artificial flavors (i.e. vanilla extract to ice cream) are added to foods for pleasantness reasons. Mixing two or more odors, may result in lowering (compensation), masking (neutralization), or increasing (synergism) the final odor perception.

### 10.2.6 Taste

Taste is an important sensorial attribute, which, along with odor, is a determinant for food acceptance. The pleasant eating sensations arise primarily from taste rather than from odor. People value foods for their taste; only few would be satisfied just by their smell before swallowing them. Moreover, they generally like sweet-tasting foods.

Taste refers only to certain modalities perceived in the mouth, almost exclusively on the tongue. The historical development of the taste modalities, emanating from Aristotle to the current view, concerns four taste qualities. It is generally accepted that the primary taste qualities or basic (true) tastes are sweet, salty, sour, and bitter (Table 10.1). All taste qualities are significant for some foods, while several of them can be found in the same food [2, 7].

Taste is sensed by the taste chemoreceptors named buds; the majority of taste buds are located in groups in the papillae on the tongue. Chemoreceptors are also located on the palate, pharynx, larynx, and other areas of the oral cavity, and their stimulation results in taste. However, the morphological structure and the sensitivity to the taste stimulus of the tongue have been well studied.

Taste buds include distinct papillae structures, which are regionally arranged on the tongue and present different taste sensitivities. Fungiform papillae occur at the anterior of the tongue, foliate papillae occupy the sides of the tongue, and circumvallate papillae, which form a chevron, are located on the posterior dorsal surface of the tongue. Filiform papillae, which are widely distributed over the dorsal surface, have no role in taste but are sensitive to touch. The central surface of the tongue contains no taste buds. The taste buds comprise a number of cells called pores in which the saliva collects. Each taste-sensing (gustatory) cell terminates in a microvillus. Cells constituting the taste buds degenerate and are regenerated over a seven-day period. Tastants from food must be primarily dissolved in the saliva, transported to the receptor centers, and interact with the microvilli. Once activated, electrical signals are sent via the nerve fibers that transmit taste sensations to the processing regions of the brain [10, 16].

Certain mechanisms have been proposed to explain the interaction between the taste substances and the taste receptors; in the initial step, the formation of a weak complex appears by the stimulus–receptor reaction that results in the initiation of the nerve impulse. According to taste stimulation by electrolytes, the taste molecules are absorbed creating a disturbance in the molecular geography of the sense organ's surface thus allowing an interchange of ions across the surface; these changes are followed by an electrical depolarization that initiates a nerve impulse. According to chemical stimulation, the taste compounds contact the taste cells and depolarize the receptor membrane activating the voltage-dependent calcium channels, while the influx of calcium triggers the transmitter release [17]. More especially, the taste compounds interact with specific proteins in the receptor cells, and sweet- and bitter-sensitive proteins have been reported. Molecules that taste sweet have functional groups that comprise an AH-B system (AH: proton donor, B: proton acceptor), and the two components are appropriately positioned so as to form hydrogen bonds to a similar system at a receptor site on a taste bud, while the receptor's membrane is depolarized, initiating a nerve impulse that results in the sensation of sweetness [16]. A relationship between the chemical structure of a compound and its taste has been established. Minor changes in the chemical structure (i.e. in isomers) may change the taste of a compound, for example, from sweet to bitter or tasteless [17].

A relationship between the concentration of the stimulus and neural response has been proposed: as the stimulus concentration increases, the response increases at a decreasing rate until a point is reached where further increase in its concentration does not produce further increase in the response. The stimulus concentration can be increased in steps according to JND indicating a difference large enough to elicit an increase in the response. The concentration of a substance required for the identification (threshold) of each of the four primary tastes is not the same for all individuals. A substance to produce a taste must be water-soluble, while pH and temperature may also affect the taste responses [17].

Regarding the sensitivity, taste buds are not equally sensitive to all taste stimuli. There is a spatial dispersion of specific receptors of the primary taste qualities on the tongue, so that regions of different sensitivities are present; taste buds near the tip are more sensitive to sweet, those on the sides to sour, those on the sides and the tip to salty, and those near the back to bitter. For each of these taste qualities, a set of stimuli is identified that arouses these sensations.

The sensation of sour is associated with hydrogen ions supplied by acids (vinegar, fruits, and vegetable) and acid salts. The intensity of the sour sensation produced by an acid depends more upon the hydrogen-ion concentration than upon the total acidity. Acids present different tastes; their sourness depends on the nature of the acid group, pH, titratable acidity, buffering effects, and the presence of other compounds (i.e. sugars). The relative sourness of most common acids found in foods has been determined based on their concentration [18].

Salty taste is due to ions of low-molecular-weight salts, most commonly of sodium chloride. The taste of salts depends on the nature of both cation and anion [7]. Sodium chloride is proved to enhance mouthfeel, sweetness, flavor, balance, and saltiness, while it masks or decreases off-notes. However, due to the current trend of reducing sodium intake in the diet, salt substitutes based on potassium chloride are suggested.

Substances that elicit the sweet sensation are primarily organic compounds as alcohols, certain amino acids, aldehydes, and glycerol. However, sugars are the main source of sweetness in foods. Sugars with their oxygen of the glycol groups form complexes with sweet-sensitive proteins on the taste buds at the receptors. The stereo-chemical position of the –OH groups influences the ability of the molecules to elicit sweet sensation. Sugars are ranked in order of sweetness using sucrose as the reference sweetener. However, the relative sweetness depends on whether sugars are in a crystal form or in solution as well as on the concentration, age, mutarotation extent, and temperature of their solutions. In food systems, their relative sweetness is influenced by the presence of acids, salts, and other constituents. Other substances with sweet taste, in some cases many times sweeter than sucrose, are used as alternative and/or noncaloric sweeteners (i.e. dipeptide of aspartame, dihydrochalcones, cyclamates, etc.).

A variety of organic or inorganic compounds of foods taste bitter, such as alkaloids (caffeine, theobromine in coffee bean, quinine in soft drinks), glucosides of phenolic compounds (naringin in grapefruit), certain amino acids, and other groups of substances (bile salts and salts of magnesium).

Although bitterness by itself is usually considered unpleasant, bitter components may be found in many foods, for example, bitter peptides are formed by enzymic hydrolysis of proteins during ripening of cheese. In addition, some alternative sweeteners present bitter aftertaste [16, 17].

It is generally agreed that the four primary taste qualities cannot describe taste completely, and therefore, their interrelationship (i.e. between sweet and sour that affects the flavor of fruits and wines) as well as other important sensations have also been suggested. The existence of a fifth taste quality called *umami* is recognized by some researchers, which is sensed in different receptors in the mouth than those for the primary tastes. It is described as delicious or savory and most commonly associated with the taste of monosodium glutamate, but it can also be elicited by certain l-amino acids [2]. Glutamate is considered a flavor enhancer having not so meaty or chicken taste but a unique flavor. In high-protein foods, it appears as a balanced taste and a mouth-filling sensation.

Alkaline taste has been attributed to the hydroxyl ion and astringency, which is considered a flavor feature, to borax and tannins (in tea), respectively. Coolness is characteristic of menthol (mint flavor), while hotness (also referred to as pungency) is characteristic of spices (piperine in black pepper, capsicum in red pepper). Metallic taste can be generated by salts of metals, such as iron or copper, and is observed in canned foods as well as an aftertaste. Enhancement or modification of the taste sensation can be obtained by some substances, such as gymnemagenin that suppresses the ability to taste sweetness and the proteins of miracle fruit that change the perception of sour to sweet [17].

Pungency, astringency, and metallic taste are part of the common chemical sense or chemesthesis (Table 10.1). The chemosensory system, distinct from taste and smell, mediates sensations of pain and/or irritation caused by chemicals in contact with the skin or mucous membrane. The oral and nasal mucosa have specialized sensory receptors that respond to both heat and cold evoking thermal and pain sensations. The oro-nasal nerve fibers are not independent sensory systems, but a component of the pain and temperature fibers that occur throughout the skin. Oro-nasal chemical irritation has long latency of sensation as compared to taste or smell; however, it is influenced by the taste and smell coelicited with the irritation, thus adding complexity to flavor perception. A number of chemicals are capable of activating irritant sensations described as burn from chili pepper, warmth from ethanol, tingle from $CO_2$, pungency from wasabi, etc. [10].

In some food products, a specific taste quality is particularly important for them, for example, sweetness in soft drinks, fruits and fruit juices, honey, and many baked products; sourness in pickles, fruits, and fruit juices and wines; saltiness in meat products or bitterness in beer.

As foods contain mixtures of substances that elicit all four taste sensations, their interaction is of great interest to food technologists, and therefore, it has been extensively investigated. Thus, subthreshold concentrations of salt reduce the sourness of acids and increase the sweetness of sucrose, while subthreshold concentrations of acids intensify the saltiness of sodium chloride. Sugar at subthreshold concentration reduces the saltiness of sodium chloride, the sourness of acids, and the bitterness of caffeine. The effect of suprathreshold concentrations of sweet, salt, sour, and bitter has also been reported [7, 16].

### 10.2.7 Texture

Texture is an important aspect of food quality, sometimes even more important than flavor and color. The texture refers to the visual, tactile and, in some cases, auditory attributes of foods, with the textural (tactile) attributes of foods being perceived with either the fingers, the tongue, the palate, or the teeth. It represents the perception that originates from the structure and the behavior of foodstuff when handled and eaten, thus generating a human experience. The textural properties vary widely in foods with hardness, cohesiveness, and moisture content being the most commonly cited characteristics. However, for some products like crispy and crunchy, the texture may be the most important sensorial attribute for consumers [13, 17].

The texture of a food product is determined by its rheological, mechanical, and structural (geometric and surface) attributes perceived by mechanical, tactile, and, where appropriate, visual, and auditory receptors [13]. Food texture can be defined as the way in which various constituents and structural elements are arranged and combined into a micro- and macro-structure, which is expressed in terms of flow and deformation [17]. Foods, due to the relationship between chemical composition and physicochemical structure, present widely different physical and mechanical properties covering a range from fluid, semi-solid to solid systems.

By handling and consuming a food or drink, the following texture qualities are perceived sequentially by the senses of sight, sound, and touch: (i) visual texture, (ii) auditory texture, (iii) tactile texture (handfeel) through touch (direct) or using a tool, such as a knife, fork, or spoon (indirect), and (iv) oral tactile texture on the tongue including kinesthetic, mouthfeel, and phase change [13, 19]. A single one or a combination of these senses may be used to perceive the texture of various food products, i.e. crispness of potato chips as tactile in the mouth and auditory texture, thickness of milkshake as visual by stirring and tactile in the mouth.

Visual texture refers to surface characteristics and other characteristics evaluated visually by slicing or pouring. Some of these may be assessed as appearance characteristics, like glossy and shine; however, they affect the perceived texture of the product. There is a correlation between visual and oral-tactile characteristics, for example, oral-tactile moistness and visual pore sizes of cake [13].

Hearing (audition) is the perceptual experience that results from the stimulation of the acoustic receptors by sound produced during biting and mastication of foods. Auditory attributes are also perceived by sampling or slicing of foods or by pouring of drinks [6]. Distinctive sounds are associated with the consumption of specific foods (i.e. crackers, snacks, etc.) affecting (psychoacoustic) significantly their quality perception and acceptability. The auditory texture of foods can contribute to the perception of food quality, while any deviation from expectations can decrease the liking of a product. Food texture studies for the assessment of crispness and crunchiness are based on the analysis of the physical or perceptual sounds emitted during mastication [2].

The sense of touch (tactile hand-feel) helps to evaluate the texture attributes of foods. When touching and manipulating a food with fingers, sensory sensations are generated either as surface feel (pressure, pain, and temperature) or as internal sense in the deeper tissues (muscles). Touch is often underestimated; however, it is as an important factor in the appreciation of certain foods (i.e. graininess in nuts or pears, crumbliness in cake, sogginess in wet bread) [7].

Kinesthesis (receptors in deep tissues, that is, muscles, tendons, and joints) and somesthesis (receptors in skin) are responsible for all nongustatory oral perceptions involved in the perception of food texture [2]. Kinesthetic perception results from the stress exerted by jaw or tongue muscles and the sensation of strain developed within the food being masticated. The perceived textural characteristics may be similar to those assessed by touch; however, kinesthetic assessment differs greatly in sensitivity from simple touch. In addition, some foods may exhibit textural changes once inside the mouth due to mixing with saliva and mouth enzymes, as in the softening of crunchy foods (i.e. biscuits). Texture changes are also associated with the melting behavior of foods in the mouth. Many foods undergo a phase change in the mouth due to fat melting at the elevated temperatures inside the oral cavity. Fat melting is considered responsible for the palatability of chocolate or ice cream and further changes in texture during mastication responsible for the enjoyment of certain foods. Mouthfeel refers to tactile, pain, and temperature sensations; tactile attributes included in the mouthfeel tend to change less dynamically than other oral texture characteristics. Mouthfeel plays an important role in the sensory pleasure of eating. A food product may be rejected on the basis of mouthfeel even though the rest sensorial characteristics are acceptable [13].

The understanding of the textural properties of foods often requires the study of their physical structure, which includes the flow and deformation properties and their macro- and micro-structure. A classification system of textural characteristics constituting the so-called "texture profile" of foods has been developed by Szczesniak, and many descriptive terms for the texture properties of foods have been widely accepted. The textural characteristics are categorized into three groups containing primary, secondary, and tertiary parameters as follows (Table 10.1) [12, 13, 20]:

1) *Mechanical characteristics* that contain five basic parameters: hardness, cohesiveness, viscosity/consistency, elasticity, and adhesiveness, and the secondary parameters: brittleness, chewiness, and gumminess.
2) *Geometrical characteristics* that include two groups: those related to particle size and shape and their arrangement, and those related to particle size and orientation.
3) *Other characteristics* related to moisture content and fat content.

The improved system for the assessment of the entire texture profile of foods suggested by Sherman relates analytical characteristics with sensory responses and groups them into three categories (primary, secondary, and tertiary) regardless of whether they are analytical, mechanical, or geometrical. The characteristics (primary, secondary, and tertiary) as they are successively perceived during texture evaluation (step 1: initial perception and step 2: initial perception on palate, initial bite, mastication, and residual masticatory impression) are as shown in Table 10.1 [17, 20].

Tertiary terms are common descriptive terms used by consumers. The tertiary terms are specified for fluid, semi-liquid, and solid products and are suggested for certain food products [13, 17]. Food texture can be evaluated by mechanical (instrumental) or sensory (physiological/psychological) methods. Texture testing methods mainly determine rheological properties based on the resistance in flow and deformation (by stress and strain action), such as compression, shearing, shear-pressure, cutting or tensile strength, while specialized test instruments exist for such measurements (succulometer, tenderometer, penetrometer, gelometer). None of these, however, duplicates exactly what occurs in the mouth. The textural parameters are evaluated by a sensory panel using the "Texture profile analysis" test. Many attempts have been made to establish objective criteria for texture measurement and to relate the results obtained by instrumental techniques of measurements with the responses provided by sensory analysis.

Although texture in foods varies widely, consumers expect food products with certain textural characteristics and any divergence from their expectations is considered a quality defect.

Temperature may affect markedly the consistency of foods (i.e. syrups), or different mechanical properties of the same product may be required depending on its intended purpose, i.e. viscous cream (when whipped) or runny [10]. Finally, sometimes a texture contrast within a food on the plate or across food products in a meal is pursued. The measurement of the mechanical properties and the food texture studies provide useful information for food processing as well as the design of texture measuring instruments for the food industry to satisfy consumers' expectations [14].

### 10.2.8 Flavor

Flavor of foods is generally the combined sensory impression of odor, taste, temperature, pain, and touch. In flavor, the sum of the sensorial attributes evaluated primarily by the senses of taste and smell are included, although the sense of feel in the form of touch and temperature may be involved (Table 10.1) [7, 12]. More simply, flavor is referred to as the combination of both taste and odor including the taste sensations perceived by the tongue and odors perceived by the nose. Flavor is considered a key factor in the quality perception of foods. The terms flavor and odor (aroma) often are used interchangeably [14]. The flavor of a given food is determined by the combination of both the basic tastes and the numerous compounds, which give foods characteristic aromas. The flavor of a food is quite complex and has not been completely described for most foods [15].

The perceived flavor of a food is derived when three independent sensory systems are activated: taste, olfaction, and oro-nasal somatic sensations (irritation, thermal, and tactile). Food components, exhibiting taste or smell, interact with the respective organs receptors, and the produced signals are carried out to the central nervous system, thus creating flavor sensation, while the sensory information is further relayed to the brain. Flavor release, texture, and taste can all be altered by mouth pH, saliva levels, enzyme activity, food temperature, and chewing rate. Since other qualities (i.e. roughness or smoothness, granularity, viscosity, hotness or coolness, and metallic taste) contribute to the overall sensation, numerous sensory inputs are processed by the brain to result in flavor perception.

Cognitive processes occurring in the brain vary depending on individual impressions [10, 17, 21]. Food flavor and aroma are subjective and therefore difficult to measure. People differ in their ability and sensitivity to detect tastes and odors. The same food is often perceived differently by different individuals, for example, an extremely bitter and unpleasant flavor for one may be perceived as slightly bitter and pleasant by another. People also differ in their preference due to cultural, social, and biological factors, for example, strong smelling fish is desirable in some cultures, whereas in other cultures, such fish would be undesirable; garlic for some people is quite pleasant, while for others, it is extremely unpleasant. Consumers' responses can lead to acceptance, rejection, and quality ratings of specific foods.

The perceived flavor of foods is often influenced by their color. For example, the flavor of cherry, raspberry, and strawberry is associated with red color; that of orange with orange color; that of lime with green color; that of chicken with yellow color; and that of beef with brown color. In butter and margarine colored with carotenes, the more yellow products are expected to have a stronger butter flavor. It is difficult for consumers to distinguish the right flavor in desserts prepared without color or with an alternative color. In addition, texture can equally influence the flavor of foods. For example, in sauces thickened with a tasteless starch or gum, the thicker product is expected to have a richer flavor. However, some texturizing agents can influence foods' taste and flavor by affecting the solubility and volatility of flavor compounds [15, 22].

Special interest has been given to the study of the flavor of particular food products. In a group of foods like bread, meat, and cheese, the flavor cannot be attributed to one or few outstanding flavor notes, but their flavor is the result of the complex interaction of a variety of taste and odor components. In other groups of foods like fruits, vegetables, and spices, the flavor can be related to one or few easily recognized components [17]. Such contributory flavor compounds are investigated, for example, fruity flavors are esters, alcohols, ethers, and ketones, and many of these are volatiles associated with common acids (citric, isocitric, malic, oxalic, tartaric, and succinic) [14]. Some compounds contribute to a pleasant flavor in some foods, but unpleasant in others. For example, volatile fatty acids function as flavor contributors in many cheese flavors, while in very low concentrations, they cause an unpleasant rancid off-flavor in milk and other dairy products. Acetaldehyde naturally occurring in fruits is essential for imparting the taste of freshness; however, it is responsible for a very unpleasant oxidized flavor in wine [17]. Undesirable flavors (off-flavors) can arise in foods as a result of contamination (taints), such as from packaging materials, cleaning agents, etc., by distortion of the normal processing or by deterioration during storage.

Many food products are specially flavored like carbonated drinks and biscuits. Natural or synthetic flavorings may be added to foods or drinks for flavor enhancement purposes. Added flavors should be safe for consumers and maintain the desirable sensory character of foods after incorporation and processing. Protection of flavor substances is attained by encapsulation techniques.

Consumers are exposed to novel foods or flavors, thereby changing their flavor experiences and expectations [7, 10].

The study of food flavors includes the instrumental analysis of specific components and the identification of those responsible for the aroma and flavor compounds. This information can be useful in developing new products using the component as a flavoring, or in product identification/authentication by detecting specific substances attributed to a certain variety or geographical origin as well as in the correlation of individual substances with flavor acceptability [12].

Information on volatile flavor compounds in a variety of food products has been obtained by the development of powerful analytical techniques, such as gas chromatography, infrared spectroscopy, or mass spectrometry) for the identification and quantification of flavor compounds.

Electronic noses (e-noses) and electronic tongues (e-tongues) have also been developed in an attempt to mimic the human smell and taste sensors and their communication with the human brain. Electronic noses comprise a vapor sampling system, an array of sensors (i.e. metal oxide, conducting polymers, optical, piezoelectric, electrochemical, field effect, or olfactory biosensors), and a method of signal processing that leads to the classification of the responses. Electronic tongues function in a similar way to the electronic nose. A sensor array produces signals that can be correlated to certain features or qualities of the sample [23].

Sensomics is another appealing approach, which combines both chemical analysis and the human olfactory or gustatory system. It involves the separation of the flavor compounds present in an extract by gas chromatography. As each compound elutes from the column, it is sniffed by a trained individual who assesses both the aroma and its intensity, while the odor-active compounds are identified by GC–MS. This procedure can be repeated on a set of serial dilutions until only the most potent aroma compounds that are more likely to contribute to the aroma of the food are detected. Each aroma is given a flavor dilution factor (FD), which is the highest dilution of the original extract at which it is still smelled by GC-olfactometry (GC-O) and is the basis of the aroma extract dilution analysis (AEDA). Quantification of the most persistent aroma compounds is ideally carried out using stable isotope dilution analysis, and the odor activity value (OAV) is defined as the concentration of the aroma compounds in the extract divided by its odor threshold. Sensomics approach is completed with the recombination of the individual odor-active aroma compounds in a suitable matrix and the comparison against the original food, preferably using a trained sensory panel [24].

In the case where flavors are evaluated by human senses, the flavor profile method has been developed to describe flavor qualities including taste, aroma, feeling, and aftertaste [17]. Flavor profile analysis is based on the identifiable taste, odor, and chemical feeling factors plus the underlying complex of sensory impressions not separately identifiable. By using this method, it is possible to define parameters such as the overall aroma and flavor impression (amplitude), the identification of perceptible aroma and flavor, the intensity of each character, the order of perception, and aftertaste [22].

## 10.3 Sensory Evaluation – Perception – Acceptance of Foods

*Sensory evaluation* has been defined by the Sensory Evaluation Division of the Institute of Food Technologists as "a scientific discipline used to evoke, measure, analyze and interpret reactions to those characteristics of foods and materials as they are perceived by the senses of sight, smell, taste, touch and hearing" [25].

*Sensory perception* is a three-step process (brain stimulus sensation → perception → response). The stimulus hits the sense organ, and it is converted to a nerve signal, which travels to the brain; the brain then organizes and integrates the incoming sensations into perceptions; lastly, a response is formulated based on the subject's perceptions. Humans often yield varied responses to the same stimulus. Sensations are not controlled by the mind, and differences in sensations between subjects cannot be influenced but must be treated as a component of the internal error as in any experiment or test. By contrast, perception and response are open to the analyst's influence. Through training and the use of reference samples, trainees can achieve to show the same response to a given stimulus [26, 27].

Sensory evaluation principles have their origin in physiology and psychology. It is of interest to relate the response with the stimulus. Since sensation cannot be measured directly, it is necessary to measure sensitivity by means of differential changes. By determining the detectable amount of difference between two stimuli (JND), one can establish a unit of sensation [25].

Fechner considers that each JND is equivalent to a unit of sensation and the JNDs are equal:

$$S = k \log R\left(\text{Weber's or Weber–Fechner or Psychophysical Law}\right),$$

where:

$S$ is the magnitude of sensation

$k$ is a constant

$R$ is the magnitude of stimulus

Stevens proposed that equal stimulus ratios result in equal sensation ratios rather than equal sensation differences:

$$\log R = k \log S^n \left(\text{Psychophysical Power Law}\right),$$

where:

$R$ is the response

$k$ is a constant

$S$ is the stimulus concentration

$n$ is the function slope

Weber noted that the perception of the difference between two products was a constant, related to the ratio of the difference:

$$K = \frac{\Delta R}{R}$$

where:

$R$ is the magnitude of the stimulus

$K$ is the constant of the JND

Mathematical relationships can be used to model the connection between *acceptance and perceived sensorial properties*. Thus, food acceptance can be predicted by a polynomial function of the weighted perceived intensities of various sensorial attributes as:

$$\begin{aligned}\text{Acceptance} = {} & k_1\left(S_1\right) + k_2\left(S_2\right) + \cdots + k_n\left(S_n\right)\left(\text{simple linear model}\right)\\ & + k_{12}\left(S_1{}^{*}S_2\right)\ \left(\text{with interaction terms}\right)\\ & + k_{1m}\left(S_1\right)_m^a\ \left(\text{high order polynomial}\right)\end{aligned}$$

where:

$S_n$ is the perceived intensity of an attribute

$n$ is the intensity

$S_n$ can also be predicted by equations such as:

$$S_n = K_n I_n^b$$

$b$ is the characteristic of the power function exponent

$I$ is the stimulus intensity in physical units.

Weighting coefficients are positive for desirable attributes but negative for product defects, and the more or less intense the sensorial attribute, the more or less acceptable is the product considered, respectively. Higher order terms may be added to approach the most desirable level. Interaction terms may take into account the simultaneous balancing of two or more attributes. Perceived intensity, in turn, can be modeled on the basis of ingredient concentrations, using relationships such as Steven's power law or alternative functions [28].

Understanding sensory thresholds (minimum perceivable levels), psychophysical functions (dose–response curves), and sensory interactions as well the weighting coefficients for sensory contributions to the overall acceptance can help the product formulator in engineering acceptable products. The relationship between the acceptance and the intensity of various sensorial characteristics provides useful information to understand and quantify the sensorial attributes of food products that determine their eventual success ([9, 29, 30]).

### 10.3.1 Sensory Evaluation Tests

They are divided into: (i) affective (or subjective), aiming to evaluate products' preference and/or acceptance and (ii) analytical (or objective) that evaluate differences or similarities between products, or the quality, and/or quantity of products' sensorial characteristics. In affective tests, the spontaneous, personal reaction of the panelist is evoked to determine preference or acceptance. In analytical tests, certain aspects of the sensorial quality of the product are of interest and not the personal reaction of the panelist. The panel is the analytical tool. In any type of analytical work, there is an obvious need for control and standardization. This is especially necessary in sensory evaluation, since it is based on the psychological evaluation of physiological sensations [31]. The International Organization for Standardization (ISO) defines an objective method as "a method in which the effects of personal influence are minimized". Specific procedures have been developed in analytical sensory tests in an effort to control or minimize the effect that psychological and physical conditions can have on the panelists' reactions. When these procedures are followed, analytical sensory tests meet the requirements to be considered "objective."

## 10.4 Sensory Control of Foods – Methodology

To provide reliable results, the sensory evaluation of foods should satisfy certain prerequisites such as sensory laboratory, trained assessors (panelists), sensory methods and standard sensory tests, data recording and results evaluation by using appropriate statistical methods. Fundamental to the successful implementation of any sensory evaluation test is the clear understanding and statement of the objective of the study [3, 26, 27, 32].

### 10.4.1 Sensory Laboratory

The sensory laboratory is fundamental for sensory analyses of foods and general specifications for the installation of a test room with the booths, and the additional premises are described by the Standard ISO 8589: 2007 "Sensory analysis; general guidance for the design of test rooms". The test room is designed to provide panelists, participating in the sensory tests, with a suitable, comfortable, and standardized environment, which facilitates work and helps to improve the repeatability and reproducibility of the results. The sensory analysis booths should be sufficiently large and comfortable.

Each booth should contain a table and a seat for the assessor, while it should have individual lighting and a sink with running, drinking water. The booths should be placed alongside one another in the premises. They should be separated by partitions, which must be sufficiently high and wide enough to isolate the tasters when seated. The booths and premises should be pleasant, meet certain specifications, and provide appropriate conditions (light colored – easy-clean construction materials, quiet – undisturbed area, controlled lighting – adequate illumination, controlled air circulation, controlled environmental conditions: 20–22 °C, 60–70% RH) to avoid affecting the results of the sensory analysis. Additional premises for samples preparation and shelves for containers or discussion rooms should also be provided.

### 10.4.2 Assessors/Panelists – Training

As mentioned before, the panel is the analytical tool in sensory evaluation. The value of this tool depends on the objectivity, precision, and reproducibility of the judgments of the panelists. Before a panel can be used with confidence, the ability of the panelists to reproduce judgments must be determined. The number of the panelists used will influence the statistical significance of the results obtained. The reliability and validity of the results will vary according to the selection, training, and instructions that the panelists are given. The rules which are essential for the panel supervisor/leader to select, train, and monitor the performance of the experts and assessors/tasters (sensory panel) are described by the Standards: ISO 8586: 2012 "Sensory analysis – General guidance for the selection, training and monitoring of assessors and expert sensory assessors", ISO 11132: 2012 "Guidelines for monitoring the performance of a quantitative sensory panel" and ISO 13300: 2006 "Sensory analysis – General guidance for the staff of a sensory evaluation laboratory, Part 1: Staff responsibilities, Part 2: Recruitment and training of panel leaders".

The purpose of training is to develop familiarity with the product and its characteristics, to develop a common language to describe these characteristics, and to improve the panel's ability to make consistent judgments. General terms and physiological terms of the sensory analysis are described in ISO 5492: 2008 "Sensory analysis – Vocabulary", while general guidance for the methodology is provided in ISO 6658: 2005 "Sensory analysis; Methodology; General guidance". Special standards for assessors' training in individual sensorial characteristics (taste and odor) are shown in Table 10.2 [3].

During training, the panelist learns to disregard personal preferences, making the evaluations more objective. Panelists who have been selected for their ability to discriminate are more likely to give the same results on a replication of the test. They will be able to detect smaller differences than randomly selected panelists.

Specific instructions and recommendations are given to the tasters, which should be followed before and during the sensory analysis, such as not to smoke at least

**Table 10.2** Standards for training on the individual sensory characteristics.

| | |
|---|---|
| ISO 6658: 2005 | Sensory analysis – Methodology – General guidance |
| ISO 3972: 2011 | Sensory analysis – Methodology – Method of investigating sensitivity of taste |
| ISO 5496: 2006 | Sensory analysis – Methodology – Initiation and training of assessors in the detection and recognition of odors |
| ISO 13301: 20 02 | Sensory analysis – Methodology – General guidance for measuring odor, flavor, and taste detection thresholds by a three-alternative forced-choice (3-AFC) procedure |

30 minutes before the test, not use any perfumes or cosmetics, fast at least one hour before tasting, and avoid tasting if not feeling physically well or being under any psychological effect that may influence their perception.

### 10.4.3 Samples

Several parameters, inherent to samples' characteristics, may affect sensory analysis, such as sample preparation, dilution, or mixing with carriers; serving temperature (preferably that proposed for the consumption of the product); serving containers (masked for color, covered for odor); quantity of samples, coding (using random numbers); and order of samples presentation or mouth rinsing between samples (with water, crackers, apples, celery, or bread). Therefore, all the above must be carefully designed and taken into account for the successful performance of a sensory test.

### 10.4.4 Sensory Tests and Methods

Several sensory methods can be used as analytical tests; the best can be selected after considering the test objective. The tests used in sensory analysis are shown in Table 10.3.

**Table 10.3** Sensory methods – sensory tests.

| | |
|---|---|
| *Discriminant tests* | |
| ISO 4120: 2004 | Sensory analysis – Methodology – Triangle test |
| ISO 10399: 2004 | Sensory analysis – Methodology – Duo–trio test |
| ISO 5495: 2005 | Sensory analysis – Methodology – Paired comparison test |
| ISO 8588: 1987 | Sensory analysis – Methodology – Test "A" no "A" |
| *Descriptive methods* | |
| ISO 6564: 1985 | Sensory analysis – Methodology – Flavor profile methods |
| ISO 11036: 1994 | Sensory analysis – Methodology – Texture profile |
| ISO 11035: 1994 | Sensory analysis – Methodology – Identification and selection of descriptors for establishing a sensory profile by a multidimensional approach |
| ISO 13299: 2003 | Sensory analysis – Methodology – General guidance for establishing a sensory profile |
| ISO 11037: 2011 | Sensory analysis – Methodology – General guidance and test method for assessment of the color of foods |
| ISO 11056: 1999 (2013 amd) | Sensory analysis – Methodology – Magnitude estimation method |
| ISO 8587: 2006 (2013 amd) | Sensory analysis – Methodology – Ranking |
| ISO 4121: 2003 | Sensory analysis – Methodology – Guidelines for the use of quantitative response scales |

They are categorized as follows: (i) discriminant tests that are used to determine differences between samples in a certain sensorial characteristic or in total and/or in their preference, (ii) descriptive tests that are used to describe the sensorial characteristics, especially the texture and flavor, through descriptive attribute terms (descriptors), while they can be applied to quantify the perceived intensity of the sensory attributes (quantitative descriptive analysis – QDA and free choice profiling – FCP) and/or the degree of liking (DOL) using scaling score. Other tests are also used for specific purposes, such as time intensity test, while preference tests are conducted to evaluate consumers' acceptance or preference [33, 34]. The above tests can be a valuable tool for the needs of the quality control in the food industry as sensory control supports input from instrumental methods and is necessary to achieve total quality control of the food products [3, 35–38].

The most important sensory tests used are cited below:

Discrimination tests are designed to discriminate differences between two or more samples and can also be used for the selection, training, and monitoring or the performance of sensory panelists. These include the following tests:

- *Paired comparison test*: It is used to determine the direction of the differences between two test samples for a specified attribute (for example, more or less sweet).
- *Triangle test*: It is used to determine whether a perceptible difference results (triangle testing for difference) or does not result (triangle testing for similarity) when, for example, a change is made in ingredients, processing, packaging, handling, or storage of samples. In this case, three samples are presented simultaneously: two samples are alike and one is different. Panelists are asked to indicate the odd sample.
- *Duo/Trio test*: It is based on the same principle as the triangle test and is also used to determine whether a sensory difference exists between samples. In this case, a reference sample is presented first and is then followed by two other samples, one of which is the same as the reference. The assessor is asked to identify which of the last two samples is the same as the reference.
- *Multiple comparison tests*: When they are applied as true difference tests, the panelist is required to separate the sample into two groups of like samples. When they are applied as directional tests, the panelist is asked to identify the groups of higher or lower intensity of a given criterion.

The above tests can also be used for preference purposes.

Descriptive tests are used to describe and evaluate more complex sensorial characteristics of foods (i.e. texture and flavor) and further provide a full description of the profile of certain food products. They can also be used to quantify primary and secondary sensorial characteristics in foods and score their DOL.

- *Descriptive Tests (Quantitative Descriptive Analysis (QDA) and Free Choice Profiling)*: They provide quantitative descriptions of the sensory attributes of a product by using a similarity scale, a graphic scale (smiley face), or a grading scale (i.e. DOL scale 1–9, 0–10, 1–7, 0–5), by taking into account all sensation that are perceived (visual, auditory, olfactory, gustatory, kinesthetic, etc.) in the order of their perception. They involve relatively few judges, who have been screened, selected, and specially trained for the particular product category to develop a common descriptive language.

- *Ranking test*: It is used to determine how several samples differ on the basis of a single characteristic. Two or more samples are provided to the panelists who are asked to arrange them in an ascending or descending order according to the intensity of the specified attribute.

Time intensity tests are used to measure the intensity of a sensation in relation to the time of its perception. They are mainly related to taste or flavor attributes that change during mastication in the mouth. The evaluation begins as soon as the sample is put in the mouth and stops when the attribute under investigation is no longer detected.

Acceptance/preference tests are usually performed at consumers' level and are used to measure liking or preference for a product. Preference can be measured directly by comparison of two or more products with each other or indirectly by determining which product has scored significantly higher rating in a multiproduct test. The two methods most frequently used to directly measure preference and acceptance are the paired comparison test and a 9-point hedonic scale. The hedonic scale extends from "dislike extremely" (i), through "dislike very much" (ii), "dislike moderately" (iii), "dislike a little" (iv), "neither not like nor dislike" (v), "like a little" (vi), "like moderately" (vii), "like very much" (viii) to "like extremely" (ix).

### 10.4.5 Presentation of Sensory Analyses Results – Correlation to Objective Analyses

Assessors, immediately after tasting, record their scores in suitable sheets, according to the sensory test requirements, using symbols (+, √) or a numbering scale (1–5, 1–9, 1–10). The sensory sheets should be easy for the tasters to complete and for the supervisor to collect and evaluate the results according to the instructions provided in the ISO standards for each individual test. The sensory data may be presented using network diagrams, etc. and processed with proper statistical methods such as Analysis of Variance (ANOVA). The data analysis method applied may influence the accuracy of the results.

In most instances, the objective of the study can be stated in the form of a null hypothesis, which will be accepted or rejected depending on the results of the statistical analysis of the data collected. A type I error consists of rejecting the null hypothesis when it is true, and a type II error consists of accepting the null hypothesis when the alternative is true. Type II errors are minimized by using acute reliable judges and/or by increasing the number of observations in which the conclusion is based. The probability of type I error is the level of significance selected (5%, 1%, or 0.1%) by the experimenter before the test. The sensory results can be correlated to the objective results (i.e. results from chemical analyses for flavor compounds, instrumental measurements of texture) using multivariate techniques such as Principal Component Analysis (PCA) [32, 39, 40].

## 10.5 Conclusions

Sensorial properties perceived by flavor, texture, odor, palatability, etc., are of paramount importance in the food industry as they are directly associated with food products' acceptability. Sensory evaluation should be incorporated in the QC program of the food industry to comply with the specifications and requirements set by the consumers. An integrated QC/sensory program involves the determination of specifications or quality standards and the selection of the most appropriate testing method. Methodologies such as descriptive analysis, difference from control, quality ratings, and in/out methods are used for evaluating food quality. Sensory schemes are developed for existing or new products by determining the individual contribution of each sensorial characteristic as a percentage to the overall sensory quality. Finally, sensory studies can be a powerful tool for the needs of R&D in improving and/or optimizing of the sensory quality and determining the shelf life of the already existing food products as well as in designing and developing new products.

## References

1 Luning, P.A., Marcelis, W.J., and Jongen, W.M.F. (2006). *Food Quality Management: A Techno-Managerial Approach*. Wageningen, the Netherlands: Wageningen Academic Publishers.
2 Cardello, A. (1998). Chapter 1 – Perception of food quality. In: *Food Storage Stability* (ed. I.A. Taub and R.P. Singh), 1–37. Boca Raton, FL: CRC Press LLC.
3 Tzia, C., Giannou, V., Lignou, S., and Lebesi, D. (2015). Chapter 2 – Sensory evaluation of foods. In: *Handbook of Food Processing: Food Safety, Quality, and Manufacturing Processes* (ed. T. Varzakas and C. Tzia), 42–71. Boca Raton, FL: CRC Press/Taylor & Francis Group, LLC.
4 Amerine, M.A., Pangborn, R., and Roesler, E.B. (1965). Principles of sensory evaluation of food. In: *Food Science and Technology Monographs*, 338–339. New York: Academic Press.
5 Jellinek, G. (1985). *Sensory Evaluation of Food—Theory and Practice*. Hemel Hempstead, U.K.: Ellis Horwood.
6 Meilgaard, M., Civille, G.V., and Carr, B.T. (1999). *Sensory Evaluation Techniques*. Boca Raton, FL: CRC Press LLC.
7 Stewart, G.F. and Amerine, M.A. (1982). Chapter 6 – Human nutrition and food science and technology. In: *Introduction to Food Science and Technology*, 2e (ed. G.F. Stewart and M.A. Amerine), 129–175. New York: Academic Press, Inc.
8 Hough, G. (2010). Chapter 2 – Principles of sensory evaluation. In: *Sensory Shelf Life Estimation of Food Products* (ed. G. Hough), 23–60. Boca Raton, FL: Taylor & Francis Group, LLC.
9 Thomson, D.M.H. (1988). *Food Acceptability*. Essex, U.K.: Elsevier Applied Science.

10 Keast, R.S.J. (2010). Chapter 3 – Food quality perception. In: *Processing Effects on Safety and Quality of Foods, Contemporary Food Engineering Series*, D.W. Sun Series Editor (ed. E. Ortega-Rivas), 67–83. Boca Raton, FL: CRC Press/Taylor & Francis Group, LLC.

11 Hutchings, J. (2002). Chapter 2 – The perception and sensory assessment of colour. In: *Colour in Food, Improving Quality* (ed. D.B. MacDougall), 7–31. Boca Raton, FL: Woodhead Publishing Limited and CRC Press, LLC.

12 Kramer, A. and Twigg, B.A. (1970). *Quality Control for the Food Industry*, 3e, vol. 1, 10–154. Westport, CT: The AVI Publishing Company, Inc.

13 Lawless, H.T. and Heymann, H. (1998). Texture evaluation (Chapter 11), Color and appearance (Chapter 12). In: *Sensory Evaluation of Food, Principles and Practices.* New York: Chapman & Hall pp. 379–405, 406–429.

14 Parker, R. (2003). Nutrition and digestion (Chapter 4), Food composition (Chapter 5), Quality factors in foods (Chapter 6). In: *Introduction to Food Science* (ed. R. Parker), 63–108. New York: Delmar, Thomson Learning.

15 Potter, N.N. and Hotchkiss, J.H. (1995). Constituents of foods: properties and significance (Chapter 3), Nutritive aspects of food constituents (Chapter 4), Quality factors in foods (Chapter 6), Food safety, risks and hazards (Chapter 23). In: *Food Science*, 5e. New York: Chapman & Hall pp. 24–68, 90–112, 532–558.

16 Charley, H. (1982). Chapter 1 – Evaluation of foods. In: *Food Science*, 2e (ed. H. Charley), 1–26. New York: Wiley.

17 deMan, J.M. (1990). Color (Chapter 6), Flavor (Chapter 7), Texture (Chapter 8). In: *Principles of Food Chemistry*, 2e, 203–333. New York: Van Nostrand Reinhold (AVI).

18 Noble, A., Philbrick, K.C., and Boulton, R.B. (2007). Comparison of sourness of organic acid anions at equal pH and equal titratable acidity. *J. Sens. Stud.* 1 (1): 1–8.

19 Heath, M.R. and Prinz, J.F. (1999). Chapter 2 – Oral processing of foods and the sensory evaluation of texture. In: *Texture Measurement and Perception* (ed. A.J. Rosenthal), 18–29. Gaithersburg, MD: Aspen Publishers, Inc.

20 Brennan, J.G. (1988). Texture perception and measurement. In: *Sensory Analysis of Foods*, 2e (ed. J.R. Piggott), 69–101. London, U.K.: Elsevier.

21 Reineccius, G. (ed.) (2006). Chapter 1 – An overview of flavour perception. In: *Flavor Chemistry and Technology*, 2e, 3–21. Boca Raton, FL, CRC Press/Taylor & Francis Group, LLC.

22 Vieira, E.R. (ed.) (1996). Chapter 2 – Composition and nutritional value of foods. In: *Elementary Food Science*, 4e, 11–46. New York: Chapman & Hall.

23 Preedy, V. and Rodríguez Méndez, M.L. (2016). *Electronic Noses and Tongues in Food Science.* Oxford, UK: Elsevier Inc.

24 Lignou, S. and Parker, J.K. (2015). Chapter 19 – Flavor production. In: *Handbook of Food Processing: Food Safety, Quality, and Manufacturing Processes* (ed. T. Varzakas and C. Tzia), 616–640. Boca Raton, FL: CRC Press/Taylor & Francis Group, LLC.

25 Stone, H. and Sidel, J.L. (2004). *Sensory Evaluation Practices*, 3e. London, U.K.: Elsevier Academic Press.

26 Civille, G.V. and Carr, B.T. (1987). *Sensory Evaluation Techniques*, 2e. London, U.K.: CRC Press, Inc.

27 Piggott, J.R. (1988). *Sensory Analysis of Foods.* Essex, U.K.: Elsevier Applied Science.

28 Lawless, H.T. and Klein, B.P. (1991). *Sensory Science Theory and Applications in Foods*. New York: Marcel Dekker, Inc.
29 Gacula, J.R.M.C. (1993). *Design and Analysis of Sensory Optimization*. Trumbull, CT: Food & Nutrition Press.
30 Gacula, J.R.M.C. and Singh, J. (1984). *Statistical Methods in Food and Consumer Research*. New York: Academic Press.
31 Kapsalis, J.G. (1987). *Objective Methods in Food Quality Assessment*. London, UK: CRC Press, Inc.
32 Lyon, D.H., Francombe, M.A., Hasdell, T.A., and Lawson, K. (1992). *Guidelines for Sensory Analysis in Food Product Development and Quality Control*. London, U.K.: Chapman & Hall.
33 Lawless, H.T. and Heymann, H. (2010). *Sensory Evaluation of Food: Principles and Practices*, 2e. New York: Springer Science & Business Media, LLC.
34 Zawirska-Wojtasiak, R. (2012). Chapter 19 – Methods for sensory analysis. In: *Food Flavors: Chemical, Sensory and Technological Properties* (ed. H. Jelen), 440–456. Boca Raton, FL, CRC Press/Taylor & Francis Group, LLC.
35 MacFie, H.J.H. and Thomson, D.M.H. (1994). *Measurement of Food Preferences*. London, U.K.: Blackie Academic & Professional.
36 Meiselman, H.L. and MacFie, H.J.H. (1996). *Food Choice, Acceptance and Consumption*. London, U.K.: Blackie Academic & Professional.
37 Moskowitz, H.R. and Beckley, J.H. (2012). *Sensory and Consumer Research in Food Product Design and Development*. Oxford, U.K.: IFT/Wiley/Blackwell Publishing.
38 Munoz, A.M., Civille, G.V., and Carr, B.T. (1992). *Sensory Evaluation in Quality Control*. London, U.K.: Van Nostrand Reinhold.
39 Burgard, D.R. and Kuznicki, J.T. (1990). *Chemometrics: Chemical and Sensory Data*. London, U.K.: CRC Press, Inc.
40 Dijksterhuis, G.B. (1997). *Multivariate Data Analysis in Sensory and Consumer Science*. Trumbull, CT: Food & Nutrition Press, Inc.

# 11

# Challenges of Sensory Science: Retention and Release

*Shradha Soni[1], Roshin Thankachan[2], Józef T. Haponiuk[3], and Sreeraj Gopi[4]*

[1]*Plant Lipids (P) Ltd., Department of Technical Services, Kolenchery, Cochin, Kerala, 682311, India*
[2]*R&D Centre, Tropical flavors (P) Ltd., Kolenchery, Cochin, Kerala, 682311, India*
[3]*Gdansk University of Technology, Department of Polymer Technology, G. Narutowicza 11/12, 80-233, Gdańsk, Poland*
[4]*R&D Centre, Aurea Biolabs (P) Ltd., Kolenchery, Cochin, Kerala, 682311, India*

## 11.1 Introduction

Food, indispensable to human beings, deals with provision of essential nutrients necessary to support human life and good health [1]. Even so, an individual's acceptability of any food item peculiarly depends on chemical senses [2]. Henceforth, human biological senses act as a safeguard to find the variance between deteriorated or unsuitable items and nutritious or healthy items; the senses also help to provide explicit and reliable qualification of food items [3]. As a result, sensory science and consumer perception toward any edibles has been a great challenge and concern for the food industry sector in terms of embracing and bewitching food products [4]. In other words, sensory science elucidates an integrative realm that helps to induce, measure, interpret, and analyze an individual's behavior or response toward product characteristics perceived by the sensory apparatus of vision, smell, taste, touch, and sound [5].

In particular, sensory appraisal imparts dexterity to food technologists, scientists, and processors to enhance food items by amendment of the ingredients. Further, sensory evaluation not only remarks on the immense change in product characteristics but also helps to disclose minuscule differences in the product's properties to meet consumer needs and satisfaction [6]. With the ramification of time, the discern vigor of sensory attributes can change. The potency of a sensory assay based on progressive measurement of intuitive variation can turn out sensory attributes over time. Incessant time-based modes rely on different parameters and aid in scrutinizing ample profane etiquette of an individual's taste stimuli like salty, bitter, sweet, and sour [3]. Sensory studies succeed in exploiting novel merchandize, and attuning to patron's preference

*Natural Flavours, Fragrances, and Perfumes: Chemistry, Production, and Sensory Approach*, First Edition. Edited by Sreeraj Gopi, Nimisha Pulikkal Sukumaran, Joby Jacob, and Sabu Thomas.

can turn out to reduce strife between the food producer and the consumer [7]. Generally, sensory evaluation operating procedures or strategies mutually pivot on food science, physics, computation, engineering, humanities, sociology, ergonomics, psychology, and further sciences [8]. Thus, a sensory evaluation study not only helps in the marketing, evolution, and modification of products, but it also portrays a consequential characteristic to form safe and qualitative products [3].

As a matter of fact, the creation of standards compatible with quality and safety has a glaring concern toward consumer acceptance. Abundant interior (hue, appearance, form, dimension, structure, fragrance, and appetite) and extraneous (cost, tag, trade name, product promotion, nutritional facts, sustenance information, emergence, business unit name, ease, and satisfaction) sensorial features ascribe to perceive food commodities [9]. Moreover, an individual's perception and acceptance of any product are hinged upon its potentiality of chemical senses. Human chemical senses such as sight, sniff, sound, tang, and touch exert neural stimulators jointly configure sensory apparatus and triggers to acquire central tastes (salt, sweet, sour, bitter, and umami). Chemistry beyond the sensory studies elucidates in terms of interpersonal chemistry linking customer perception and market exigency. Thus, sensory chemistry assists consumer consciousness, deportment, and discernment concerning edible products [10]. Consequently, this chapter presents an insight into the sensory science and overview on the sensorium organs. Also, the chapter briefs about the challenges faced and the different factors involved in the flavor release and retention.

## 11.2 Bottlenecks and Novel Insights of Sensory Science

In the ancient era, sensory analysis was based on the solitary authority that was responsible for entire chores from manufacturing to modification of the product. However, later on, sensory determination by a single authority or expert became a snag for the food enterprise because solitary analyzers might face various issues such as physiological and psychological illness, retirement, travel, etc., along with reduced consumer acceptability of the product [11]. To conquer these sagas and overcome hurdles, sensory evaluation modus operandi evolved into different methods such as discriminatory, descriptive, and effective methods [12]. These analytical (discriminatory and descriptive) and affective tests are based on genuine demonstrations or trials followed by maintaining appropriate documents [5].

The discriminatory test, also known as the difference test, is further stratified into a triangle, duo-trios, paired comparison, ranking, and multicomparison tests. These analytical tests conducted by trained and untrained panels generally rely on aggregate periodicity of tactful variability in the samples [13]. On the other hand, descriptive tests are much more expensive, extensive, and factual and are only performed by a perfectly trained panel. It works on four models, namely sensory spectrum, flavor profile, texture profile, and quantitative analysis [14]. The effective sensory method of consumer testing is centered on the scale of liking and disliking a specific product through the patron's preference or acceptability. This straightforward assessment is executed with the aid of nine numeral hedonic scales, ranking, and paired preference methods through an unskilled group of people [9]. These specified analytical and affective sensory techniques are carried out by the *Homo sapiens* sensorial panel through appropriate methods. The sensory analyzers are thus required to attend formal training. Alongside, only particular number of members can participate in tasting for effectual outcomes. The involvement of a specified number of analyzers in a panel and lengthy tasting procedures erratically lead to sensory fatigue [15].

Even though the organoleptic sensory analysis remains one of the most pre-eminent test methods in the food industry, it is still associated with several stumbling blocks such as intuitiveness, inconsistency, infections, and diseases and increased exposure to toxic compounds. To overcome these barriers, scientists and researchers have developed a novel technique that involves the use of modern electronic pieces of machinery that can mimic human senses [16]. This novel human mimic sense, also known as electronic senses, includes various analytical devices such as e-eye [17], texture analyzers [18], e-nose, e-tongue [16], and aural devices [19]. In this modern era, these mechanical senses have made a mark in terms of perfection, efficiency, effectiveness, and continuity [16].

## 11.3 Sensorium Organs

With growing interest in personalized foods, the theme of chemosensory is progressing among purchasers and food manufacturers [20]. The neural-perceptible network model includes sensory receptors, neural avenue, and part of the brain that plays a salient role in the perception of humans [21]. Sensorium organs carry various somatic, olfactory, taste, light, and sound receptors that help transfer stimulus from specific organs to the neural system. Generally, sensory receptors are classified as photo, thermal, mechanical, and chemical receptors, which act as an actuator that transform stimulation into effectors' potential [22].

### 11.3.1 Sensory of Sight

The human eye can detect light only from 380 to 700 nm wavelengths. Hence, this wavelength range is known as the visible light electromagnetic spectrum [23]. Photo or vision receptors carry chromophores that accompany the mechanism of electromagnetic effectors, G-protein, phosphor-diastase, and cGMP barrier ion passage. The retina, which is near the optic nerve, is the part that collects illumination[24, 25]. Optic retinal cone and rod photoreceptors convert luminous energy into electrical signals, which are forwarded to the brain as neural signals. The number of cylindrical rod cells in the retina is high, in contrast to conical cone cells. Approximately 92 million rod cells and 4.6 million cone cells are localized on the retinal membrane [26]. Luminous potency and optical perspicuity of rod cells are higher, whereas cone cells work vice-versa. Rod cells possess highly light-delicate achromous pigment that is capable of night vision. On the other hand, retinal cone cells induce color vision with the aid of M, S, and L cone cells, which accelerate vibrant green, blue, and red hues [27, 28].

The phenomenon of optical illusion transduction is initiated when the human eye pupil fetches the visible electromagnetic beam. Constrained light by the pupil streams on the peripheral retinal part and further impinges on the medial retinal bipolar receptors, where the receptor signals emitted to intramural retinal ganglion cells that mutually form ocular nerves, which assist to configure the brainiac concept. Later, light stimulation on receptors causes the hyperpolarization of pigment receptor (rhodopsin or iodopsin), which leads to the change in pigment configuration. Rhodopsin (R) binds to 11-cis retinal isoforms in the presence of a photon, and it is converted to all-trans retinals [25, 29]. Opsin forms metarhodopsin 11 (M11) by atomic spatial arrangement. Further M11 pinion to transducin or G-protein transducin, which segregates α, β, and γ trinity from GDP-bound transducin ($G\alpha_t$-GDP-$G\beta_{\gamma t}$). Activation of transducin by M11 on $G\alpha_t$ position generates a change from bound GDP to GTP through swift segregation of $G\beta_{\gamma t}$ from $G\alpha_t$-GTP and detachment of transducin from M11. After successive activation of transducin, α-GTP binds to one of two tetramer cGMP phosphodiesterase enzymes (PDE6 or PDE) lead to the activation of PDE and hydrolysis of cGMP. Trimming of the cGMP cytoplasmic mass results in termination of sodium transport, which further carries on the inflow of potassium ion and potential drop of calcium ion passage because of hyperpolarization. Simultaneously, the declining level of glutamate releases ions from photoreceptor cells due to the closure of the CNG (cyclic nucleotide-gated) channel, which leads to the hyperpolarization of bipolar off-centered cone cells and depolarization of bipolar on-centered cells. Hence, reduced glutamate excites the nerves due to lowering of the obstruction of the retinal nerve [28, 30, 31].

A favorable ending of the phototransduction cascade is necessary for the persistent recovery of light response or for new signals. Consequently, the inactivation of major activated proteins (M11, $G\alpha_t$-GTP, and PDE6) and cytoplasm cGMP is induced by the enzyme guanylyl cyclase [32]. Calcium–guanylate cyclase-activating protein (GCAP) plays a prominent role in the recovery of discharged cGMP amount by the alienation of $Ca^{2+}$ and GCMP, which resumes the transport of the cation through the cGMP barrier. Activated transducin is transformed into a deactivated form through GAP. Ultimately, the rhodopsin kinase enzyme depletes holding empathy of transducin. Arrestins, a group of proteins, ends the phototransduction process by inactivating M11 Phosphorylation. Hence, this results in the opening of the cationic channels, in the plasma membrane, and the influx of calcium [33].

### 11.3.2 Sensory of Olfaction

The human olfactory system consists of a stereotyped anatomy structure, which plays an evolutionary role in social and psychic behaviour. Long back, the sensory olfaction system was believed to be the unimportant sensory system in humans. But recent research shows that, even with the weak characterization of the olfactory receptors, the cortical region of integration of the olfactory sensation has a vital affinity with language, memory, and neuron-vegetative areas where it affects the *H. sapiens* to perceive food by cherishing its flavour and it shows beyond doubt a very essential economic factor [34, 35]. Chemosensory of human odorant or olfactory system is a transduction vehicle which is initiated by sniffing in the odorant molecules from the air followed by the incorporation of the bony skeletal nose, nares, nasal chamber that overhead to olfaction mucoid epithelium pellicle for olfactory perception, bowman's glands, olfactory sensory or receptor, olfactory bulb, neurons, and nerve fibers [36].

The initiation of olfaction sensory triggers in the nasal cavity includes three zones. The primary zone is the nasal vestibule, which appears in the interior section of the nostrils. The secondary zone is the respiratory region found over the nasal vestibule, which aids in confining foreign particles and humidifying the air through the mucus. The tertiary zone is a distinctive sheet of epithelial tissues, also known as the olfactory epithelium zone, which carries three types of cells, namely basal cells, supporting cells, and olfaction sensorial receptors [34]. Functionally, basal cells act as stem cells that have the potential of formation, cleavage, and differentiation of cells into olfactory or supporting tissues. The supporting cells are stratified into non-neural microvillar and sustentacular cells along with chemoreceptors [37]. Microvillar cells produce cKIT protein, also known as stem cell growth factor receptor (SCFR), whereas sustentacular cells help in clarifying harmful chemicals and provide physical and metabolic support to olfactory epithelium [38]. The olfactory chemoreceptor cells bind with bipolar neurons that manifest olfaction effectors towards immobile bristles in the vicinity of dendrite node termination where they produce transducer sites for olfaction impetus [39].

The lamina propria connective tissue is placed underneath the sheet of olfactory epithelium, which accommodates Bowman's glands or olfactory glands. These glands secrete mucus that protects the olfactory epithelium realm. The bipolar neurons that are present on the olfactory chemoreceptor have two ledges. The olfactory cilia or hairs capture the odorants that are further imprisoned by the mucus [36]. The second receptor ledge performs as an axon that exploits fusion with additional chemoreceptor axons. The fusion of axons is mutually named cranial nerve 1 or olfactory nerve. This minute cranial nerve 1 or olfactory nerve elapses through the cribriform plate. The cribriform plate is a scrap of the ethmoid bone that gives rise to the olfactory area. The tapered and intense furrowed cribriform plate creates a passage with olfactory foramina for permitting cranial nerve 1 to the interior of the olfactory bulb. The olfactory bulb communicates information to the temporal lobe in the region of the olfactory cortex by the second series of neurons through olfactory transit [35].

When odor stimulation occurs in the epithelium layer of cilia, the G-protein-coupled receptors capture the odorants. At the same time, various odorant molecules bind with the receptors. Among thousand receptors, only a specific odorant molecule imprisons the specific receptor. Consequently, the G-olfactory receptor activates, which leads to action transduction. The activation of a couple of receptors triggers the opening of sodium and calcium routes that permit the stream of ions within the cell. Simultaneously, depolarization of the cell membrane occurs due to the inflow of positive calcium ions and outflow of negative chloride ions, which leads to bombardment in neurons. On receiving the indication by the olfactory bulb, it produces a juncture with the olfactory nerve with mitral cells (2nd order neurons). After turning on the receptors, mitral cells are depolarized with the aid of released glutamate, which further gives rise to the olfactory tract [40]. This tract is divided into medial tractor stria and lateral tractor stria. The lateral tractor is the top spot of olfaction that flicks to the piriform cortex of the temporal lobe. The filaments coming from the lateral area can proceed to the limbic system. The limbic system is the zone of emotions and memories [36]. This system exists beneath the cerebrum on the pair of the thalamus, which contrives into the hippocampus, amygdala, and hypothalamus. Besides, the medial tractor bears the axons that stretch across from the piriform cortex contralateral olfactory region, which spreads the olfaction to the couple portion of the brain [36, 41]. After an interval of rhythm, the odor gets adapted, the odorants initiate inducing lesser effects which causes a final drop in perceiving the odors [42].

### 11.3.3 Sense of Touch

The sense of touch or tactile, also known as somatosensory, relies on a straight connection between two physical forms. According to neurology, somatosensory is a peculiar system where the body can perceive touch, pressure, pain, and vibration concerning the responsive mind. Being the largest organ of the body, skin plays a vital role in somatosensory transduction [43]. It carries an immense quantity of thermoreceptors, nociceptors, mechanoreceptors, and chemoreceptors. *H. sapiens* sense of tactile evolves extremely in hands where it receives the information about the intrinsic and extrinsic environment. There are four varieties of mechanoreceptors found in glabrous myelinated skin, which cause somatic transduction. These receptors work based on the different scheme that includes tactile cells or Merkel Ranvier cells, Pacinian corpuscles, Meissner's corpuscles, and Ruffini corpuscles [44].

Merkel cells are oval and round-shaped mechanoreceptors that are responsible for slow or light touch perceptions. These cells are widely present on the human fingertips, and they reciprocate the shape, form, and pressure of the material they touch. Pacinian corpuscles, called lamellar corpuscles, are present as long cylindrical oval-shaped under extensive layers of skin. These receptors are unusual nerve endings in the skin that stimulate the texture of the objects concerning vibrate sensation [45]. Meissner's corpuscles, or tactile corpuscles, are broadly apportioned at the area of lips and fingers rather than other portions of the skin. These enveloped and unmyelinated nerve fiber cells are found among neurolemmocytes, especially inside the dermis papillae and underneath the epidermis. Generally, tactile corpuscles give swift stimulation by the elicitation of touch with an object [46]. Ruffini corpuscles are spotted as a long capsulated form at the intense sheath of skin. Usually, these receptors are abysmally stimulated in human skin and play a prominent role in the self-motion of the body, such as finger actions, cuticle stretch, etc. [47].

Along with mechanoreceptors, other receptors such as thermoreceptors, nociceptors, and chemoreceptors, play an important role in somatosensation. Among them, thermoreceptors differentiate the various ranges of temperature stimulation on the skin [48]. Nociceptors signify the pain sensations in tissues, such as acute pressure, chemical burn, or vigorous temperature, whereas chemoreceptors recognize the variation in the chemical environment and generate action potential by a stimulant. Regarding food perception, texture plays a prominent role in terms of somatosensory. In somatosensation, not only eyes but fingers, palms, and hands touch also reveal the form, state, or texture of the food product [49]. Oral somatosensation also has a considerable part in sensation. Oral somatosensation commences exertion by stimulation. The attribute of the food item immediate in the oral inlet is passed on to the brain through trigeminal nerve ganglia followed by the trigeminal nerve where it collects various sensations from different receptors. Consequently, it emanates and proceeds to the tertiary stage of neurons in the thalamus [50]. The aggregated structure of axons emerges from the thalamus and passes the information through the primary somatic sensation cortex to the cortical area of the brain. The cortical area forms an illustration of the product with the aid of various cumulative indicators [48].

### 11.3.4 Sensory of Taste

Taste modality imparts an essential function in an individual's day-to-day life regarding emotions, satiety, preferences, safety, and qualities of food. Moreover, taste perception depends on various factors, including genetic factors, body mass index, hunger, smoking, health conditions and pregnancy, age, gender, and environmental and biological factors. The human brain experiences various appetites, like sweetness that fulfills the desire for energy and carbohydrates, and food with salt makes a healthy balance of fluid, muscle, and nerve activity in the body [51]. It also enhances the flavor of food. Sourness in meals indicates digestion and also stabilizes the excess sweet or fatty taste. Savory or umami indicates the pleasant protein in food. Bitter is just an enigma that signals health in terms of detoxification [52]. Along with five tastes, fatty appetite and temperature also come under consideration, where fat signals the mellow, richness, luxurious or creamy appetite, and temperature generates a spirited relish in the food. Eventually, taste cells or receptors associated with taste buds of the gustatory system encompass the tongue, epiglottis, palate, esophagus, pharynx, and larynx [53]. Different receptors in the oral cavity signal various tastes such as bitterness, sweetness, umami, sourness, and saltiness to the gustatory cortex. The G-protein-coupled receptor T2Rs signal bitterness, T1R2 and T1R3 indicate sweetness, and T1R1 and T1R3 trigger umami taste. The PKD2L1 and Car4 express sourness and saltiness that is signaled by ENaC (epithelial sodium channel) [54]. The action potential of taste begins with the stimulation of receptors by the food. The gustatory receptors are coupled with primary sensory axons that are cranial nerves VII, IX, and X (vagus nerve). After receiving the signal, information of taste flows through the cranial nerves, cranial nerve VII routes into the chorda tympani, cranial nerve IX runs through the inferior ganglion, and cranial nerve X comes from the receptors present on the pharynx, epiglottis, and soft palate, which proceeds through nodose ganglion [55]. These primary nerves enter the medulla and meet the gustatory nucleus or nucleus of the solitary tract. The fibers from the portion of the nucleus transmit the signals to the ventral posterior thalamus, where it further send the fibers to insula, operculum, and other parts of the brain, such as the hippocampus (responsible for retaining the memory of taste) and hypothalamus where it gives the perception of taste [56].

### 11.3.5 Sense of Hear

The acoustical sense manifests, devouring something with ears. Although aural sensorial is a neglected flavor, it is considered during the phase of sensory evaluation in terms of crunchiness, crankiness, crispiness, carbonated, and squeaky. The sound of food comes after biting and mastication, which imparts numerous hearing cues where it makes it easier to understand about textural characteristics of food. The sound accords enjoyment and also acts as a foremost sensory in the title of quality while rating the range of pleasant crisp of product [57]. Indeed, food sound depends on the makeup and arrangement of product cells. For instance, some juicy products like apples and pears that contain sap lead to the wet crisp sound whereas, products like biscuits, chips, grain flakes, and cheese balls contain air in their cells known as dry crisp. The carbonated or fizzy products are suspected to give a pop and sparkle sound, whereas creamy products raise the friction to feel with the oral area, and that ultimately produces slightly sonic cues. Products like halloumi cheese produce squeaky sounds with a stick–sneak biting motion. Therefore, sonic cues perception phenomena initiate either from food or by an individual's oral cavity [58].

The mechanical sonic vibrations or waves pass through the passage of the outer, middle and inner area of the ear to the primary auditory cortex of the brain, where it perceives the structure of sound. While biting and chewing the food, the sound vibrations generate and travel through the medium of air to the outer part of the ear. The visible auricle or pinna leads the canal-like pathway, where the myringa or tympanic membrane partitions the outer area of the ear from the middle area of the ear [59]. The refined sound waves oscillate by the eardrum, and in the area of the eardrum, sebaceous and ceruminous glands secrete ear wax, which plays a role in defending the tympanic nerve. The ossicle bones, namely incus, mallus, and stapes, aid in the transmission of sonic waves from the air to the fluid-filled area (cochlea) [60]. The tensor tympani and stapedius muscles locale in the middle ear behaves as a safeguard against acoustic reflex of sound and further, it passes on the waves to the inner area of the ear. The cochlea inside the interior ear has an organ of Corti, which carries receptors and is responsible for neural transduction [61]. The organ of Corti carries two types of hair cells the inner hair cells, known as the hearing receptors (which hold the sensitive stereocilia cells ), and the outer hair cells aid in tuning the cochlea. The movement of stereocilia cells or the basilar layer causes depolarization. After depolarization, the signal goes to the posterior area of the brain (brainstem), where sound is perceived [62].

## 11.4 Factors Affecting Flavor Retention and Release

Flavor retention and release are willfully planned to design healthier products without conceding on traditional product acceptability, functional beverages, and beverages with exotic features. For instance, in skim milk, reduction of hydrophobic flavors challenges consumer's acceptability compared to that of high-fat milk, instead, the potential health benefit of soymilk suffers from a beany off-flavor originating from lipoxygenase activity. Accordingly, flavor release or retention is largely affected by the intrinsic chemical properties of the flavor (hydrophobicity, hydrophilicity [log *P* value], and volatility), the composition of the medium (lipid, protein, salt, sugar, etc.), and finally environmental conditions (temperature, pH). On the contrary, the interaction between flavor compounds and other food ingredients under given environmental conditions determines the intensity of flavor retention or release from a product [63].

The release of aroma compounds from foods is determined by the partition coefficient between the air phase and food matrix and, in the retronasal case, by the partition coefficient between the water phase (saliva) and the food matrix. If an aroma compound is added to the water matrix in a closed system and allowed to reach equilibrium, it will distribute between the air and water phases according to its air-to-water partition coefficient [64]. For instance, lipids present in the form of an emulsion are the strongest volatile retainers due to the lipophilic nature of most of the volatile flavors. Proteins also have flavor retention properties, whereas carbohydrates hardly have a retention effect in beverages. Smaller components such as sugars and salts can change the water activity, thereby facilitating flavor release. Alternatively, salts can also indirectly affect the binding sites of proteins, leading to release (e.g. NaCl and $Na_2SO_4$) or retention (NaCSN and $Cl_3CCOONa$) of flavors [65].

### 11.4.1 Flavor Binding and Entrapment

The perception of flavors is contributed by aroma, flavor, texture, and mouthfeel, which depends on the physicochemical interactions of flavors with other molecules. For instance, in food industries, flavor retention and release are on the whole studied to develop more healthy products (low-fat milk, low-alcoholic beer, etc.) without compromising on traditional product acceptability, practical liquids (consisting of drinkable meal replacers or recreation dietary supplements), and liquids with unusual capabilities (distinct fruit tastes, cocktails, fusions, etc.) [63]. Yet, the loss of free volatile compounds due to unsuccessful binding of flavor compounds is known to have an effect on the purchaser acceptability. Thus, flavor binding studies, binding mechanisms, and factors affecting the interaction involved in the flavor entrapment are of great interest to flavor chemists. For instance, starch, an extensively studied hydrocolloid, amylose has been shown to form complexes with aroma compounds, and the physical state of carbohydrates is one parameter influencing flavor retention. Yet, the important impact of hydrocolloids seems to be a hassle for the diffusion of aroma compounds due to changes in viscosity. Addition of fat induces sizable retention of hydrophobic-flavor compounds resulting in important effects on flavor notion [66].

### 11.4.2 Flavor–Matrix Interaction

Food is a complicated system containing mixtures of volatile compounds liable for flavor perception and non-volatile compounds. Thus, to be perceived, aroma compounds have to be released from the food matrix [67]. Yet, flavor perception has an influence on the Interactions between flavor compounds and food ingredients. For instance, proteins are known to bind flavor compounds, the most-studied example, β-lactoglobulin, has hydrophobic interactions with volatiles, due to the effect of the medium on the conformation of the protein and its ability to bind flavor compounds. Moreover, the retention of volatiles by protein is much lower than that by fat, and in emulsions, the presence of protein at the oil/water interface induces a significant effect on flavor release and flavor perception of hydrophobic flavor compounds [66].

## 11.5 Future Prospects

Food choice is a complex process inspired by a variety of things associated with the product (intrinsic and extrinsic properties), the consumer (for instance, knowledge, beliefs, attitudes), and the consumption context (for instance, occasion, cultural environment). The role of the consumer in determining the market achievement of a product is of maximum relevance. The motivation for food preference can be stimulated by an interest in health, weighs, sensory pleasure, ideological reasons, comfort, rate, or familiarity [68]. Moreover, the significance of sensory insight in terms of degree of acceptance of products is extensively appreciated by the agro-food sector, and it brings forth a challenge for experts. Various qualitative methods such as microbiological, chemical, physical, and sensorial are being utilized for sustainability of food products [69]. Earlier, sensory wings were only part of the role in dispensing the quality assurance specifics, but with the advancing scenario, it participates in the whole cycle of product. Consequently, together with the researchers team and the merchandize members anticipate the perception which is needful for evolution of the product [12]. For example, changing the fat content modifies the general perception of a mixture of taste compounds from exclusive chemical classes. The melting factor of the fat affects the solubility of flavors and accordingly the retention and release of flavors. Emulsification and size of droplets also affect flavor release and perception. Thus, more research is required on the results of actual food samples containing combinations of different flavor compounds [66].

## References


1 Sonesson, U., Davis, J., Flysjö, A. et al. (2017). Protein quality as functional unit – a methodological framework for inclusion in life cycle assessment of food. *J. Cleaner Prod.* 140, Part 2: 470–478, ISSN 0959-6526, https://doi.org/10.1016/j.jclepro.2016.06.115.

2 Sirangelo, T.M. (2019). Sensory descriptive evaluation of food products: a review. *J. Food Sci. Nutr. Res.* https://doi.org/10.26502/jfsnr.2642-11000034.

3 Carolyn, F. (2009). Ross, sensory science at the human–machine interface. *Trends Food Sci. Technol.* 20 (2): 63–72, ISSN 0924-2244, https://doi.org/10.1016/j.tifs.2008.11.004.

**4** Zeng, X., Ruan, D., and Koehl, L. (2008). Intelligent sensory evaluation: concepts, implementations, and applications. *Math. Comput. Simul.* 77 (5–6): 443–452. https://doi.org/10.1016/j.matcom.2007.11.013.

**5** Duizer, L.M. and Walker, S.B. (2016). The application of sensory science to the evaluation of grain-based foods. In: *Encyclopedia of Food Grains*, 2e (ed. C. Wrigley, H. Corke, K. Seetharaman and J. Faubion), 144–153. Academic Press, ISBN 9780123947864, https://doi.org/10.1016/B978-0-12-394437-5.00134-0.

**6** Heymann, H. (2019). A personal history of sensory science. *Food, Cult. Soc.* 22 (2): 203–223. https://doi.org/10.1080/15528014.2019.1573043.

**7** Vivek, K., Subbarao, K., Routray, W. et al. (2020). Application of fuzzy logic in sensory evaluation of food products: a comprehensive study. *Food Bioprocess Technol.* 13: 1–29. https://doi.org/10.1007/s11947-019-02337-4.

**8** Grunert, K.G. (2015). The common ground between sensory and consumer science. *Curr. Opin. Food Sci.* 3: 19–22, ISSN 2214-7993, https://doi.org/10.1016/j.cofs.2014.11.003.

**9** Story, M. and French, S. (2004). Food advertising and marketing directed at children and adolescents in the US. *Int. J. Behav. Nutr. Phys. Act.* 1 (1): 3. https://doi.org/10.1186/1479-5868-1-3.

**10** Landis, B.N. and Hummel, T. (2019). The smell and taste corner: the human chemical senses are useless, are they not? *ORL* 81: 183–184. https://doi.org/10.1159/000501972.

**11** Lawless, H. and Heymann, H. (2010). Introduction. In: *Sensory Evaluation of Food. Food Science Text Series*. New York, NY: Springer https://doi.org/10.1007/978-1-4419-6488-5_1.

**12** Stone, H. (2018). Example food: what are its sensory properties and why is that important? *NPJ Sci. Food* 2: 11. https://doi.org/10.1038/s41538-018-0019-3. PMID: 31304261; PMCID: PMC6550172.

**13** Drake, M.A. (2008). Modern sensory practices. In: *The Sensory Evaluation of Dairy Products* (ed. S. Clark, M. Costello, M. Drake and F. Bodyfelt). New York, NY: Springer https://doi.org/10.1007/978-0-387-77408-4_17.

**14** Civille, G.V. and Oftedal, K.N. (2012). Sensory evaluation techniques – make "good for you" taste "good". *Physiol. Behav.* 107 (4): 598–605. https://doi.org/10.1016/j.physbeh.2012.04.015. PMID: 22554616.

**15** Abdullah, A.H., Rahim, N.A., Masnan, M.J. et al. (2016). Chapter 11 – Rice and the electronic nose. In: *Electronic Noses and Tongues in Food Science* (ed. M.L.R. Méndez), 103–113. Academic Press, ISBN 9780128002438, https://doi.org/10.1016/B978-0-12-800243-8.00011-1.

**16** Njoman, M.F., Nugroho, G., Chandra, S.D.P. et al. (2017). The vulnerability of human sensory evaluation and the promising senses instrumentation. *Br. Food J.* 119 (10): 2145–2160. https://doi.org/10.1108/BFJ-10-2016-0505.

**17** Apetrei, C., Apetrei, I.M., Villanueva, S. et al. (2010). Combination of an e-nose, an e-tongue and an e-eye for the characterisation of olive oils with different degree of bitterness. *Anal. Chim. Acta* 663 (1): 91–97, ISSN 0003-2670, https://doi.org/10.1016/j.aca.2010.01.034.

**18** Liu, Y.-X., Cao, M.-J., and Liu, G.-M. (2019). Chapter 17 – Texture analyzers for food quality evaluation. In: *Woodhead Publishing Series in Food Science, Technology and Nutrition, Evaluation Technologies for Food Quality* (ed. J. Zhong and X. Wang), 441–463. Woodhead Publishing, ISBN 9780128142172, https://doi.org/10.1016/B978-0-12-814217-2.00017-2.

**19** Dias-Faceto, L.S., Salvador, A., and Conti-Silva, A.C. (2020). Acoustic settings combination as a sensory crispness indicator of dry crispy food. *J. Texture Stud.* 51: 232–241. https://doi.org/10.1111/jtxs.12485.

**20** Nolden, A.A. and Feeney, E.L. (2020). Genetic differences in taste receptors: implications for the food industry. *Annu. Rev. Food Sci. Technol.* 11 (1): 183–204.

**21** Dalesio, N.M., Barreto Ortiz Sebastian, F., Pluznick Jennifer, L., and Berkowitz Dan, E. (2018). Olfactory, taste, and photo sensory receptors in non-sensory organs: it just makes sense. *Front. Physiol.* 9: https://doi.org/10.3389/fphys.2018.01673.

**22** Moini, J., Avgeropoulos, N.G., and Samsam, M. (2021). Chapter 1 – Anatomy and physiology. In: *Epidemiology of Brain and Spinal Tumors* (ed. J. Moini, N.G. Avgeropoulos and M. Samsam), 3–40. Academic Press, ISBN 9780128217368, https://doi.org/10.1016/B978-0-12-821736-8.00002-9.

**23** Foroni, F., Pergola, G., and Rumiati, R. (2016). Food color is in the eye of the beholder: the role of human trichromatic vision in food evaluation. *Sci. Rep.* 6: 37034. https://doi.org/10.1038/srep37034.

**24** Baker, S.A. and Kerov, V. (2013). Chapter 7 – Photoreceptor inner and outer segments. In: *Current Topics in Membranes*, vol. 72 (ed. V. Bennett), 231–265, ISSN 1063-5823. Academic Press, ISBN 9780124170278, https://doi.org/10.1016/B978-0-12-417027-8.00007-6.

**25** Salesse, C. (2017). Physiology of the visual retinal signal: from phototransduction to the visual cycle. *J. Fr. Ophtalmol.* 40 (3): 239–250, ISSN 0181-5512, https://doi.org/10.1016/j.jfo.2016.12.006.

**26** Grossniklaus, H.E., Geisert, E.E., and Nickerson, J.M. (2015). Chapter 22 – Introduction to the retina. In: *Progress in Molecular Biology and Translational Science*, vol. 134 (ed. J.F. Hejtmancik and J.M. Nickerson), 383–396, ISSN 1877-1173. Academic Press, ISBN 9780128010594, https://doi.org/10.1016/bs.pmbts.2015.06.001.

**27** Mustafi, D., Engel, A.H., and Palczewski, K. (2009). Structure of cone photoreceptors. *Prog. Retinal Eye Res.* 28 (4): 289–302, ISSN 1350-9462, https://doi.org/10.1016/j.preteyeres.2009.05.003.

**28** Weiss, E.R., Ducceschi, M.H., Horner, T.J. et al. (2001). Species-specific differences in expression of G-protein-coupled receptor kinase (GRK) 7 and GRK1 in mammalian cone photoreceptor cells: implications for cone cell phototransduction. *J. Neurosci.* 21 (23): 9175–9184. https://doi.org/10.1523/JNEUROSCI.21-23-09175.2001.

**29** Phan, N.M., Cheng, M.F., Bessarab, D.A., and Krivitsky, L.A. (2014). Interaction of fixed number of photons with retinal rod cells. *Phys. Rev. Lett.* 112: 213601. https://doi.org/10.1103/PhysRevLett.112.213601.

**30** Rieke, F. and Baylor, D.A. (1998). Single-photon detection by rod cells of the retina. *Rev. Mod. Phys.* 70: 1027. https://doi.org/10.1103/RevModPhys.70.1027.

31 Arshavsky, V.Y., Lamb, T.D., and Pugh, E.N. Jr. (2002). G proteins and phototransduction. *Annu. Rev. Physiol.* 64 (1): 153–187. https://doi.org/10.1146/annurev.physiol.64.082701.102229.
32 Gray-Keller, M.P. and Detwiler, P.B. (1994). The calcium feedback signal in the phototransduction cascade of vertebrate rods. *Neuron* 13 (4): 849–861, ISSN 0896-6273, https://doi.org/10.1016/0896-6273(94)90251-8.
33 Scott, K. and Zuker, C. (1997). Lights out: deactivation of the phototransduction cascade. *Trends Biochem. Sci* 22 (9): 350–354, ISSN 0968-0004, https://doi.org/10.1016/S0968-0004(97)01100-6.
34 Sela, L. and Sobel, N. (2010). Human olfaction: a constant state of change-blindness. *Exp. Brain Res.* 205 (1): 13–29. https://doi.org/10.1007/s00221-010-2348-6. . PMID: 20603708; PMCID: PMC2908748.
35 Huart, C., Rombaux, P., and Hummel, T. (2013). Plasticity of the human olfactory system: the olfactory bulb. *Molecules* 18 (9): 11586–11600. https://doi.org/10.3390/molecules180911586.
36 Silva Teixeira, C.S., Cerqueira, N.M.F.S.A., and Silva Ferreira, A.C. (2016). Unravelling the olfactory sense: from the gene to odor perception. *Chem. Senses* 41 (2): 105–121. https://doi.org/10.1093/chemse/bjv075.
37 Elsaesser, R. and Paysan, J. (2007). The sense of smell, its signalling pathways, and the dichotomy of cilia and microvilli in olfactory sensory cells. *BMC Neurosci.* 8: S1. https://doi.org/10.1186/1471-2202-8-S3-S1.
38 Genovese, F. and Tizzano, M. (2018). Microvillous cells in the olfactory epithelium express elements of the solitary chemosensory cell transduction signaling cascade. *PLoS One* 13 (9): e0202754. https://doi.org/10.1371/journal.pone.0202754.
39 Maßberg, D. and Hatt, H. (2018). Human olfactory receptors: novel cellular functions outside of the nose. *Physiol. Rev.* 98 (3): 1739–1763. https://doi.org/10.1152/physrev.00013.2017.
40 Antunes, G., Sebastião, A.M., and Simoes de Souza, F.M. (2014). Mechanisms of regulation of olfactory transduction and adaptation in the olfactory cilium. *PLoS One* 9 (8): e105531. https://doi.org/10.1371/journal.pone.0105531.
41 Sosulski, D.L., Bloom, M.L., Cutforth, T. et al. (2011). Distinct representations of olfactory information in different cortical centres. *Nature* 472 (7342): 213–216. https://doi.org/10.1038/nature09868.
42 Firestein, S.J., Margolskee, R.F., and Kinnamon, S. (1999). Olfaction. In: *Basic Neurochemistry: Molecular, Cellular and Medical Aspects*, 6e (ed. G.J. Siegel, B.W. Agranoff, R.W. Albers, et al.). Philadelphia: Lippincott-Raven https://www.ncbi.nlm.nih.gov/books/NBK28226.
43 ten Donkelaar, H.J., Broman, J., and van Domburg, P. (2020). The somatosensory system. In: *Clinical Neuroanatomy*. Cham: Springer https://doi.org/10.1007/978-3-030-41878-6_4.
44 Haggard, P. and de Boer, L. (2014). Oral somatosensory awareness. *Neurosci. Biobehav. Rev.* 47: 469–484, ISSN 0149-7634, https://doi.org/10.1016/j.neubiorev.2014.09.015.
45 Germann, C., Sutter, R., and Nanz, D. (2021). Novel observations of Pacinian corpuscle distribution in the hands and feet based on high-resolution 7-T MRI in

healthy volunteers. *Skeletal Radiol.* 50 (6): 1249–1255. https://doi.org/10.1007/s00256-020-03667-7.

**46** Bonny, J.M., Sinding, C., and Thomas-Danguin, T. (2017). Functional MRI and sensory perception of food. In: *Modern Magnetic Resonance* (ed. G. Webb). Cham: Springer https://doi.org/10.1007/978-3-319-28275-6_132-1.

**47** Purves, D., Augustine, G.J., Fitzpatrick, D. et al. (ed.) *Neuroscience*, 2e. Sunderland, MA: Sinauer Associates; 2001. Mechanoreceptors Specialized to Receive Tactile Information. https://www.ncbi.nlm.nih.gov/books/NBK10895.

**48** Piqueras-Fiszman, B. and Spence, C. (2012). The influence of the feel of product packaging on the perception of the oral-somatosensory texture of food. *Food Qual. Preference* 26 (1): 67–73, ISSN 0950-3293, https://doi.org/10.1016/j.foodqual.2012.04.002.

**49** Spence, C., Hobkinson, C., Gallace, A. et al. (2013). A touch of gastronomy. *Flavour* 2: 14. https://doi.org/10.1186/2044-7248-2-14.

**50** Wang, Q.J., Mielby, L.A., Junge, J.Y. et al. (2019). The role of intrinsic and extrinsic sensory factors in sweetness perception of food and beverages: a review. *Foods* 8 (6): 211. https://doi.org/10.3390/foods8060211. PMID: 31208021; PMCID: PMC6617395.

**51** Kikut-Ligaj, D. and Trzcielińska-Lorych, J. (2015). How taste works: cells, receptors and gustatory perception. *Cell. Mol. Biol. Lett.* 20 (5): 699–716. https://doi.org/10.1515/cmble-2015-0042.

**52** Noel, C. and Dando, R. (2015). The effect of emotional state on taste perception. *Appetite* 95: 89–95, ISSN 0195-6663, https://doi.org/10.1016/j.appet.2015.06.003.

**53** Maehashi, K. and Huang, L. (2009). Bitter peptides and bitter taste receptors. *Cell. Mol. Life Sci.* 66 (10): 1661–1671. https://doi.org/10.1007/s00018-009-8755-9. PMID: 19153652.

**54** Lee, A.A. and Owyang, C. (2019). Chapter 16 – Sugars, sweet taste receptors, and brain responses. In: *Molecular Nutrition: Carbohydrates* (ed. V.B. Patel), 265–283. Academic Press, ISBN 9780128498866, https://doi.org/10.1016/B978-0-12-849886-6.00020-3.

**55** Da Conceicao Neta, E.R., Johanningsmeier, S.D., and McFeeters, R.F. (2007). The chemistry and physiology of sour taste—a review. *J. Food Sci.* 72: R33–R38. https://doi.org/10.1111/j.1750-3841.2007.00282.x.

**56** Kim, U.-K., Breslin, P.A.S., Reed, D., and Drayna, D. (2004). Genetics of human taste perception. *J. Dent. Res.* 83 (6): 448–453. https://doi.org/10.1177/154405910408300603.

**57** Tunick, M.H., Onwulata, C.I., Thomas, A.E. et al. (2013). Critical evaluation of crispy and crunchy textures: a review. *Int. J. Food Prop.* 16 (5): 949–963. https://doi.org/10.1080/10942912.2011.573116.

**58** Zdunek, A., Konopacka, D., and Jesionkowska, K. (2010). Crispness and crunchiness judgment of apples based on contact acoustic emission. *J. Texture Stud.* 41: 75–91. https://doi.org/10.1111/j.1745-4603.2009.00214.x.

**59** Van Der Bilt, A., de Liz Pocztaruk, R., and Abbink, J.H. (2010). Skull vibration during chewing of crispy food. *J. Texture Stud.* 41: 774–788. https://doi.org/10.1111/j.1745-4603.2010.00254.x.

**60** Casale, J., Kandle, P.F., Murray, I. et al. (2022). Physiology, cochlear function. In: *Stat Pearls*. Treasure Island, FL: Stat Pearls Publishing https://www.ncbi.nlm.nih.gov/books/NBK531483.

**61** Brown, M.C. (2009). Auditory system: efferent systems to the auditory periphery. In: *Encyclopedia of Neuroscience* (ed. L.R. Squire), 753–757. Academic Press, ISBN 9780080450469, https://doi.org/10.1016/B978-008045046-9.00248-5.

**62** Kozlov, A.S., Risler, T., Hinterwirth, A.J., and Hudspeth, A.J. (2012). Relative stereociliary motion in a hair bundle opposes amplification at distortion frequencies. *J. Physiol.* 590 (2): 301–308. https://doi.org/10.1113/jphysiol.2011.218362.

**63** Ammari, A. and Schroen, K. (2018). Flavor retention and release from beverages: a kinetic and thermodynamic perspective. *J. Agric. Food. Chem.* 66 (38): 9869–9881. https://doi.org/10.1021/acs.jafc.8b04459.

**64** De Roos, K.B. (2000). Physicochemical models of flavour release from foods. *Flavour Release* 126–141.

**65** Naknean, P. and Meenune, M. (2010). Factors affecting retention and release of flavour compounds in food carbohydrates. *Int. Food Res. J.* 17: 23–34.

**66** Guichard, E. (2002). Interactions between flavor compounds and food ingredients and their influence on flavor perception. *Food Rev. Int.* 18 (1): 49–70. https://doi.org/10.1081/FRI-120003417.

**67** Paravisini, L. and Guichard, E. (2016). Interactions between aroma compounds and food matrix. In: *Flavour* (ed. E. Guichard, C. Salles, M. Morzel and A.-M. Le Bon). Wiley https://doi.org/10.1002/9781118929384.ch9.

**68** Kaya, I. (2016). Motivation factors of consumers' food choice. *Food Nutr. Sci.* 7: 149–154. https://doi.org/10.4236/fns.2016.73016.

**69** Kemp, S.E. (2008). Application of sensory evaluation in food research. *Int. J. Food Sci. Technol.* 43: 1507–1511. https://doi.org/10.1111/j.1365-2621.2008.01780.x.

# 12

# Virtual Screening: An *In Silico* Approach to Flavor Compounds

*Nirosha Pulikkal[1], Nimisha P. Sukumaran[2], and Dhanesh Haridas[1,3]*

[1]*Epixel Pvt Ltd., R&D Centre, 15/159(5), Palakkad, Kerala, 678001, India*
[2]*Aurea Biolabs (P) Ltd., R&D Centre, Aurea Research Centre, Kolenchery, Cochin, Kerala, 682311, India*
[3]*Epixel Solutions, Department of IT and Engineering, 3651 Lindell Rd., Suite D1050, Las Vegas, NV, 89103, USA*

## 12.1 Introduction

Flavor compounds are naturally present in foods or are a consequence of physiological or enzymatic process as well as generated by microorganisms during the fermentation process. These compounds can be produced or modified through chemical, biochemical, or microbial changes during their extraction, processing, and storage, thus affecting the overall food quality and sensory profile [1]. Flavor compounds are naturally present in plants as secondary metabolites, and they are a diverse series of low-molecular-weight compounds. Besides, these chemical diversities in these metabolites make them important candidate for discovery of new flavors and "signature" molecules [2]. Thus, being a fundamental sense in humans, olfaction makes flavor and fragrance (F&F) an indispensable human–environment interaction. Consequently, F&F industry research highlights this perspective as a vast and attractive field, which aims to decipher the mechanism of action of flavor ligands in activating the receptors, leading to advancements in flavor research field [3].

Flavor can be broadly classified into taste and odors, where taste is divided into five primary tastes – sweet, sour, salty, bitter, and umami. On the other hand, odors are much more diverse and difficult to classify, which include spicy, flowery, fruity, resinous or balsamic, burnt, and foul [4]. For instance, flavor-enhanced foods and beverages is found to be more appealing as chemosensory changes arise [5]. The scientific community is now witnessing a newer, faster, and sophisticated approach to screen flavor diversity to improve the aroma and flavors with the aid of groundbreaking opportunities in food biotechnology. For instance, beer being the most popular beverages in the world, and its flavor profiling is important for brewers to optimize beer production and to guarantee odor quality and taste stability of the final products. Tatiana et al. developed method to identify marker compounds for

*Natural Flavours, Fragrances, and Perfumes: Chemistry, Production, and Sensory Approach,*
First Edition. Edited by Sreeraj Gopi, Nimisha Pulikkal Sukumaran, Joby Jacob, and Sabu Thomas.

beer flavor profiling to achieve beer quality optimization and quality control in routine laboratories [6]. Similarly, Yang et al., distinguished different metabolites phenotypes in Chinese liquor by determining the relationship between physico-chemical properties and their metabolic profiles. They identified 647 compounds of non-volatile compounds and volatile compounds which were caused by different metabolic pathways in its starter culture (Daqu) [7].

*In silico* approaches are gaining ground as a fast and cheap tool to predict the generation, characteristics, and structure taste relationship of sensory peptides in foods prior to complex *in vitro* and *in vivo* studies by using instrumental and sensory analysis [8]. *In silico* analysis of chemical structure can give guidance to chemists in terms of identification of known substructure that are present in known toxic compounds [9]. Computational biology can be aided informatics infrastructure to provide the basis for *in silico* studies that no longer require the generation of data and instead facilitate the collection, organization, and analysis of existing datasets that can drive discovery [10]. Thus, applying scientific methods to explicate the characteristics of flavor molecules deepens the understanding of the property theory and provides a foundation for discovery and industrial application of new flavor molecules. Moreover, chemoinformatics has been established as an important scientific discipline, bringing about a paradigm shift combining with bioinformatics. However, its applicability in the field of food science is less appreciated, despite the recognized potential for a significant contribution to this field [11]. Moreover, it is considered as an interface between chemistry and informatics or as a collection of methods that comprise a plethora of computational techniques to organize, excavation, visualize, and analyze the diversity and coverage of the chemical space of compound collections [12]. This chapter deals with recent advancements in flavor bioinformatics for the *in silico* prediction, characterization and decipher of flavor-forming pathways, which will certainly spread its use at artisanal and industrial levels.

## 12.2 Flavor Bioinformatics

### 12.2.1 Comparative Genomics

As a fundamental tool of genome analysis, comparative genomics is an area of research in which comparison of genome sequences between different species are performed to understand their difference at the molecular level [13]. It also provides a powerful tool to help in identifying genes that are preserved or common among species, in addition to genes that give each organism its unique characteristics like flavor [14]. Thus, by applying comparative genomics approach, functional annotation of the key enzymes in the formation of flavor compounds from amino acids can be developed by combining phylogeny, gene context, and experimental evidence. For instance, with over 20 genomes fully sequenced for lactic acid bacteria, the available genomic information provides new opportunities to study the flavor-forming potential of nonstarter cultures in the dairy industry [15]. Thus, to produce diverse flavors, gene manipulation at starter or nonstarter culture is used as a prime tool. This gene manipulation is achieved by the availability of the whole genome sequences of the microbes involved in the fermentation process and the new possibilities of "omics" technologies. Consequently, with comparative genomics, genomic traits to phenotypic outputs of strains can be linked, which helps to dig deep into metabolic diversity of these starter cultures, thereby analyze the metabolic routes to flavor compound formation, identify the strains with flavor-forming potential, and select them for possible commercial application [16].

### 12.2.2 Omics Technologies

With a surge in the demand for plant metabolites (primary and secondary), the advanced "omics technologies" are most sought after for a faster research and better characterization of the natural products. Besides, with the arrival of the advanced bioinformatics, genomics, and proteomics and the synergy between combinatorial chemistry and structure-based molecule design, the process of characterizing secondary metabolites has been revolutionized [17]. Omics approach includes high-throughput next-generation sequencing (NGS)-based methods, such as genomics/metagenomics and metatranscriptomics targeting DNA and RNA (Figure 12.1) [18]. For instance, metagenomic sequencing used to analyze the metabolic potential of the microbial populations on the surfaces of the test cheeses revealed a high relative abundance of metagenomic clusters associated with flavor development [19].

### 12.2.3 Bioactive Peptides

Bioactive peptides are defined as specific protein fragments that have a positive effect on body functions or conditions and may influence health [20]. They are also responsible for the taste of food. These diverse peptides can be synthesized, and the data obtained are deposited in databases for future use as shown in Figure 12.2 [21]. At present, more than 1500 diverse bioactive peptides have been reported in a

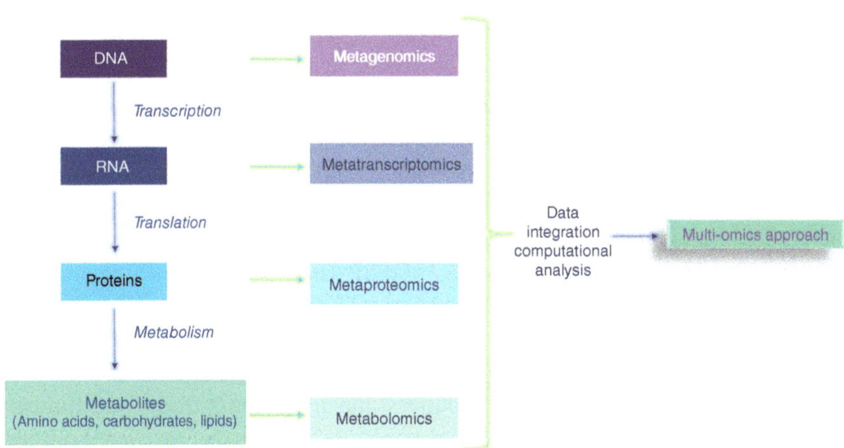


**Figure 12.1** Integrating multi-omics approaches for the study of flavor development.

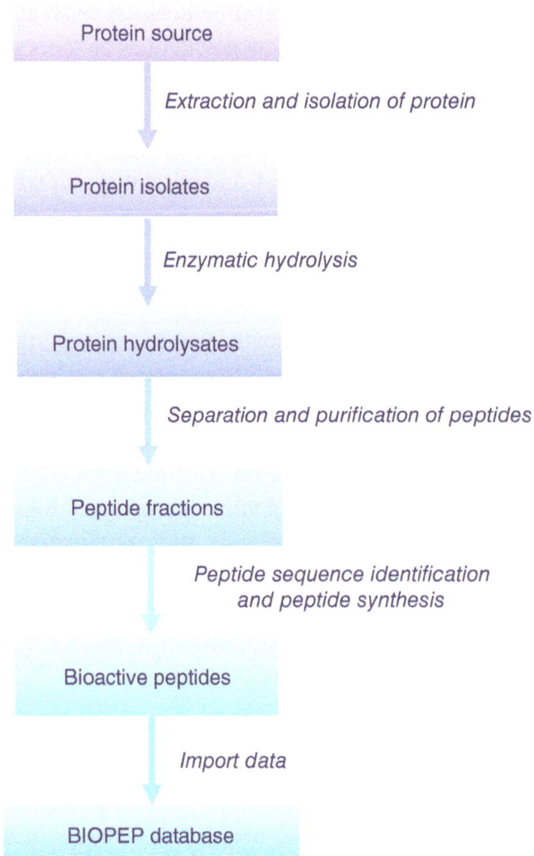


**Figure 12.2** Schematic representation of biopeptide biosynthesis and database depository.

database named "BIOPEP". For instance, the fractionation of brown rice protein into peptides, when evaluated for flavor characteristics using BIOPEP database demonstrated had high occurrence frequencies of umami peptides (ESDVVSDL, GSGVGGAK, and SSVGGGSAG) suggesting that these peptides may be used as a fortifying health ingredient with good taste [22]. In the same way, bioactive peptides in dry-cured ham is described as a taste-active compounds, which are released during the dry-cured ham processing as dipeptides and tripeptides in large amounts. The potential of dipeptides and tripeptides to impart taste characteristics to dry-cured ham has been evaluated using the BIOPEP database [23].

Starter cultures release biopeptides during milk fermentation in dairy industry, for instance, casein-derived peptides formed in cheese industry are associated with cheese ripening and cheese flavor formation [24]. Moreover, there is a growing demand in developing novel food produces and ingredients from food industry waste to improve the sustainability. For example, defatted seeds from major oilseeds (i.e. soybean, rapeseed, cottonseed, sunflower seed, flaxseed, and peanut) are considered waste by the industry, but are a rich source of protein [25].

## 12.3 Computational Strategies

### 12.3.1 Homology Modeling

Homology modeling is the most accurate computational method to create reliable structural models and is commonly used in many biological applications. Homology modeling predicts the 3D structure of a query protein through sequence alignment of template proteins. Generally, the process of homology modeling involves four steps [26]: target identification, sequence alignment, model building, and model refinement. The first step is to identify protein structure(s) to act as template(s). Secondly, the sequence of the protein of known structure is aligned with the protein to be modeled (the target sequence). Thirdly, the alignment is used to guide how the target sequence is overlaid on the 3D coordinates of the template structures to generate the initial model. Finally, the model is optimized using structural, stereochemical, and energy calculation techniques (Figure 12.3) [27]. Often, this process is repeated until a suitable model is obtained. The main difference between the various modeling methods is how the 3D model is calculated from the augment [28].

The procedure is as follows: A previously unknown protein structure is fitted according to its sequence (target) into a known structure (template), given a certain level of sequence homology between the target and the template. First, the sequences

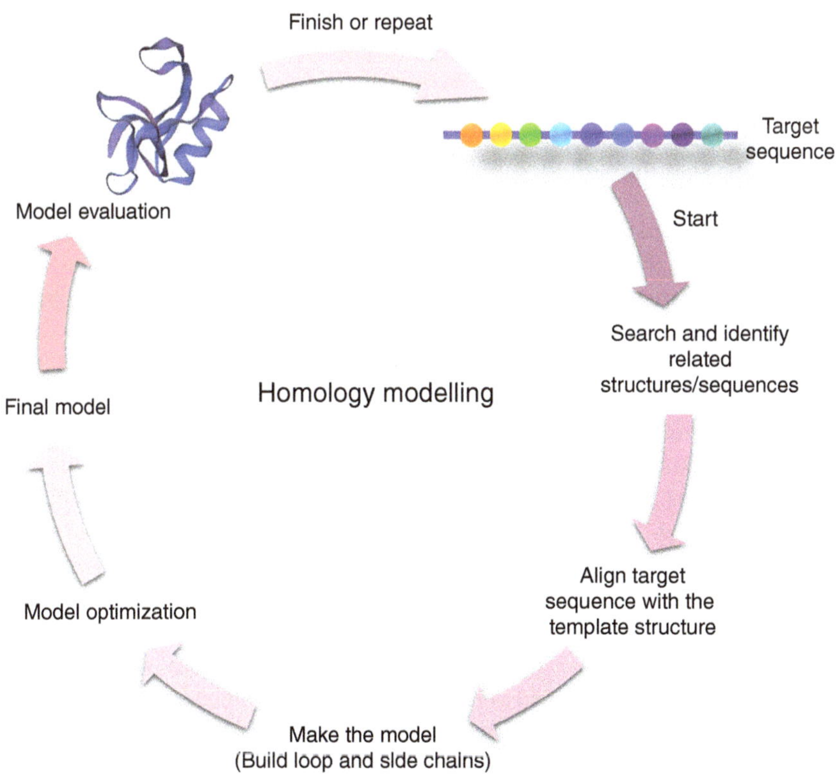


**Figure 12.3** Steps involved in homology modeling.

of the template structure(s) should be retrieved using multiple alignments. Several multiple-sequence alignment (MSA) software applications and webservers, namely MUSCLE, Clustal Omega, BLAST, PSI-Search, and FASTA, can be used for this task. After finding the sequences with high homology to the query model, the ones with available 3D structures must be filtered. Some available methods/webservers (e.g. MODELLER, SWISS-MODEL, FoldX, HHpred, PRIME, and ROBETTA, automatically search for the structural database, yielding templates with resolved structures, and their respective Protein Data Bank (PDB) ids (PDBids). The corresponding coordinates of the template GPCR can be downloaded directly from the PDB (http://www.rcsb.org). There are a few online databases that provide specific template suggestions and homology modeling of the helical regions of GPCRs, which can be quite useful as an initial guess [28].

Homology modeling has proven to be the most successful approach for protein structure prediction, but it requires that a template exist in PDB, such that better the template the more accurate the prediction. This approach is greatly benefited by the fact that structure of at least one member almost all the families of proteins have been deduced experimentally, which can be used as a template for modeling the other members of that particular family [29]. Homological methods play an increasing role in structure-based drug discovery. Although computational structure prediction methods provide a cost-effective alternative in the absence of experimental structures, developing accurate enough models remains a big challenge. [30].

### 12.3.2 Synthetic Ligands for Taste Receptors

Until recently, it was believed that there was just a single receptor or binding site for sweet taste sensations; however, it was later revealed that few strongly sweet plant proteins and superpotent guanidino compounds could evoke electrophysiological responses independent from those evoked by small sweet-tasting carbohydrate and related compounds; this prompted the thought that various receptors or binding sites might exist. The receptors for sweet taste appear to have the exact stereochemical requirements or recognition units as exhibited by the way that generally few compounds elicit a sweet taste sensation [31].

Taste receptors are proteins that recognize taste stimuli of various types, consequently functioning as the initial component in the process of sensing and discriminating ingested materials [32]. There are two principal classes of taste receptors for the five basic tastes of vertebrates: particle channels and G protein-coupled receptors (GPCRs); ion channels responsible for salty and sour taste, intervened by the epithelial sodium channel (ENaC) and acid sensing ion channels (ASICs) proteins that identify Na+ and H+ ions. GPCRs intercede the sensation of bitter, sweet, and umami taste. GPCRs are transmembrane proteins that change conformation when activated by an extracellular ligand, setting off an intracellular signaling cascade [33]. The taste receptor type 1 (T1r) family perceives "palatable" tastes in nutrients, like sugars and L-amino acids. In humans and rodents, the T1r2–T1r3 heterodimer perceives sweet substances like sugars, though the T1r1–T1r3 heterodimer tests umami (savory tastes) of L-amino acids including glutamate. The physiology of taste perception is embodied in the characteristics of T1r function. Numerous T1r receptors have broad ligand specificity: the human T1r2–T1r3 receptor responds to mono-to oligosaccharides, artificial sugars without saccharide groups, some D-amino acids, and even proteins [34].

Two distinct techniques have been developed to recognize taste molecules. For salty and sour tastes, it is generally accepted that ion channels act as receptors. Here H+ (sour) and Na+ (salty) ions are supposed to move through the channels into the cell. For sweet and umami, a group of three GPCRs, named T1R1, T1R2, and T1R3, act two by two (T1R1 + T1R3 for umami, and T1R2 + T1R3 for sweet) to recognize molecules bestowing these taste characteristics. The bitter receptors, the T2Rs, include a considerably bigger group of GPCRs, with around 25 individuals [35].

Taste buds are populated by various kinds of cells. A portion of the cells has properties proposing that they are supporting or glial-like cells. Specifically, these cells take up or degrade neurotransmitters released during taste excitation. They additionally function as spatial buffers for potassium particles delivered into the tight intercellular spaces during taste bud activity. Different cells in the taste bud are specific receptor cells for sweet, sour, salty, bitter, or umami taste stimuli. These chemosensory cells express particular membrane receptors or ion channels specific for those stimuli. Taste bud chemosensory receptor cells discharge synapses when they are stimulated by gustatory mixtures and these transmitters excite tangible afferent fibers that innervate the taste buds [36].

### 12.3.3 Molecular Docking of Flavor Compounds

Molecular docking is considered as the molecular modelling tool that is liked for the prediction of the ligand-receptor interaction when both the particles are bound together to form a stable complex [37]. Docking is a significant method to uncover counter actions between the chemicals. These studies were considered an imperative tool and showed up at one bio molecule GST with an affinity towards numerous harmful agrochemicals [38]. Molecular docking is the process by which two molecules fit together in a 3D space, which is a vital device in the structural biology and computer-aided drug design. To acquire an exact docking, a steady standard dock was utilized as a default standard dock setting on which the boundaries were set. The best ligands were chosen based on their best conformation that permits the lowest free binding energy. The outcome acquired by the docking software was analyzed to study the binding energy and the interactions of the docked structure. For most docking concentrate on instruments, users typically need to set up the construction of the binding site and the ligand. The selected software produces a straightforward strategy to derive the binding site from the bounded ligand, and it can likewise naturally consider the impact of hydrogen atom on the binding side [39].

While performing molecular docking default settings can be utilized for all the calculations. The interaction between the receptor protein and the ligands is studied in Pymol. The limiting energy (kcal/mol) with hydrogen bonds, number of hydrogen bonds, hydrogen bond length, and amino acid residues interactions are also recognized [40]. Autodock 4.0 is a graphical UI program used to prepare, run, and examine the docking simulations. The tubulin protein model was prepared for docking simulation by allotting partial charges, solvation parameters, and hydrogen to the receptor molecule. Water molecules were eliminated from the protein to make it a free receptor. Since ligands are not peptides, Gasteiger's charge was assigned, and nonpolar hydrogen was then merged [41].

### 12.3.4 Virtual Screening Tools for Flavor Compounds

Virtual screening is a computational method that utilizes computer programs to look through likely hits from virtual piece libraries. There are two generally utilized approaches: (i) Structure-based strategies (e.g. docking) and (ii) ligand-based strategies. As opposed to experimental techniques, virtual screening is rapid and economical but has moderately low accuracy of predictions and rapid accumulation of errors [42]. Virtual screening has become well known as a tool; one can use alternative strategies that could assist with picking a suitable set of compounds, eliminate superfluous costs with resources, and minimize production of unwanted compounds. Virtual screening has several advantages over traditional ways of screening, which incorporates a higher hit rate than customary screening techniques, and plays a novel part in accomplishing the assignment of examining the interaction of all existing natural products with all possible targets. Moreover, virtual screening is not important to collect and assay several natural compounds [43].

The ligand-based approach accentuates the chemical features of compounds significant for effective protein binding. The structure-based approach depends on the structural information of the binding pocket on the target protein [44]. The best performing pharmacophore and shape-based models as well as the docking work process were utilized for the virtual screening of the in-house data set. The action of highest level hits was likewise anticipated with additional external *in silico* profiling tools, which were applied in the lineup as autonomous grouping programming. All forecasts of every program were summed up in a prediction matrix. After the biological testing, the hit records were analyzed, and the performance of the applied tools were assessed and analyzed [45] (Figure 12.4).

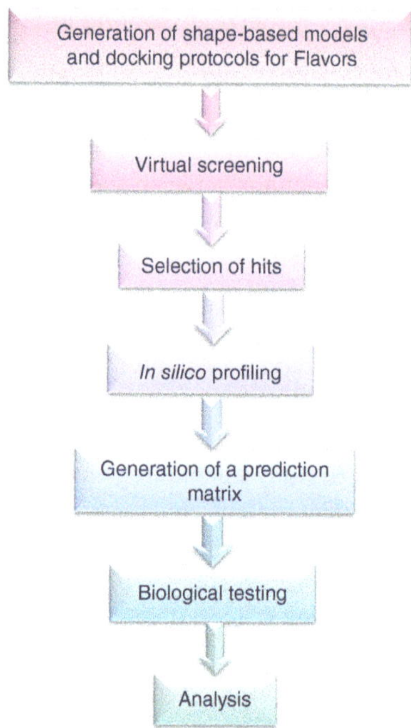


**Figure 12.4** Virtual screening and prediction of matrix.

With a known target structure, virtual screening can promptly distinguish compounds that have shape complementarity proper for binding, usually in the range of 5–20% of an overall chemical library, contingent upon the general imperatives of the binding site, possibly decreasing the quantity of compounds to be at first screened or to be followed up after the initial screen. Virtual screening can quickly foresee the ligand-restricting power and estimate the binding affinity by force field-based, empirical or knowledge-based methods [46].

#### 12.3.4.1 QSAR-Based Virtual Screening for Flavor Compounds

The utilization of molecular modelling techniques, for example, QSAR and molecular docking have become vital techniques. QSAR (quantitative structure–activity relationship) is a strong strategy developed by utilizing statistical techniques. To develop these models, multiple linear regression (MLR) investigations, multiple nonlinear regression (MNLR), and artificial neural networks (ANNs) are utilized. The predictive capacity of created models is tried by a few approval procedures that are internal and external validations and Y-randomization techniques [47]. QSAR is a methodology that utilizes numerical models to correlate the biological activity (e.g. LD50) and descriptive parameters (descriptors) related to the structure of a molecule. QSARs are regression models that are widely applied to biological and chemical sciences. In the field of the classification of hazardous chemicals, QSAR models can be helpful to foresee the hazardous nature of various synthetic compounds in situations where experimental data is not accessible. The target of a QSAR model is to relate a group of predictor variables with the potency of the responsive variable. The predictors contain the data that describes theoretical molecular descriptors or physicochemical properties of chemicals; the responsive variables include different properties, for example, organic action, that is utilized for the classification of hazardous chemicals. The development of descriptors from molecular structure and correlating molecular descriptors with responsive activities through multivariate analysis are two steps in layout QSAR models [48].

A blend of various methodologies (2D and 3D QSAR) prompts appropriate QSAR models. This joint treatment ought to assist while giving powerful expectations. In line with the principal objective of QSAR investigation, which is to predict new molecules of potential interest, novel structures with high anticipated biological activity are recognized based on their outcomes [49]. A QSAR model for virtual screening of potential inhibitors of a protein has been created using an original computational system. The principal objective of this procedure is to acquire exact QSAR models coordinated by a minimum number of molecular descriptors since models with a huge number of descriptors endure poor generalizability and complex interpretability QSAR approaches based on machine learning use the feature selection method for picking the most instructive subset of molecular descriptors; however, this calculation investigates only a small amount of the entire combinatorial of a possible subset [50].

#### 12.3.4.2 Model Validation

Mechanistic models can be a useful tool to predict aroma release and thus serve as a key step in understanding, individual experience with the perceived flavor. To test the validity of models, the elements of oral processing parameters and aroma release parameters at different stages of mastication were estimated [51]. The validation is done to assess the predictive ability of the acquired QSAR model. There are two sorts of validation methods: internal and external. Interior validation is done utilizing the leave one out (LOO) technique. The best model was chosen based on different statistical parameters like the square of the correlation coefficient ($r^2$), and the quality of each model was assessed from the cross-validated squared correlation coefficient. Leave one out cross validation (LOO CV) is perhaps the best technique for approval of a model with a small training data set. This technique was used to determine the predictive power of the model [52]. Statistical validation is a vital process of robust QSAR model development. Therefore, different sorts of statistical validations were acquired, and the best model was chosen by applying these three distinct statistical parameters: (i) coefficient of assurance ($r^2$) ought to be more noteworthy than 0.7; (ii) correlation coefficient ($q^2$) ought to be more noteworthy than 0.5; and (iii) correlation coefficient of the outside approval (pred_$r^2$) set ought to be more noteworthy than 0.5. For an acceptable QSAR model, the value of average $rm^{2\prime}$ greater than 0.5 and "Delta $rm^{2}$" should be under 0.2, degree of freedom ought to be higher, $F$ – test for statistical significance of the model is to be higher $Z$ score calculated by randomization test ought to be higher and the applicability domain characterized as "0" and "1" [53].

There is a requirement for validated methods to improve detection limits and simultaneously measure polycyclic flavor compounds (both parent and alkylated homologous) in biota by gas chromatography due to their environmental significance. The validation of the technique was acted in accordance with the Eurachem manual for quality in analytical chemistry [54]. A semi-quantitative high-throughput technique has been validated in a range of various media. It is quick and economical and will track down the application in metabolomic studies as it allows one to limit the initial search field prior to employing the costlier and tedious traditional quantitative approach [55]. Partial least square regression (PLSR) was preferred to correlate the instrumental data with the inclination results. PLSR was utilized to make outside inclination guides to decide the connections between instrumental volatile data and acceptability data. Thus, identified flavor compounds by the PLSR model are chosen and added to food for the palatability test [56].

#### 12.3.4.3 Docking Setups

Docking has turned into a pillar in virtual screening of large chemical databases and lately has been applied to a lot more extensive issues, for example, predicting the potential protein targets of a particular molecule. Protein–ligand docking techniques have two primary parts: a search algorithm to generate a plausible 3D configuration of a ligand bound to a protein and a scoring function to evaluate the quality of protein–ligand interaction. Generally, docking is firmly incorporated with a wide scope of other molecular modeling and cheminformatics methods. For instance, docking strategies require at least one representative model of the protein receptor, which has been properly prepared, ordinarily from an X-ray crystal structure. Moreover, ligand structures are frequently extracted from a two-dimensional (2D) electronic database and require a few stages of cheminformatic processing and filtering to yield a drug-like arrangement of 3D models for docking. Following docking, high-scoring poses might be investigated and rescored utilizing an assortment of simulation and data mining techniques [57]. Autodock is one of the most well-known software packages for docking; however, its computerization is not trivial for an assignment like (i) virtual screening of a library of ligands against a bunch of potential receptors, (ii) the utilization of receptor adaptability, and (iii) making a blind docking experiment regarding the entire receptor surface [58].

A docking study as a rule begins with the meaning of a binding site, that is, a restricted region of the protein. Autodock and Vina need receptor and ligand representation in a file format called pdbqt, which is a modified protein data bank format containing atomic charges, atom type definitions, and ligand's topological data. Ligands for subsequent docking runs can be set one up by one through pyMOL choices or by specifying a directory containing a library of ligands to be docked. Both Autodock and Vina take into account the adaptability of predefined side chains during docking. Here the plugging facilitates the determination of a flexible side chain. Sidechains inside the docking box can be envisioned straightforwardly, and pyMOL selections can be translated into an adaptable receptor definition [59]. Docking can be used to virtually screen new compounds comparably to experimental high-throughput screening as well as offering atomistic level knowledge to work with structure-based design to track down stable binding affirmations between a ligand and a receptor. Briefly, atomic docking strategies use information-based scoring capacities [60]. Molecular docking comprises two major tasks for which separate calculations are utilized. The examining calculation predicts the numerous confirmations, alluded to as poses, which the ligand can assume within the binding or active pocket. A scoring function then predicts the binding energies between the ligand and the receptors for each predicted pose. The created binding poses are then ranked in light of their binding energies, where the highest level pose ought to compare to the correct confirmation of the ligand. Scoring functions are accordingly fit for separating through and positioning large databases of compounds in virtual screening, where the most noteworthy positioned binding energies should correspond to a potential lead [61].

### 12.3.5 Structural Motifs in Flavor Compounds

In a chain-like biological molecule, like a protein or nucleic acid, a structural motif is a common 3D structure that shows up in a wide range of evolutionarily unrelated molecules [62]. A structural motif does not need to be related to a sequence motif; it tends to be addressed by various and unrelated sequences in different proteins or RNA. The structural motifs characterized above treat every part of the biological network as a unique and anonymous entity, ignoring any other useful biological information possibly known about them [63]. For instance, the presence of bitter-tasting motifs is theoretically estimated as bitter-tasting indicators to study the bitterness of soybean proteins. The analysis covered the hydrolysis of five major soybean-originating protein sequences using bromelain, ficin, papain, and proteinase K, which was verified using in vitro (RP-HPLC for monitoring the proteolysis, and identification of peptides using RP-HPLC-MS/MS). Thus, this fragmentomic idea of research might provide a supportive method for predicting the bitterness of hydrolysates [64]. Recently, an extra vestibular binding site in taste receptor 2 (TAS2R) (in TAS2R46) was proposed. This is especially significant because the TAS2R family is the G protein-coupled receptors that facilitate the detection of countless structurally diverse bitter substances. Thus, the human TAS2R gene family, which includes ~25 member-like odorant receptors, has different ligand profiles. Subsequently, these proposed TAS2R46 binding sites might help to pre-filter ligands from the bulk of irrelevant compounds present in complex food matrices [65].

## 12.4 Quality and Safety of Flavor Compounds

Current advancements in high-throughput "omics" innovations allow growing more rational approaches to deal with fermentation processes both from the food functionality as well as from the food safety viewpoint [66]. Regular bioinformatics methods and approaches are utilized to work on different aspects of the microbial production of fermented food products and safety. These include genomics-based functional predictions, developing genome-scale metabolic models, and prediction of complex boggling food properties like taste and texture and properties of complicated fermentations. Thus, these methods can be utilized to work on the microbial production of fermented food products [67].

## 12.5 Conclusion

Metabolically engineered organisms and their metabolic libraries are becoming the driving force behind a variety of evolutionary and revolutionary changes in the flavoring, cosmetic, and perfumery industries. Consequently, laying out the biosynthetic routes in these organisms can completely replace any natural source for flavor and fragrance molecules. With recent scientific advances, it will be increasingly possible to design taste receptor targeted or multifunctional flavor compounds to develop commercially significant products. While both genome sequencing and *in silico* approaches are under development, omics technologies and synthetic biology will likely continue to have the greatest impact for the foreseeable future. In addition, bioengineering has a decisive impact on future feasibility of innovations in the fields of functional and convenience foods and thus needs to expand research activities in a targeted manner. These are the exciting times for *in silico* research, and the pace of scientific discovery in this area is gaining momentum.

## References

1 Tylewicz, U., Inchingolo, R., and Rodriguez-Estrada, M.T. (2022). Chapter 9 – Food aroma compounds. In: *Nutraceutical and Functional Food Components*, 2e (ed. C.M. Galanakis), 363–409. Academic Press, ISBN 9780323850520, https://doi.org/10.1016/B978-0-323-85052-0.00002-7.

**2** Alseekh, S., de Souza, L.P., Benina, M., and Fernie, A.R. (2020). The style and substance of plant flavonoid decoration; towards defining both structure and function. *Phytochemistry* 174: 112347, ISSN 0031-9422, https://doi.org/10.1016/j.phytochem.2020.112347.

**3** Ishikawa, M., Warita, Y., Takahisa, E., and Ohkubo, Y. (2010). Chapter 4.15 – Human–environment interactions (1): flavor and fragrance. In: *Comprehensive Natural Products II* (ed. H.-W. (Ben) Liu and L. Mander), 595–629. Elsevier, ISBN 9780080453828, https://doi.org/10.1016/B978-008045382-8.00107-6.

**4** Barrett, D.M., Beaulieu, J.C., and Shewfelt, R. (2010). Color, flavor, texture, and nutritional quality of fresh-cut fruits and vegetables: desirable levels, instrumental and sensory measurement, and the effects of processing. *Crit. Rev. Food Sci. Nutr.* 50 (5): 369–389. https://doi.org/10.1080/10408391003626322.

**5** Marcus, J.B. (2019). Chapter 6 – Flavor enhancement ingredients. In: *Aging, Nutrition and Taste* (ed. J.B. Marcus), 173–206. Academic Press, ISBN 9780128135273, https://doi.org/10.1016/B978-0-12-813527-3.00006-5.

**6** Cucu, T., David, F., Devos, C., and Sandra, P. (2021). Untargeted flavor profiling of lager beers by stir bar sorptive extraction–capillary gas chromatography – time-of-flight mass spectrometry: high analytical performance with a green touch. *J. Chromatogr. A* 1647: 462164, ISSN 0021-9673, https://doi.org/10.1016/j.chroma.2021.462164.

**7** Yang, L., Fan, W., and Xu, Y. (2021). GC × GC-TOF/MS and UPLC-Q-TOF/MS based untargeted metabolomics coupled with physicochemical properties to reveal the characteristics of different type daqus for making soy sauce aroma and flavor type baijiu. *LWT* 146: 111416, ISSN 0023-6438, https://doi.org/10.1016/j.lwt.2021.111416.

**8** Gallego, M., Toldrá, F., and Mora, L. (2022). Quantification and in silico analysis of taste dipeptides generated during dry-cured ham processing. *Food Chem.* 370: 130977, ISSN 0308-8146, https://doi.org/10.1016/j.foodchem.2021.130977.

**9** Kenakin, T.P. (2012). Chapter 10 – Safety pharmacology. In: *Pharmacology in Drug Discovery* (ed. T.P. Kenakin), 211–227, ISBN 9780123848567. Academic Press https://doi.org/10.1016/B978-0-12-384856-7.00010-0.

**10** Kalbfleisch, T.S., Rempala, G.A., and Ramos, K.S. (2010). Chapter 2.33 – Genomics, bioinformatics, and computational biology. In: *Comprehensive Toxicology*, 2e (ed. C.A. McQueen), 641–661. Elsevier, ISBN 9780080468846, https://doi.org/10.1016/B978-0-08-046884-6.00236-0.

**11** Holton, T.A., Vijayakumar, V., and Khaldi, N. (2013). Bioinformatics: current perspectives and future directions for food and nutritional research facilitated by a Food-Wiki database. *Trends Food Sci. Technol.* 34 (1): 5–17, ISSN 0924-2244, https://doi.org/10.1016/j.tifs.2013.08.009.

**12** Gómez-Verjan, J.C., Rodríguez-Hernández, K.D., and Reyes-Chilpa, R. (2017). Chapter 8 – Bioactive coumarins and xanthones from calophyllum genus and analysis of their druglikeness and toxicological properties. In: *Studies in Natural Products Chemistry*, vol. 53 (ed. Atta-ur-Rahman), 277–307, ISSN 1572-5995. Elsevier, ISBN 9780444639301, https://doi.org/10.1016/B978-0-444-63930-1.00008-9.

**13** Kawashima, T. (2019). Comparative and evolutionary genomics. In: *Encyclopedia of Bioinformatics and Computational Biology* (ed. S. Ranganathan, M. Gribskov, K. Nakai and C. Schönbach), 257–267. Academic Press, ISBN 9780128114322, https://doi.org/10.1016/B978-0-12-809633-8.20236-7.

**14** Hardison, R.C. (2003). Comparative genomics. *PLoS Biol.* 1 (2): E58. https://doi.org/10.1371/journal.pbio.0000058.

**15** Liu, M., Nauta, A., Francke, C., and Siezen, R.J. (2008). Comparative genomics of enzymes in flavor-forming pathways from amino acids in lactic acid bacteria. *Appl. Environ. Microbiol.* 74 (15): 4590–4600. https://doi.org/10.1128/AEM.00150-08.

**16** McAuliffe, O., Kilcawley, K., and Stefanovic, E. (2019). Symposium review: genomic investigations of flavor formation by dairy microbiota. *J. Dairy Sci.* 102 (1): 909–922, ISSN 0022-0302, https://doi.org/10.3168/jds.2018-15385.

**17** Sabitha, R.A., Neelima, G., Mukhopadhyay, R. et al. (2018). In silico characterization of plant secondary metabolites. In: *Silico Approach for Sustainable Agriculture* (ed. D. Choudhary, M. Kumar, R. Prasad and V. Kumar). Singapore: Springer https://doi.org/10.1007/978-981-13-0347-0_15.

**18** Anastasiou, R., Kazou, M., Georgalaki, M. et al. (2022). Omics approaches to assess flavor development in cheese. *Foods* 11 (2): 188. https://doi.org/10.3390/foods11020188.

**19** Bertuzzi, A.S., Walsh, A.M., Sheehan, J.J. et al. (2018). Omics-based insights into flavor development and microbial succession within surface-ripened cheese. *mSystems* 3 (1): e00211–e00217. https://doi.org/10.1128/mSystems.00211-17.

**20** Sánchez, A. and Vázquez, A. (2017). Bioactive peptides: a review. *Food Qual. Saf.* 1 (1): 29–46. https://doi.org/10.1093/fqsafe/fyx006.

**21** Jakubczyk, A., Karaś, M., Rybczyńska-Tkaczyk, K. et al. (2020). Current trends of bioactive peptides—new sources and therapeutic effect. *Foods* 9 (7): 846. https://doi.org/10.3390/foods9070846.

**22** Selamassakul, O., Laohakunjit, N., Kerdchoechuen, O. et al. (2020). Bioactive peptides from brown rice protein hydrolyzed by bromelain: relationship between biofunctional activities and flavor characteristics. *J. Food Sci.* 85: 707–717. https://doi.org/10.1111/1750-3841.15052.

**23** Gallego, M., Mora, L., and Toldrá, F. (2019). The relevance of dipeptides and tripeptides in the bioactivity and taste of dry-cured ham. *Food Prod Process. Nutr.* 1: 2. https://doi.org/10.1186/s43014-019-0002-7.

**24** Li-Chan, E.C.Y. (2015). Bioactive peptides and protein hydrolysates: research trends and challenges for application as nutraceuticals and functional food ingredients. *Curr. Opin. Food Sci.* 1: 28–37, ISSN 2214-7993, https://doi.org/10.1016/j.cofs.2014.09.005.

**25** Xin, Y., Dominic, A., Chibuike, U. et al. (2021). Manufacturing of plant-based bioactive peptides using enzymatic methods to meet health and sustainability targets of the sustainable development goals. *Front. Sustainable Food Syst.* 5: https://www.frontiersin.org/article/10.3389/fsufs.2021.769028.

**26** Skariyachan, S. and Garka, S. (2018). Chapter 1 – Exploring the binding potential of carbon nanotubes and fullerene towards major drug targets of multidrug resistant bacterial pathogens and their utility as novel therapeutic agents. In: *Fullerens,*

*Graphenes and Nanotubes* (ed. A.M. Grumezescu), 1–29. William Andrew Publishing, ISBN 9780128136911, https://doi.org/10.1016/B978-0-12-813691-1.00001-4.

**27** Pitman, M.R. and Menz, R.I. (2006). Chapter 2 – Methods for protein homology modelling. In: *Applied Mycology and Biotechnology*, vol. 6 (ed. D.K. Arora, R.M. Berka and G.B. Singh), 37–59, ISSN 1874-5334. Elsevier, ISBN 9780444518071, https://doi.org/10.1016/S1874-5334(06)80005-5.

**28** Sensoy, O., Almeida, J.G., Shabbir, J. et al. (2017). Chapter 16 – Computational studies of G protein-coupled receptor complexes: structure and dynamics. In: *Methods in Cell Biology*, vol. 142 (ed. A.K. Shukla), 205–245. Academic Press, ISSN 0091-679X, ISBN 9780128133194, https://doi.org/10.1016/bs.mcb.2017.07.011.

**29** Jabeen, A., Mohamedali, A., and Ranganathan, S. (2019). Protocol for protein structure modelling. In: *Encyclopedia of Bioinformatics and Computational Biology: ABC of Bioinformatics*, vol. 3 (ed. M. Cannataro, B. Gaeta and M. Asif Khan), 252–272. Elsevier https://doi.org/10.1016/B978-0-12-809633-8.20477-9.

**30** Cavasotto, C.N. and Phatak, S.S. (2009). Homology modeling in drug discovery: current trends and applications. *Drug Discovery Today* 14, 13–14: 676–683, ISSN 1359-6446, https://doi.org/10.1016/j.drudis.2009.04.006.

**31** Sulikowski, M.M., Kim, K., Sulikowski, G.A. et al. (1995). Ligands for the study of sweet taste receptors: synthesis of fluorescent N,N′-disubstituted guanidineacetic acids. *Bioorg. Med. Chem. Lett.* 5 (23): 2875–2878, ISSN 0960-894X, https://doi.org/10.1016/0960-894X(95)00501-J.

**32** Boughter, J.D., Munger, S.D., and Receptors, T. (2013). *Encyclopedia of Biological Chemistry*, 2e (ed. W.J. Lennarz and M.D. Lane), 366–368. Academic Press, ISBN 9780123786319, https://doi.org/10.1016/B978-0-12-378630-2.00397-2.

**33** Freund, J.R. and Lee, R.J. (2018). Taste receptors in the upper airway. *World J. Otorhinolaryngol.* 4 (1): 67–76, ISSN 2095-8811, https://doi.org/10.1016/j.wjorl.2018.02.004.

**34** Nuemket, N., Yasui, N., Kusakabe, Y. et al. (2017). Structural basis for perception of diverse chemical substances by T1r taste receptors. *Nat. Commun.* 8: 15530. https://doi.org/10.1038/ncomms15530.

**35** Mennella, J.A. (2017). Swaiman's pediatric neurology. *Taste Smell* 58–64: https://doi.org/10.1016/B978-0-323-37101-8.00009-6.

**36** Roper, S.D. (2020). *The Senses: A Comprehensive Reference, Microphysiology of Taste Buds*, 187–210. https://doi.org/10.1016/B978-0-12-809324-5.24152-7.

**37** Lakhera, S., Devlal, K., Ghosh, A., and Rana, M. (2021). In silico investigation of phytoconstituents of medicinal herb 'Piper Longum' against SARS-CoV-2 by molecular docking and molecular dynamics analysis. *Results Chem.* 3: 100199, ISSN 2211-7156, https://doi.org/10.1016/j.rechem.2021.100199.

**38** Ritika, A., Ritika, G., Nikita, J. et al. (2021). In silico prediction, characterization and molecular docking studies on Glutathione-S-transferase as a molecular sieve for toxic agrochemicals explored in survey of North Indian farmers. 7 (9): https://doi.org/10.1016/j.heliyon.2021.e07875.

**39** Kareti, S.R. and Subash, P. (2020). In silico molecular docking analysis of potential anti-Alzheimer's compounds present in chloroform extract of carissa carandas leaf using gas chromatography MS/MS. *Curr. Ther. Res.* 93: 100615, ISSN 0011-393X, https://doi.org/10.1016/j.curtheres.2020.100615.

**40** Srikanth Kumar, K., Lakshmana Rao, A., and Basaveswara Rao, M.V. (2018). Design, synthesis, biological evaluation and molecular docking studies of novel 3-substituted-5-[(indol-3-yl) methylene]-thiazolidine-2,4-dione derivatives. *Heliyon* 4 (9): e00807, ISSN 2405-8440, https://doi.org/10.1016/j.heliyon.2018.e00807.

**41** Quan, P.M., Binh, V.N., Ngan, V.T. et al. (2019). Molecular docking studies of Vinca alkaloid derivatives on Tubulin. *Vietnam J. Chem.* 57: 702–706. https://doi.org/10.1002/vjch.201900087.

**42** Chen, H., Zhou, X., Gao, Y. et al. (2017). Chapter 2.08 – Fragment-based drug design: strategic advances and lessons learned. In: *Comprehensive Medicinal Chemistry III* (ed. S. Chackalamannil, D. Rotella and S.E. Ward), 212–232, ISBN 9780128032015. Elsevier https://doi.org/10.1016/B978-0-12-409547-2.12319-4.

**43** Sarker, S.D., Nahar, L., Miron, A., and Guo, M. (2020). Chapter 2 – Anticancer natural products. In: *Annual Reports in Medicinal Chemistry*, vol. 55 (ed. S.D. Sarker and L. Nahar), 45–75. Academic Press, ISSN 0065-7743, ISBN 9780128210192, https://doi.org/10.1016/bs.armc.2020.02.001.

**44** Liu, Z., Chen, H., Wold, E.A., and Zhou, J. (2017). Chapter 2.13 – Small-molecule inhibitors of protein–protein interactions. In: *Comprehensive Medicinal Chemistry III* (ed. S. Chackalamannil, D. Rotella and S.E. Ward), 329–353. Elsevier, ISBN 9780128032015, https://doi.org/10.1016/B978-0-12-409547-2.12326-1.

**45** Kaserer, T., Temml, V., Kutil, Z. et al. (2015). Prospective performance evaluation of selected common virtual screening tools. case study: cyclooxygenase (COX) 1 and 2. *Eur. J. Med. Chem.* 96: 445–457, ISSN 0223-5234, https://doi.org/10.1016/j.ejmech.2015.04.017.

**46** Hevener, K.E., Pesavento, R., Ren, J. et al. (2018). Hit-to-lead: hit validation and assessment. *Methods Enzymol.* 610: 265–309. https://doi.org/10.1016/bs.mie.2018.09.022. Epub 2018 Oct 25. PMID: 30390802.

**47** Daoui, O., Elkhattabi, S., Chtita, S. et al. (2021). QSAR, molecular docking and ADMET properties in silico studies of novel 4,5,6,7-tetrahydrobenzo[D]-thiazol-2-Yl derivatives derived from dimedone as potent anti-tumor agents through inhibition of C-Met receptor tyrosine kinase. *Heliyon* 7 (7): e07463, ISSN 2405-8440, https://doi.org/10.1016/j.heliyon.2021.e07463.

**48** Chakrabarty, A. (2016). Multiscale modeling for process safety applications. In: *Application of Modeling for Industrial Hygiene and Â Toxicological Issues*, 397–406. Butterworth-Heinemann Elsevier Ltd. https://doi.org/10.1016/b978-0-12-396975-0.00009-7.

**49** Katritzky, A., Slavov, S., Dobchev, D., and Karelson, M. (2007). Comparison between 2D and 3D-QSAR approaches to correlate inhibitor activity for a series of indole amide hydroxamic acids. *QSAR Comb. Sci.* 26: 333–345. https://doi.org/10.1002/qsar.200630021.

**50** Ponzoni, I., Sebastián-Pérez, V., Martínez, M.J. et al. (2019). QSAR classification models for predicting the activity of inhibitors of beta-secretase (BACE1) associated with Alzheimer's disease. *Sci. Rep.* 9: 9102. https://doi.org/10.1038/s41598-019-45522-3.

51 How, M.S., Jones, J.R., Morgenstern, M.P. et al. (2021). Modelling the role of oral processing on in vivo aroma release of white rice: conceptual model and experimental validation. *LWT* 141: 110918, ISSN 0023-6438, https://doi.org/10.1016/j.lwt.2021.110918.

52 Ramalakshmi, N., Manimegalai, P., Bhandare, R.R. et al. (2021). 2D-Quantitative structure activity relationship (QSAR) modeling, docking studies, synthesis and in-vitro evaluation of 1,3,4-thiadiazole tethered coumarin derivatives as antiproliferative agents. *J. Saudi Chem. Soc.* 25 (7): 101279, ISSN 1319-6103, https://doi.org/10.1016/j.jscs.2021.101279.

53 Alam, S. and Khan, F. (2018). Virtual screening, docking, ADMET and system pharmacology studies on garcinia caged xanthone derivatives for anticancer activity. *Sci. Rep.* 8: 5524. https://doi.org/10.1038/s41598-018-23768-7.

54 Idowu, I., Francisco, O., Thomas, P.J. et al. (2018). Validation of a simultaneous method for determining polycyclic aromatic compounds and alkylated isomers in biota. *Rapid Commun. Mass Spectrom.* 32: 277–287. https://doi.org/10.1002/rcm.8035.

55 Haggerty, J., Bowyer, P., Jiranek, V., and Taylor, D. (2016). Metabolomic screening studies of wines. *Aust. J. Grape Wine Res.* 22: 3–10. https://doi.org/10.1111/ajgw.12167.

56 Chen, M., Chen, X., Nsor-Atindana, J. et al. (2017). Optimization of key aroma compounds for dog food attractant. *Anim. Feed Sci. Technol.* 225: 173–181, ISSN 0377-8401, https://doi.org/10.1016/j.anifeedsci.2016.12.005.

57 Waszkowycz, B., Clark, D.E., and Gancia, E. (2011). Outstanding challenges in protein–ligand docking and structure-based virtual screening. *WIREs Comput. Mol. Sci.* 1: 229–259. https://doi.org/10.1002/wcms.18.

58 Zhang, Y. and Sanner, M.F. (2019). AutoDock CrankPep: combining folding and docking to predict protein-peptide complexes. *Bioinformatics* 35 (24): 5121–5127. https://doi.org/10.1093/bioinformatics/btz459. PMID: 31161213; PMCID: PMC6954657.

59 Seeliger, D. and de Groot, B.L. (2010). Ligand docking and binding site analysis with PyMOL and Autodock/Vina. *J. Comput.-Aided Mol. Des.* 24: 417–422. https://doi.org/10.1007/s10822-010-9352-6.

60 Ainsley, J., Lodola, A., Mulholland, A.J. et al. (2018). Chapter 1 – Combined quantum mechanics and molecular mechanics studies of enzymatic reaction mechanisms. In: *Advances in Protein Chemistry and Structural Biology*, vol. 113 (ed. T.G. Karabencheva-Christova and C.Z. Christov), 1–32. Academic Press, ISSN 1876-1623, ISBN 9780128139165, https://doi.org/10.1016/bs.apcsb.2018.07.001.

61 Berry, M., Fielding, B., and Gamieldien, J. (2015). Chapter 27 – Practical considerations in virtual screening and molecular docking. In: *Emerging Trends in Computer Science and Applied Computing, Emerging Trends in Computational Biology, Bioinformatics, and Systems Biology* (ed. Q.N. Tran and H. Arabnia), 487–502. Morgan Kaufmann, ISBN 9780128025086, https://doi.org/10.1016/B978-0-12-802508-6.00027-2.

62 Johansson, M.U., Zoete, V., Michielin, O., and Guex, N. (2012). Defining and searching for structural motifs using DeepView/Swiss-PdbViewer. *BMC Bioinf.* 13: 173. https://doi.org/10.1186/1471-2105-13-173.

**63** Pizzuti, C. and Rombo, S.E. (2019). Algorithms for graph and network analysis: clustering and search of motifs in graphs. In: *Encyclopedia of Bioinformatics and Computational Biology* (ed. S. Ranganathan, M. Gribskov, K. Nakai and C. Schönbach), 95–101. Academic Press, ISBN 9780128114322, https://doi.org/10.1016/B978-0-12-809633-8.20486-X.

**64** Iwaniak, A., Hrynkiewicz, M., Minkiewicz, P. et al. (2020). Soybean (glycine max) protein hydrolysates as sources of peptide bitter-tasting indicators: an analysis based on hybrid and fragmentomic approaches. *Appl. Sci.* 10 (7): 2514. https://doi.org/10.3390/app10072514.

**65** Behrens, M., Briand, L., de March, C.A. et al. (2018). Structure–function relationships of olfactory and taste receptors. *Chem. Senses* 43 (2): 81–87. https://doi.org/10.1093/chemse/bjx083.

**66** Bergholz, T.M., Moreno Switt, A.I., and Wiedmann, M. (2014). Omics approaches in food safety: fulfilling the promise? *Tr. Microbiol.* 22 (5): 275–281. https://doi.org/10.1016/j.tim.2014.01.006.

**67** Alkema, W., Boekhorst, J., Wels, M., and van Hijum, S.A. (2016). Microbial bioinformatics for food safety and production. *Briefings Bioinf.* 17 (2): 283–292. https://doi.org/10.1093/bib/bbv034.

# 13

# Endpoint: A Sensory Perception of Future

*Nimisha P. Sukumaran and Sreeraj Gopi*

*Aurea Biolabs (P) Ltd, R&D Centre, Kolenchery, Cochin 682311, Kerala, India*

## 13.1 Introduction

The importance of food to humans is undeniable; there is still no way of living without eating. Consequently, this commodity is of utmost importance for the well-being of every man, woman, and child today. Therefore, flavor or taste is the inevitable impression of food or other substances, and consequently, a busy future lies ahead for flavor-enhanced foods. Also, to some extent, they play a vital role in the shelf life of products, for instance, due to their antioxidant characteristics. Moreover, flavors continue to be a driving force and provide essential characteristics to food items and provide variety, interest, and increase preferability in the market to what we consume. Further, in a competitive global market, flavor enhancers are essential to enable the food industry to meet the increasingly challenging market demands [1]. So, there is a growing interest in aroma-forming compounds in the preparation of functional and nutritional foods for providing more health benefits with less chemically synthesized food additives. The term "taste enhancer" is used in the food industry to describe a substance that enhances the sensation of food (or food ingredients) when introduced into the mouth. The use of the term "taste" is colloquial and refers to flavor (both taste and smell) because chemicals from food activate receptors in the nose and mouth [2].

Flavors or flavoring materials are volatile or aromatic substances of low molecular weight, most below 300 [3]. There are about 2500 flavoring materials used in the food industry, and they are classified as natural or synthetic. Flavoring agents are composed of large and divergent groups of materials as follows (i) Artificial substances that are unlikely to occur naturally in food; (ii) Natural materials not normally used as food, their derived products, and the equivalent nature-identical flavorings; (iii) Herbs and spices, their derived products, and the equivalent nature-identical flavorings; (iv) Natural flavoring substances (vegetable and animal products) that are normally consumed as food, irrespective of whether they are processed or not, and their

*Natural Flavours, Fragrances, and Perfumes: Chemistry, Production, and Sensory Approach,*
First Edition. Edited by Sreeraj Gopi, Nimisha Pulikkal Sukumaran, Joby Jacob, and Sabu Thomas.

synthetic equivalents [4]. Natural flavors can be extracted from volatile compounds found in different parts of plants, mostly herbs. The usage of natural flavors over synthetic flavors has several disadvantages in the food industry. First, their flavor profile is not consistent. Second, the strength of the flavor is usually not strong enough because of the presence of many other natural components. Third, natural flavors cost more because of their limited quantity in nature. Hence, synthetic flavors are preferred as an alternative to natural flavors. Moreover, they are composed of volatile flavoring materials that are mixed using the same formula as natural flavors [5].

With current trends toward "cleaner" food labels (lower salt, sugar, and fat levels in food), a growing number of consumers prefer to buy natural products. Moreover, there is an ever-demanding need to keep material costs down, and a progression towards processes that have a minimal environmental impact [6]. For instance, flavor enhancers like sweeteners are usually substances of low energy value that provide a sweet taste, but without the calories of carbohydrates or their cariogenic or glycemic effects. These compounds have been extensively used as sugar substitutes to prepare foods with restricted sugar levels, intended for diets of diabetic individuals and for healthy diets with limiting energy intake. Sugar substitutes are widely used in a variety of beverages and foods, such as soft drinks (the most popular sugar-free product), noncarbonated soft drinks, yogurts, fruit juices, chewing gums, ice creams, baked goods, and confectionery products. They are also used in pharmaceutical preparations, such as throat lozenges, cough syrups or drops, and oral hygiene products such as mouthwashes, oral strips, and toothpaste [7].

The integration and interpretation of taste and smell stimuli in the brain leading to the ultimate liking or disliking of a food product are poorly understood. The balance of flavor components in a given food product determines whether that food product is liked or disliked [8]. Interestingly, human responses to bitterness, for instance, show a remarkable diversity, ranging from absolute rejection to strong acceptance. Besides, the perception of the bitter taste changes significantly during aging. For instance, some foodstuffs that are usually rejected by infants and children become acceptable to adults when a certain degree of bitterness in food and beverages starts to become pleasant and contributes to attraction and palatability as they age. Many natural bitter compounds also have a certain degree of toxicity; therefore, the bitter taste receptors have been designed to recognize and avoid them, particularly when the organism is potentially more susceptible to intoxication, as during childhood, which signifies an evolutionary significance [9].

On the other hand, the flavor concept is the result of a complex balance. Its intensity critically and positively depends on the levels of chemosensory attributes [10]. Sensory experience and consumer perception of food products is a multi-disciplinary approach that emphasizes the role of food oral processing concerning food structure, food intake, and sensory perception [11]. Yet, sensory perception is neither uniform across the senses nor among all age groups. Thus, the use of substances to augment the original flavors of foods and beverages in flavor enhancement needs a broad and individualistic approach that should be taken for a diverse flavor and aromatic profile [12]. Consequently, the chapter highlights the different aspects of sensory perception related to other disciplines including oral food processing behavior, flavor receptors, and food choice, and psychology.

## 13.2 Sensory Perception

Sensory perception, a complex process, is influenced by many factors, such as the content of flavor components, texture, and appearance [8]. Sensory quality means different things to different people and is a term that needs to be well defined to allow effective communication between suppliers, processors, and retailers. Thus, sensory quality is the texture, flavor (taste), aroma, and visual effects [13]. Food experiences extend beyond the eating of food. It involves fantasizing about food, perceiving the venue where one can buy or consume it, seeing or smelling the food from a distance, touching its package or container, the tools one uses to prepare or cook, the cutlery one uses to eat, the way one disposes of the leftovers, and so on. In each of these stages, multiple sensory impressions conveyed by the senses of touch, hearing, smell, vision, and taste contribute to the overall experiences [14].

Sensory evaluation is the use of human senses to measure and interpret flavor and sensory characteristics of foods, beverages, other materials. All the human senses are involved in describing and measuring what can be detected by laboratory instruments. The sensory evaluation comprises a set of techniques that can provide analytical information of the sensory attribute of food or non-food products to define the technical specification of the product and the hedonic perception of that product as defined by the user/consumer [15]. There are two types of sensory methods used: (i) analytical ones comprising of difference/discriminative tests and descriptive tests and (ii) affective, divided into acceptance/rating and preference/choice-based tests. Analytical sensory tests define food profiles and sample variability of food attributes but do not take into account any liking considerations. Discrimination tests are considered "user friendly" since they easily distinguish differences between two or more samples by using reference standards and binomial statistical distributions [16].

Color is an integral part of the human daily sensory experience; a particular foodstuff has to be of a specific color quality to be edible. Food triggers physiological and psychological reactions conditioned by experience, tradition, education, and environment. The impression of colorlessness is often unconsciously associated with ill health or poor quality. On the other hand, natural, bright colors give the sensory impression of high quality, healthy, nutritious food; thus, berry and plant extracts such as saffron, paprika, annatto, carrots, and peppers have been used worldwide to refine foods [17]. By and large, during the consumption of food, all the senses are involved in the perception of its sensory properties. Even before consuming food, judgments are made based on appearance and smell. Perception of food texture begins with the first bite and continues throughout mastication and even after swallowing. Throughout the oral processing of food, the senses of taste and smell, as well as the trigeminal senses are important in contributing the flavor perception. The sense of taste (gustation) provides the perception of the five basic tastes (sweet, salty, sour, bitter, and umami), while the sense of smell (olfaction) is triggered by the volatile compounds. Trigeminal sensations are responsible for the perception of temperature, feelings associated with heating and cooling, as well as the pain sensation associated with spicy foods. Older adults over the years' experience many changes that can alter the way food sensory properties are perceived [18].

## 13.3 Flavor Perception

Flavor perception is defined as the sensation arising from the integration or interplay of signals produced as a consequence of sensing chemical substances by smell, taste, and irritation stimuli from food or beverage. In the mouth, five basic differences like sweet, bitter, salt, acid, and umami regarding flavor are sensed by taste-receptor cells and cold and hot sensations. In the nose, many different receptors (3–400) are present and can respond to a large variety of volatile flavor components [8].

### 13.3.1 Flavor Receptors

Many different analytical tools to identify the physical, chemical, and sensory characteristics are in use. This is because, when eating and tasting, the stimulation of a substance to the human tongue varies with its concentrations. A taste curve can be established to determine the correlation between the concentration of the substance and the intensity of the taste. Therefore, an objective and rapid taste evaluation method is vital for industries that intend to quickly narrow down the range of flavor enhancer content in a food product [19]. Moreover, high-throughput screening methods developed to search for novel molecules that modulate the activity of taste receptors could have a profound impact on the flavor ingredients industry and on the wider food and beverage industry [20]. This enables food producers to radically alter the nutritional profile of their products without having any impact on taste.

Our taste perception system recognizes five distinct taste sensations sweet, sour, bitter, salt and umami, a savory taste imparted by monosodium glutamate (MSG) that are triggered by binding interactions between tastant ligands and taste receptors expressed on the surface of taste cells [21]. The taste receptors are evolved to enable humans and other organisms to find carbohydrates, proteins, and other nutrients, and to avoid harmful substances. From a nutritional standpoint, taste receptors have received great interest in designing new artificial sweeteners, taste enhancers, synthetic flavors, and in general, detailed knowledge of taste receptors should allow the rational design of new ingredients (ranging from foods stabilizers to drug excipients) with enhanced palatability that avoids (or masking) bitter and sour tastes [22].

### 13.3.2 Food Oral Processing

The manipulation and processing of food in the mouth, i.e., food oral processing, is the very beginning of food digestion. Then, through mastication by the teeth and mixing by the tongue, plus the enzymatic actions of saliva, food is broken down into smaller pieces and converted to a bolus which is then transported (swallowed) through the esophagus into the stomach for further digestion [23]. Issues included in food oral processing are as follows: (i) food provision to elderly populations and those with eating and swallowing difficulties; (ii) the use of model foods for eating and swallowing studies; (iii) food oral breakdown and compound residuals and release; (iv) food and saliva interactions and bolus formation, and (v) the dynamic textural perceptions during textural attribute evaluations [24]. Food oral processing is a dynamic process. This dynamic process is controlled by three very different mechanisms: physical/mechanical, colloidal, and biochemical/enzymatic. Size reduction and hugely increased food–saliva contact area enables the simultaneous sensation of texture and flavor release. Colloidal interaction between salivary proteins and food emulsion could lead to instant destabilization [25].

Moreover, food oral processing is an essential procedure for the consumption and digestion of foods, and also for the appreciation, food texture pleasure, and food flavor. Two key variations are considered in food oral processing and sensory perception: the individuality of human beings and the properties of food materials. The former reflects the variation of oral physiology (because of age, sex, health status, etc.), while the latter reflects the effects of food rheology and texture (such as hardness, softness, geometric dimensions, etc.). Both variations play an important role in influencing how a portion of food is orally processed and sensually perceived [26]. Modification in oral processing behavior of composite foods can be done by changing single food properties. Thus, promising approaches to influence the eating rate and thereby energy intake include adding particles, adding accompanying foods, or changing single food properties, especially mechanical properties, shape, and concentration [11]. Food structure is known to modulate food oral processing behavior, by which both food intake and sensory perception are affected.

## 13.4 Consumer Perception

"Consumer perception" of food is a very complex phenomenon that is influenced by a wide range of characteristics. The key motivation for food science and nutrition should be sensory features, cost/price balance, and consumer health (sufficient/balanced nutrition). However, there are important differences between theory and reality. Food choice is a complex process influenced by several factors related to the product, the consumer, and the consumption context. The role of the consumer in determining the market success of a product is of maximum relevance. Consumer perceptions and preferences are dynamic. Understanding and analyzing consumers' motivation factors, perceptions, and preferences are important for the food industry [27].

Food safety, human health, and environmental concern along with sensory attributes such as nutritive value, taste, freshness, and appearance influence organic food consumer preferences. Demographic variables may define organic consumers, but the correlation is not very significant. A consumer also associates organic food with the natural process, care for the environment and animal welfare, and the non-use of pesticides and fertilizers. Premium price continues to suppress organic food consumptions [28]. Studying consumer attitudes toward food produced by employing new food technologies is a prerequisite for market success, especially at an early stage of their transformation into marketable products. The most important value and cost dimensions that define customer value perception are recency and utility-related, in particular, "functional (economic) value" and "performance (taste-related) cost". Moreover, additional value and cost types of affective nature play an important role, such as "emotional value," "(dis) trust" and "(un) familiarity" [29]. The factors related to individual characteristics of consumers are (i) "connectedness" (information seeking and knowledge), (ii) sensory motive, (iii) local identity, (iv) self-identity, (v) naturalness (e.g. the use of fresh raw materials), and (vi) price perception. These factors might be the drivers of consumers' perception of product craftsmanship [30].

Moreover, the market reviews revealed four main products (or process) attributes affecting consumers' perception of craftsmanship of foods: (i) price perception, (ii) food processing system, (iii) brand humanization and personification (i.e. the human-side traits of the brand), and (iv) production scale [30]. Consumers form perceptions of various food products through their comprehension of the visual stimuli on food packaging that attract their attention. These perceptions directly affect their purchasing decision at the point of purchase, which emphasizes the importance of attention-capturing packaging attributes. The study indicated that participants mainly based their perception of food packaging on its functional (being purposive, recyclable, and informative) and physical attributes (being attractive, of high quality, and hygienic). This indicates that the visual attribute of packaging gains consumers' attention in-store [31]. The traditional versions of foods are preferred over the others as they are perceived as more natural, original, and authentic. In the same way, modernized versions being perceived as modern are preferred less. Consumers have the same general perception of traditional modern food; however, they have different attitudes toward traditional foods regardless of their demographic background (age, marital status, level of education, and the number of family members). Consumers have a high potential interest in traditional products; therefore, it can provide a direction for product developers in developing modernized traditional food [32].

### 13.4.1 Food Choice

Food selection, or food choice, is the study of the factors that influence choice. Several fields of resear ch have examined this relationship, including physiology, psychology, economics, and consumer behavior, to name a few. Food selection is a function of the interactive combination of the person, the product, and the situation in which a food selection is made. Several variables have been shown to influence food choice. Habitual behaviors can influence food usage, in certain instances even more so than food acceptability. Attitudes, traits, and expectations also influence acceptability and choice. Persistent negative expectations or stereotypes are important in understanding the critical evaluations of institutional foods and other foods that are regularly criticized such as airline food. Many environmental or situational variables have been shown to influence food selection. Finally, the decor, design, and visual and auditory elements of the food selection facility can influence choice [33]. The important food choice determinants are as follows: (i) Biological determinants (satiety signals [macronutrients, energy density of diets], palatability [taste, smell, texture, sound, and sight]), (ii) Economic elements (cost and income, availability), (iii) Structural determinants (access education, food variety, cooking facilities, skills, time), (iv) Social characteristics (culture, family, peers, and meals patterns), (v) attitudes, beliefs, and knowledge about foods, (vi) psychological determinants (Stress, Mood) [34].

Food choice is an extremely complex phenomenon. It is influenced by both physiological and psychosocial impulses; it is both a conscious and an unconscious process, is affected by both internal and external forces, and has been approached from a myriad of theoretical positions and disciplines – psychologists, behavioral economists, social scientists, public health researchers, and neuroscientist are all represented in the quest to better understand why we choose to eat what we do [35]. Food choices are the result of automatic and undeliberated processes. In food choices of people, senses such as visual, olfactory, and auditory factors also play a role, but often occur without full awareness, and food choices are, in large part the result of undeliberated response to contextual food cues, many of which lead to increased caloric consumption and poor dietary choices [36]. The sensory properties of foods and beverages are the most important determinants of food choice. Hedonic properties of foods are related to flavor components including, spices that seem to influence ingestion behavior. These flavor active compounds are also involved in digestive, absorptive, and metabolic processes through direct activation of signaling pathways or through neurally mediated cephalic phase responses [37].

Moreover, the gustatory system of humans represents the key sensing system involved in the selection of food and is therefore responsible for the intake of micro-, and macronutrients required for health. The preference for umami taste for instance is innate and related to the uptake of protein-rich food products [38]. MSG for many years has been the best known and most widely used flavor enhancer. MSG is normally effective in terms of a relatively few parts per thousand, but far less powerful than the newer flavor potentiators [39]. Newborn human infants have an innate taste preference for L-glutamate, which is the predominant amino acid in human breast milk [38].

### 13.4.2 Food Psychology

Food psychology is for u nderstanding eating and drinking behavior and applying this knowledge to make better and more accepted food products. The following five fallacies are frequently reported: the idea (i) that people are uniform, (ii) that they are consistent, (iii) that they make rational choices, (iv) that their perception is more important than their memory of sensory impressions, and (v) that situations are characterized by objectively measurable context variables [40]. The bidirectional influences between emotion and food consumption to find emotional factors that influence food choice, eating, and drinking behavior independently from traditional factors as liking, wanting, and appropriateness (strength and valence of emotions that influence food choice and food intake). The food may improve mood, particularly in stressed people. Most emotion measurements are unduly explicit and verbal, and the implicitly integrated emotions should be measured through remembrance [41]. Psychological insight is needed in answering many of the questions raised, and sensory and consumer science should be more attentive in adapting to changes in psychological theory. Psychology, on the other hand, has been mainly occupied with verbal and visual processes in perception and memory and has neglected the "lower" senses involved in eating and drinking behavior.

Food intake is not only regulated by metabolic needs but also affected by emotional states, motivations, and self-regulatory processes. Negative affect and increased or decreased cognitive control can override the basic maintenance of energy balance, with subgroups of individuals, reducing or increasing food intake. Thus, coping with stress and negative emotions or difficulties with emotional regulation be a transdiagnostic risk and maintenance factor for all forms of eating and weight disorders [42]. Food psychology is the science of studying to choose what to eat, when, and how much. Behavioral interventions are a common approach to weight loss and can vary greatly in the form in which they are delivered such interventions typically includes the following elements (i) An attempt to understand and control eating behavior (e.g. emotional triggers of eating) (ii) Attitudes to eat (iii) Good nutrition (iv) Seeking and utilizing social support and (v) Exercise [43].

In doing so, it has missed an important chance to get a more balanced view of the vital processes that govern our lives. Perhaps sensory consumer scientists should be more active to try and bridge the gap with academic psychology and should make more of an effort to convince them that eating and drinking behavior is a much better and more natural area to research incidental learning, implicit memory, and perception without awareness than verbal learning and memory or vision and audition [44]. For instance, the consumption of confectionery chewing gums changes smoking behavior, and it is proven that the chewing gum flavor is an important component that has contributed to the reduction of smoking. Gum and candy affect blood flow to the brain. The property such as smell and taste has also influenced the brain, and especially the sweet flavor is believed to play a role in smoking reduction treatments. Hence, the regulation of sweetener levels in smokeless tobacco products may be an effective measure to modify the attractiveness and initiation of the products [45].

## 13.5 Future of Flavors

The epidemic of overweight and obesity is giving rise to the design of functional foods for durable suppression of appetite. Adopting a multidisciplinary approach that combines principles from the sensory science of flavors, gastrointestinal physiology, ingredient technology, and texture design could bring the engineering of such food products within reach [46]. Be it confectionary or beverage, sugar-free products are also available in the market. If the flavors are enhanced, they can be consumed without fear. Flavor enhanced food and a beverage when consumed enable an age-old person to gain additional calories, which are also known to develop a positive approach toward food and beverage consumption by improving health [47]. Flavors can also be utilized to replace components that are hard to source, handle, or seasonally available. This, in turn, can reduce the cost of manufacturing by replacing high-cost ingredients [48]. For instance, soft drinks including both clear and cloudy carbonated beverages and non-carbonated drinks such as squash and cordials use flavorings for the benefits like; (i) to impart characteristic profile, (ii) to impart correct physical appearance to the product, (iii) to provide stability to heat, light, acids, and preservatives [49].

Similarly, there is no doubt that the search for organic products is already more than a trend; it is an indisputable reality. More and more people are opting for a healthier lifestyle that starts with food, which has awakened a growing interest in understanding the reasons for these purchases. The motivational attributes of consumers' decisions regarding the consumption of organic products are as follows: they are free from toxic pesticides, synthetic fertilizers, and genetically modified organisms (GMOs) [50]. The market for organic products is growing rapidly, probably attributable to the general customer perception that they are healthier foods with a better nutritional profile than conventional ones [51]. Consumer interest in organic products is growing alongside the diversification of the supply. To serve consumers' actual needs and wants regarding organic products, those involved in the market need to be informed about consumers' perceptions of organic products [52]. Therefore, the state of research with regard to consumers' perception of organic product characteristics includes flavor-enhancing compounds among other characteristics.

In the same way, emotions represent a major driver behind a consumption behavior. Thus, an emotional profiling under informed and uninformed condition may provide more important information beyond consumers' preferences [53]. For instance, preference for a particular wine style may be influenced strongly by genetic background of a wine consumer and the judgment of a winemaker may also be influenced by his or her genes, it is a challenging but vital aspect of wine making, bearing as it does on consumer perception of the product [54].

## 13.6 Conclusion

In almost all areas of the food and flavor industry, flavor plays a major role. Thus, flavoring is one of the vastly growing fields in these industries and is also a field that is widely accepted by consumers. Increased focus on the chemodiversity of secondary metabolites might aid in the successful development of sensobolome signatures. Obstacles such as consumer perception of flavor, challenges in flavoring properties, compliance with nutritional interventions, need to be addressed. By addressing these challenges and considering the perspectives discussed, exceptional flavor and aromas from secondary metabolites to create a large repertoire of versatile sensory profiles would come to dominate the consumer products industry.

## References


1 Yang, X., Jiang, Y., Yang, J. et al. (2015). Prenylated flavonoids, promising nutraceuticals with impressive biological activities. *Trends Food Sci. Technol.* 44 (1): 93–104, ISSN 0924-2244, https://doi.org/10.1016/j.tifs.2015.03.007.

2 Schiffman, S.S. (2003). Taste enhancers. In: *Encyclopedia of Food Sciences and Nutrition*, 2e (ed. B. Caballero), 5734–5737. Academic Press, ISBN 9780122270550, https://doi.org/10.1016/B0-12-227055-X/01180-9.

3 Chan, P.N.A. (2015). Chemical properties and applications of food additives: flavor, sweeteners, food colors, and texturizers. In: *Handbook of Food Chemistry* (ed. P. Cheung and B. Mehta). Berlin, Heidelberg: Springer https://doi.org/10.1007/978-3-642-36605-5_38.

4 Kaitano, T.E. (2014). Food additives: flavors and flavor enhancers. In: *Encyclopedia of Food Safety* (ed. Y. Motarjemi), 466–470. Academic Press, ISBN 9780123786135, https://doi.org/10.1016/B978-0-12-378612-8.00230-4.

5 Bautista, E.N., Tanchoco, C.C., Tajan, M.G. et al. (2013). Effect of flavor enhancers on the nutritional status of older persons. *J. Nutr. Health Aging* 17: 390–392. https://doi.org/10.1007/s12603-012-0438-9.

6 Grigorov, M.G., Schlichtherle-Cerny, H., Affolter, M., and Kochhar, S. (2003). Design of virtual libraries of umami-tasting molecules. *J. Chem. Inf. Comput. Sci.* 43 (4): 1248–1258. https://doi.org/10.1021/ci025652g.

7 Favaro-Trindade, C.S., Rocha-Selmi, G.A., and dos Santos, M.G. (2015). Chapter 17 – Microencapsulation of sweeteners. In: *Microencapsulation and Microspheres for Food Applications* (ed. L.M.C. Sagis), 333–349. Academic Press, ISBN 9780128003503, https://doi.org/10.1016/B978-0-12-800350-3.00022-4.

8 Smit, G., Smit, B.A., and Engels, W.J.M. (2005). Flavour formation by lactic acid bacteria and biochemical flavour profiling of cheese products. *FEMS Microbiol. Rev.* 29 (3): 591–610. https://doi.org/10.1016/j.fmrre.2005.04.002.

9 Bassoli, A., Borgonovo, G., and Busnelli, G. (2007). Alkaloids and the bitter taste. In: *Modern Alkaloids* (ed. E. Fattorusso and O. Taglialatela-Scafati). Wiley-VCH Verlag GmbH & Co. KGaA, https://doi.org/10.1002/9783527621071.ch3.

10 Kershaw, J.C. and Mattes, R.D. (2018). Nutrition and taste and smell dysfunction. *World J. Otorhinolaryngol. Head Neck Surg.* 4 (1): 3–10. https://doi.org/10.1016/j.wjorl.2018.02.006.

11 van Eck, A. and Stieger, M. (2020). Oral processing behavior, sensory perception and intake of composite foods. *Trends Food Sci. Technol.* 106: 219–231, ISSN 0924-2244, https://doi.org/10.1016/j.tifs.2020.10.008.

**12** Saffarionpour, S. and Ottens, M. (2018). Recent advances in techniques for flavor recovery in liquid food processing. *Food Eng. Rev.* 10: 81–94. https://doi.org/10.1007/s12393-017-9172-8.

**13** Weller, R.F., Marley, C.L., and Moorby, J.M. (2007). Chapter 7 – Effects of organic and conventional feeding regimes and husbandry methods on the quality of milk and dairy products. In: *Woodhead Publishing Series in Food Science, Technology and Nutrition, Handbook of Organic Food Safety and Quality* (ed. J. Cooper, U. Niggli and C. Leifert), 97–116. Woodhead Publishing, ISBN 9781845690106, https://doi.org/10.1533/9781845693411.2.97.

**14** Schifferstein, H.N.J. (2021). Chapter 7 – Designing food experiences: a multisensory approach. In: *Woodhead Publishing Series in Consumer Sci & Strat Market, Transdisciplinary Case Studies on Design for Food and Sustainability* (ed. S. Massari), 115–130. Woodhead Publishing, ISBN 9780128178218, https://doi.org/10.1016/B978-0-12-817821-8.00015-1.

**15** Sinesio, F. (2005). Sensory evaluation. In: *Encyclopedia of Analytical Science*, 2e (ed. P. Worsfold, A. Townshend and C. Poole), 283–290. Elsevier, ISBN 9780123693976, https://doi.org/10.1016/B0-12-369397-7/00561-6.

**16** Djekic, I., Lorenzo, J.M., Munekata, P.E.S. et al. (2021). Review on characteristics of trained sensory panels in food science. *J. Texture Stud.* 52 (4): 501–509. https://doi.org/10.1111/jtxs.12616.

**17** Amaya, E. and Nickell, D. (2015). Chapter 11 – Using feed to enhance the color quality of fish and crustaceans. In: *Woodhead Publishing Series in Food Science, Technology and Nutrition, Feed and Feeding Practices in Aquaculture* (ed. D.A. Davis), 269–298, ISBN 9780081005064. Woodhead Publishing https://doi.org/10.1016/B978-0-08-100506-4.00011-8.

**18** Field, K. and Duizer, L.M. (2016). Food sensory properties and the older adult. *J. Texture Stud.* 47: 266–276. https://doi.org/10.1111/jtxs.12197.

**19** Wang, K., Zhuang, H., Bing, F. et al. (2021). Evaluation of eight kinds of flavor enhancer of umami taste by an electronic tongue. *Food Sci. Nutr.* 9: 2095–2104. https://doi.org/10.1002/fsn3.2178.

**20** Roper, S.D. and Chaudhari, N. (2017). Taste buds: cells, signals and synapses. *Nat. Rev. Neurosci.* 18 (8): 485–497. https://doi.org/10.1038/nrn.2017.68. PMID: 28655883; PMCID: PMC5958546.

**21** Sheridan, C. (2004). A taste of the future. *Nat. Biotechnol.* 22: 1203–1205. https://doi.org/10.1038/nbt1004-1203.

**22** Beltrami, M.C., Döring, T., and Lindner, J.D.D. Sweeteners and sweet taste enhancers in the food industry. *Food Sci. Technol.* 38 (2): https://doi.org/10.1590/fst.31117.

**23** Bornhorst, G.M. and Singh, R.P. (2012). Bolus formation and disintegration during digestion of food carbohydrates. *Compr. Rev. Food Sci. Food Saf.* 11: 101–118. https://doi.org/10.1111/j.1541-4337.2011.00172.x.

**24** Wang, X. and Chen, J. (2017). Food oral processing: recent developments and challenges. *Curr. Opin. Colloid Interface Sci.* 28: 22–30, ISSN 1359-0294, https://doi.org/10.1016/j.cocis.2017.01.001.

25 Chen, J. (2015). Food oral processing: mechanisms and implications of food oral destruction. *Trends Food Sci. Technol.* 45 (2): 222–228, ISSN 0924-2244, https://doi.org/10.1016/j.tifs.2015.06.012.

26 Chen, J. (2009). Food oral processing—a review. *Food Hydrocolloids* 23 (1): 1–25, ISSN 0268-005X, https://doi.org/10.1016/j.foodhyd.2007.11.013.

27 Kaya, I. (2016). Motivation factors of consumers' food choice. *Food Nutr. Sci.* 7: 149–154. https://doi.org/10.4236/fns.2016.73016.

28 Shafie, F.A. and Rennie, D. (2012). Consumer perceptions towards organic food. *Proc. Social Behav. Sci.* 49: 360–367, ISSN 1877-0428, https://doi.org/10.1016/j.sbspro.2012.07.034.

29 Perrea, T., Grunert, K.G., and Krystallis, A. (2015). Consumer value perceptions of food products from emerging processing technologies: a cross-cultural exploration. *Food Qual. Preference* 39: 95–108, ISSN 0950-3293, https://doi.org/10.1016/j.foodqual.2014.06.009.

30 Rivaroli, S., Baldi, B., and Spadoni, R. (2020). Consumers' perception of food product craftsmanship: a review of evidence. *Food Qual. Preference* 79: 103796, ISSN 0950-3293, https://doi.org/10.1016/j.foodqual.2019.103796.

31 Venter, K., van der Merwe, D., de Beer, H. et al. (2011). Consumers' perceptions of food packaging: an exploratory investigation in Potchefstroom, South Africa. *Int. J. Consum. Stud.* 35: 273–281. https://doi.org/10.1111/j.1470-6431.2010.00936.x.

32 Fibri, D.L.N. and Frøst, M.B. (2019). Consumer perception of original and modernised traditional foods of Indonesia. *Appetite* 133: 61–69, ISSN 0195-6663, https://doi.org/10.1016/j.appet.2018.10.026.

33 Meiselman, H.L. and Bell, R. (2003). Eating habits. In: *Encyclopedia of Food Sciences and Nutrition*, 2e (ed. B. Caballero), 1963–1968. Academic Press, ISBN 9780122270550, https://doi.org/10.1016/B0-12-227055-X/00379-5.

34 Donini, L.M. (2017). Chapter 2 – Control of food intake in aging. In: *Woodhead Publishing Series in Food Science, Technology and Nutrition, Food for the Aging Population*, 2e (ed. M.M. Raats, L.C.P.G.M. de Groot and D. van Asselt), 25–55, ISBN 9780081003480. Woodhead Publishing https://doi.org/10.1016/B978-0-08-100348-0.00002-0.

35 Neacsu, M., McBey, D., and Johnstone, A.M. (2017). Chapter 22 – Meat reduction and plant-based food: replacement of meat: nutritional, health, and social aspects. In: *Sustainable Protein Sources* (ed. S.R. Nadathur, J.P.D. Wanasundara and L. Scanlin), 359–375. Academic Press, ISBN 9780128027783, https://doi.org/10.1016/B978-0-12-802778-3.00022-6.

36 Poelman, M.P. and Steenhuis, I.H.M. (2019). Chapter 7 – Food choices in context. In: *Context* (ed. H.L. Meiselman), 143–168, ISBN 9780128144954. Woodhead Publishing https://doi.org/10.1016/B978-0-12-814495-4.00007-6.

37 Donini, L.M., Poggiogalle, E., and del Balzo, V. (2016). Chapter 9 – Food preferences in the elderly: molecular basis. In: *Molecular Basis of Nutrition and Aging* (ed. M. Malavolta and E. Mocchegiani), 121–126, ISBN 9780128018163. Academic Press https://doi.org/10.1016/B978-0-12-801816-3.00009-1.

38 Suess, B., Festring, D., and Hofmann, T. (2015). Chapters 15 – Umami compounds and taste enhancers. In: *Woodhead Publishing Series in Food Science, Technology*

*and Nutrition, Flavour Development, Analysis and Perception in Food and Beverages* (ed. J.K. Parker, J.S. Elmore and L. Methven), 331–351. Woodhead Publishing, ISBN 9781782421030, https://doi.org/10.1016/B978-1-78242-103-0.00015-1.

**39** Considine, G.D. (ed.) (2006). Flavor enhancers and potentiators. In: *Van Nostrand's Scientific Encyclopedia*. John Wiley & Sons, Inc. https://doi.org/10.1002/0471743984.vse8048.

**40** Köster, E.P. (2003). The psychology of food choice: some often encountered fallacies. *Food Qual. Preference* 14 (5–6): 359–373, ISSN 0950-3293, https://doi.org/10.1016/S0950-3293(03)00017-X.

**41** Köster, E.P. and Mojet, J. (ed.) (2015). From mood to food and from food to mood: a psychological perspective on the measurement of food-related emotions in consumer research. *Food Res. Int.* 76 (Part 2): 180–191, ISSN 0963-9969, https://doi.org/10.1016/j.foodres.2015.04.006.

**42** Yau, Y.H. and Potenza, M.N. (2013). Stress and eating behaviors. *Minerva Endocrinol.* 38 (3): 255–267.

**43** Lasikiewicz, N., Myrissa, K., Hoyland, A., and Lawton, C.L. (2014). Psychological benefits of weight loss following behavioural and/or dietary weight loss interventions. A systematic research review. *Appetite* 72: 123–137, ISSN 0195-6663, https://doi.org/10.1016/j.appet.2013.09.017.

**44** Köster, E.P. (2009). Diversity in the determinants of food choice: a psychological perspective. *Food Qual. Preference* 20 (2): 70–82, ISSN 0950-3293, https://doi.org/10.1016/j.foodqual.2007.11.002.

**45** Miao, S., Beach, E.S., Sommer, T.J. et al. (2016). *Nicotin Tob. Res.* 18 (11): https://doi.org/10.1093/ntr/ntw141.

**46** Ruijschop, R.M.A.J., Burseg, K.M.M., Lambers, T.T., and Overduin, J. (2009). Chapter 24 – Designing foods to induce satiation: a flavour perspective. In: *Woodhead Publishing Series in Food Science, Technology and Nutrition, Designing Functional Foods* (ed. D.J. McClements and E.A. Decker), 623–646. Woodhead Publishing, ISBN 9781845694326, https://doi.org/10.1533/9781845696603.3.623.

**47** Jacqueline, B., Marcus, M.S., and RDN, LDN, CNS, FADA, FAND (2019). Chapter 6 – Flavour enhancing ingredients. In: *Aging, Nutrition and Taste* (ed. J.B. Marcus), 173–206. Academic Press ISBN 9780128135273, https://doi.org/10.1016/B978-0-12-813527-3.00006-5.

**48** Heath, H.B. and Allen, B.B. (2019). *The Application of Flavor in Food Processing*. London, England: Industry News, R&D, Flavor & Fragrance Industry.

**49** Sayed, A. and Abdellatif, A. (2018). The beverages. *Agri. Res. Tech Open Access J.* 14 (5): 555933. https://doi.org/10.19080/ARTOAJ.2018.14.555933.

**50** Nunes, F., Madureira, T., and Veiga, J. (2021). The organic food choice pattern: are organic consumers becoming more alike? *Foods* 10 (5): 983. https://doi.org/10.3390/foods10050983. PMID: 33946339; PMCID: PMC8145405.

**51** Dall'Asta, M., Angelino, D., Pellegrini, N., and Martini, D. (2020). The nutritional quality of organic and conventional food products sold in Italy: results from the food labelling of italian products (FLIP) study. *Nutrients* 12 (5): 1273. https://doi.org/10.3390/nu12051273. PMID: 32365788; PMCID: PMC7282013.

**52** Schleenbecker, R. and Hamm, U. (2013 Dec). Consumers' perception of organic product characteristics. A review. *Appetite* 71: 420–429. https://doi.org/10.1016/j.appet.2013.08.020. PMID: 24012637.

**53** Ismael, D. and Ploeger, A. (2020). Consumers' emotion attitudes towards organic and conventional food: a comparison study of emotional profiling and self-reported method. *Foods* 9 (1): 79. https://doi.org/10.3390/foods9010079. PMID: 31936840; PMCID: PMC7023274.

**54** Considine, J.A. and Frankish, E. (2014). Chapter 2 – Flavors and aromas in foods and beverages. In: *A Complete Guide to Quality in Small-Scale Wine Making* (ed. J.A. Considine and E. Frankish), 11–21. Academic Press, ISBN 9780124080812, https://doi.org/10.1016/B978-0-12-408081-2.00002-0.

# Index



*Natural Flavours, Fragrances, and Perfumes: Chemistry, Production, and Sensory Approach*,
First Edition. Edited by Sreeraj Gopi, Nimisha Pulikkal Sukumaran, Joby Jacob, and Sabu Thomas.

## g



## h



## i

www.ingramcontent.com/pod-product-compliance
Ingram Content Group UK Ltd.
Pitfield, Milton Keynes, MK11 3LW, UK
UKHW050842101026
15624UKWH00018B/442

9 783527 347087